The Lost Land of Lemuria

A

BOOK

The Philip E. Lilienthal imprint
honors special books
in commemoration of a man whose work
at the University of California Press from 1954 to 1979
was marked by dedication to young authors
and to high standards in the field of Asian Studies.
Friends, family, authors, and foundations have together
endowed the Lilienthal Fund, which enables the Press
to publish under this imprint selected books
in a way that reflects the taste and judgment
of a great and beloved editor.

The Lost Land of Lemuria

Fabulous Geographies, Catastrophic Histories

Sumathi Ramaswamy

UNIVERSITY OF CALIFORNIA PRESS
Berkeley Los Angeles London

University of California Press
Berkeley and Los Angeles, California

University of California Press, Ltd.
London, England

Library of Congress Cataloging-in-Publication Data

Ramaswamy, Sumathi.
The lost land of Lemuria : fabulous geographies, catastrophic histories / Sumathi Ramaswamy.
p. cm.
Includes bibliographical references and index.
ISBN 0-520-24032-4 (cloth : alk. paper)—ISBN 0-520-24440-0 (pbk : alk. paper)
1. Lost continents. 2. Tamil (Indic people). 3. Tamil (Indic people)—India—History. 4. Civilization, Ancient. 5. Lemuria. 6. Tamil Nadu (India)—Civilization. I. Title.
GN751.R26 2004
001.94—dc22 2003022825

Manufactured in the United States of America

13 12 11 10 09 08 07 06 05 04
10 9 8 7 6 5 4 3 2 1

The paper used in this publication is both acid-free and totally chlorine-free (TCF). It meets the minimum requirements of ANSI/NISO Z39.48–1992 (R 1997) (*Permanence of Paper*).

For my father,
in the shadow of whose loss I wrote this book

A map of the world that does not include Utopia
is not worth even glancing at, for it leaves out the one country
at which Humanity is always landing.
And when Humanity lands there, it looks out,
and seeing a better country, sets sail.
Progress is the realisation of Utopias.

OSCAR WILDE, *The Soul of Man Under Socialism*

CONTENTS

ILLUSTRATIONS

ACKNOWLEDGMENTS

This has not been the easiest thing to do, to write a scholarly book on a lost continent. Over the past six years I have been engaged in this enterprise, I have encountered some skeptics among my academic peers, who wondered about the relevance of my work and queried me about my reasons for undertaking it. To these good folks, one and all, I owe my first expression of gratitude, for their skepticism has provoked me to be as analytically rigorous as possible, even while their incredulity has reminded me about the importance of scholarly accountability for the projects we undertake.

But I have also had the good fortune of receiving the encouragement and support of many. Foremost among these, I must mention David Gilmartin, who over numerous cups of tea (or glasses of beer!)—be it in Washington, Seattle, or Raleigh—patiently heard my ideas out, vigorously argued with me, gently urged me to consider avenues not taken, and then, in a rigorous reading of the penultimate version of this manuscript, offered concrete suggestions for improvement and finesse. I thank him for his mentorship, his friendship, and above all, for his own quirky sense of imagination which encouraged mine to flourish! Among others who read earlier versions of different chapters and offered me their advice, I wish to thank E. Annamalai, David Arnold, Dipesh Chakrabarty, Mathew Edney, Thomas Laquer, Tom Metcalf, Indira Peterson, Tom Trautmann, and Marcia Yonemoto. Over the past few years, I have spoken about Lemuria at the University of Pune, University of Chicago, University of Michigan, Penn State University, University of North Carolina at Chapel Hill, University of Colorado at Boulder, the School of Oriental and African Studies in London, Johns Hopkins University, and at the American Historical Association annual meetings, among others. I thank the audiences at these venues for their thoughtful suggestions and critiques. My colleagues, friends, and students in Ann Arbor gamely heard many of my ideas and helped me think through them. In particular, I wish to thank Lee Schlesinger for helping to keep me intellectually honest, and for emailing me the Oscar Wilde epigraph when I was most in need of something like this to spur me on to finish this book!

In Tamilnadu, numerous individuals were gracious with their time and resources. In particular, I want to thank R. Muthukumaraswamy, R. Mathivanan, N. Kasinathan, "Cilampoli" Chellapan, I. Mahadevan, Kodumudi Shanmugham, S. Padmanabhan of Nagerkovil, M. V. Chockalingam of the Tamilnadu Textbook Society, and Dorothy and B. Krishnamoorthy in Pudukottai. Without the help of the following, I could not have located many of the sources in Tamil that have gone into this study: Murugan at the Maraimalai Adigal Library, Shankaralingam and Sundar at the Roja Muthiah Research Library, Sunderarajan at the Perasiriyar K. Anbazhagan Library, plus Sivakumar, Kannan, Sairam, Prince, Namasivayam, and Rajendran at the Tamilnadu State Archives; I thank them one and all. Nalini Persad at the India Office Library helped me with sources in the British Library that I was unable to find in India, and Mary Rader tracked down many a recalcitrant reference from her resourceful office at the University of Michigan's library.

A special thanks to Theodore Baskaran for offering a helpful ear, and especially for introducing me to his brother, the geologist Christopher Jayakaran, who completed his own book on Lemuria while I was writing mine. And far away from India, but not in spirit, Irina Glushkova in Moscow helped with me information I needed on an elusive Russian publication on Lemuria, and at final crunch-time, worked hard to put me in touch with the publishers of the book, from whom I needed permission for the reproduction of the map that appears on the cover.

This book would also not have been possible but for the very generous financial and intellectual support I received from the following organizations: the National Endowment for the Humanities, the American Institute of Indian Studies, the Social Science Research Council, and the University of Michigan, Ann Arbor. A J. B. Harley Research Fellowship in the History of Cartography enabled me to spend some delightful months in the British Library's Map Room, and I thank Tony Campbell and Catherine Delano-Smith for their support and encouragement. At the University of California Press, I wish to thank Sheila Levine, my original commissioning editor, and especially Reed Malcolm, who took over this project when it was near completion and gave it his unwavering support. I also want to express my appreciation to my copy editor Ruth Steinberg, and to Kate Warne and her production team for shepherding the manuscript through the final stages to print.

My mother, in Chennai, housed me and fed me great meals over the many summers I did research for this book, and most recently, when I was writing parts of it. In addition, she lent a critical ear, asking questions that pushed me to amend and clarify. Across the world, in Tampa, Florida, my mother-in-law has always relentlessly encouraged me to pursue my ideas and imagination, however "wacky" they might seem to the outside world, and her friend Elaine Johnson gifted me with crucial occult books on Lemuria when I most needed them. Over the many years I have been married to him, Rich Freeman has been my best friend and critic, allowing me to generously borrow from his own areas of expertise in anthropology and Indology, keeping a check on my wilder flights of imagination, and demanding precision in language and thought. At a time when he himself was pushed for time, and when my own patience had run out, he read every word of this manuscript, helping me to finesse my arguments, adding a crucial sentence here, deleting an embarrassing one there. As always, I owe him enormous gratitude for sharing my life, both professional and personal.

I began my research on Lemuria in 1996 when my father was still alive. A civil engineer by training, he prided himself (as many Nehruvians of his generation in India still do) on his scientific rationality and on his rigorous querying of received traditions. Yet he was open to the very idea of Lemuria, always reminding me to rein in my skepticism and my own sense of incredulity as I would come home from a day at the archives having encountered one more "incredible" and "fantastic" notion. In retrospect, his gentle doubts about my own skepticism pushed me to pursue the line of thinking I eventually adopt in this book, more so perhaps than many a fashionable postcolonial critique of the politics of knowledge production. To the memory of my father—and in memory of the many lovely conversations and spirited arguments I had with him about Lemuria over a cup of South Indian coffee—I dedicate this book.

New Delhi
May 2003

NOTE ON TRANSLITERATION

For the sake of readability, I have kept the use of diacritical marks for Tamil words to a minimum. Tamil terms and the titles of Tamil texts have been transliterated following the University of Madras *Tamil Lexicon.* Tamil place-names directly pertaining to Lemuria alone have been transliterated. For place-names in contemporary usage, and for proper names of individuals, deities, castes, and institutions, I have followed the most recognizable Anglicized form. Unless indicated otherwise, all translations from Tamil are mine.

Chapter 1

Placing Loss

The most fascinating terrae incognitae of all are those that lie within the minds and hearts of men.[1]

PLACE-MAKING IN THE SHADOW OF LOSS

This is a book about place-making and imagining. It is about a place that once (might have) existed, but no more. No one involved in its making has ever seen it, much less lived there, and likely they never will. Its primary identity in their place-making labors is that it once was, but has since long vanished into the deep mists of time. It is a lost place from a lost time.

This is also, therefore, a book about loss. I am interested in the preoccupation with loss as this manifests itself in the fascination with vanished homelands, hidden civilizations, and forgotten peoples and their ignored pasts that ranges from the passionate disinterest of the scholar to the melancholic yearning of the nostalgic. I propose that this preoccupation with the lost—and with the vanished, the disappeared, the hidden and the forgotten—is an inevitable, even irresistible, condition of modernity, but that its poetics and politics vary across life-worlds, ideologies, disciplines. I suggest that high modernity has not been merely preoccupied with progress and advance, but also with loss and disappearance. Correspondingly, loss is good to think in regard to what it means to be modern.

And so, this is as well a book about disenchantment. Or more correctly, it is about the refusal to give up (entirely) on enchantment in a late-modern world where disenchantment is offered as the desirable norm to which we should all aspire as true moderns. As the rationalizations and intellectualizations of our physical and human sciences disavow truths that once mattered and discard wonders that had once captivated, the world is leached of magic, mystery, and marvel. How is a lost place made in the wake of such disenchantment? This, too, is one of my concerns.

But mostly, this is a book about Lemuria, a land that is declared to have once existed but that is no more. In following the discursive adventures of this vanished land across the globe, I consider the dilemmas confronting Lemuria's place-makers as they seek to gain recognition for it on an earth from which it has catastrophically disappeared. I follow their struggles as they usher in a tangible presence out of sheer absence. And I explore their trials and victories as they allow themselves to be caught in the play of the fabulous in an intellectual climate that is fiercely impatient with things and matters that smack of magic and marvel. In doing all this, I argue for the politics and pleasures but also the pathos of making a lost land in an age and time that has been famously pronounced disenchanted.

THE LOST LAND OF LEMURIA

This is not the first time that Lemuria has been subjected to scholarly scrutiny, although this certainly is the first attempt to do so across cultures, transnationally and theoretically. Like Atlantis and Mu, Lemuria is a staple among those freelance scholars in Europe and the United States who write about Earth's prehistory under the sign of "lost continent." Indeed, the very category of "lost continent"—once widely deployed in the natural sciences—has today a purchase largely in this world of popular scholarship which caters to the interested educated reader who is, however, neither a specialist nor academically disciplined. The majority of these publications are clearly caught up in the wonder of it all as they crack the mystery of the Easter Island statues, or the ruins of Machu Picchu, by resorting to the lost continents of Atlantis, Lemuria, or Mu. Their very titles (*Timeless Earth, Worlds before Our Own, Mysterious Places, Lands of Legend,* and so on), formulaic and unsurprising though they might be, are also revealing of the fascination with disappeared realms, lost cultures, and vanished races that continues to fuel popular interest in places like Lemuria in our time.[2] As the British archaeologist Glyn Daniel once observed tongue-in-cheek,

> We know only too well that all over the world, from wayward undergraduate to B.B.C. producer to publisher's reader there are people, otherwise sensible and sane, people who would not believe in six-headed cats and blood-curdling spectral monsters, who yet read some folly about Noah's ark or Atlantis or cataclysmic world-tides, and say, with a contented sigh, "There may be something in it, you know."[3]

Yet it is not captivation and enchantment alone that drives this market of undisciplined scholarship. Every now and then, the dubious and the disbelieving speak out as well, seeking to disenchant a reading public that might otherwise choose to remain captivated by it all.[4]

In contrast, lost continents have rarely attracted the attention of professional scholars or career academics like myself in the metropole. It is almost as if we wonder with the Harvard archaeologist Stephen Williams, "Why study this fringe area . . . ? If one considers it worthless, lacking in veracity, won't discussing it give it more credibility than it deserves? If we ignore it, perhaps it will go away."[5] In the course of the last half century, a few of my disciplined peers have chosen not to ignore "this fringe area," in the belief that it will "just go away." Some among them—the empiricist and positivist—patiently sift "fact" from "fancy," in order to extract the kernel of "truth" that might lie at the core of the "fantastic" accounts of lost continents, sunken cities, or disappeared civilizations, that every century—and culture—seems to have produced since Plato's ruminations on Atlantis.[6]

But more often than not, metropolitan academics are convinced that lost continents are the work of fakes and frauds who have misappropriated the painstaking labors of professional scholarship. They complain about the seemingly never-ending popular appetite for "the most fantastic schemes when these are presented in the richly exciting idiom of Lost Continents and Lost Tribes."[7] In these circumstances, it is the solemn duty of the scholar-in-discipline to debunk these "fraudulent" and "wild" writers "on the lunatic fringe" who have seduced a gullible readership, for "their extremely popular writings persuade so many intellectually unwary people that research is simply a process of manipulating facts, intuition, and imagination in equal parts."[8] In a moment of self-reflection, the anthropologist Robert Wauchope admitted in 1962 that such writings are a source of such anxiety to his peers because "it disturbs them to realize that . . . no amount of scientific protestation seems capable of shaking the world-wide conviction of the former existence . . . of Atlantis."[9] In the face of this anxiety, professional scholars take refuge in their disciplines, in the conviction that these will enable them to either get to the real truth of the matter and set the record straight, or at least expose the charlatans for what they are.

But neither the freelance scholar nor the disciplined academic explores what it means to categorize a place such as Lemuria as "lost." What is a lost place? What symbolic capital does a lost place command that an available place does not, or cannot? Nor is the scholarship on Lemuria (or on lost continents more generally) concerned with the technologies of making a lost place. How does one remember a lost place? What trajectories does imagining (have to) take to summon into existence a world that is no longer available, and perhaps has never been? Not least, these scholars are quite sanguine about the preoccupation with loss that generates a place such as Lemuria in the first place. Why this fascination with a lost place and its fate and fortunes? Under what circumstances does the past return to haunt the present as loss—as the disappeared, the vanished, the submerged, and the hidden? Are we not always already losing the past?

These, then, are the questions that are at the heart of this book, and that distinguish it from other scholarly works on Lemuria, or for that matter, on other places declared lost. To anticipate what follows, I consider Lemuria as a *place-world* that is the product of varied *labors of loss* underwritten by place-making imaginations that I characterize as *fabulous* and *catastrophic.* The history that I document is necessarily *off-modern, eccentric,* and *oppositional,* as it foregrounds matters that have been marginal in their own time in their own place, or have been deemed unworthy of the professional scholar's attention. I chart, therefore, what Foucault called an "insurrection of subjugated knowledges."[10] I do not see this, however, as a project of recuperation or rehabilitation, but rather one of recognition, for Lemuria is the work of imaginations that have been discarded or disavowed by the disenchanted disciplines within which we function as professional scholars. Walter Benjamin, we are told by Adorno, was attracted to "everything that has slipped through the conventional conceptual net or . . . things which have been esteemed too trivial by the prevailing spirit for them to have left any traces other than those of hasty judgement."[11] Lemuria, it seems to me, is one of these things, dismissed quite hastily as too trivial to bother with. And so, taking heart from Benjamin, I write about it in order to learn more about what it means to be an imagining (and imaginative) modern preoccupied with loss.

PLACE-MAKING A LOST WORLD

Whatever else it might be to others who have written about it, for me, Lemuria is a place-world which, following Keith Basso, I consider as "a particular universe of objects and events . . . wherein portions of the past are brought into being."[12] Place-worlds are summoned into existence through the power of imagination and the politics of narration, through acts of place-making, "retrospective world-building."[13] For Basso, as indeed for the phenomenologist Edward Casey, to whom he is indebted in this regard, place-making "involves multiple acts of remembering and imagining which inform each other in complex ways."[14] Place-making, Basso suggests, "is a way of constructing the past, a venerable means of *doing* human history." It is also, he writes, "a way of constructing social traditions and, in the process, personal and social identities. We *are,* in a sense, the place-worlds we imagine."[15]

Taking my cue from Basso's remarkable ethnography of place-making among the Western Apache, I consider Lemuria to be the work of its "place-makers," a motley crowd whose "main objective is to speak the past into being, to summon it with words and give it dramatic form, to *produce* experience by forging ancestral worlds in which others can participate and readily lose themselves."[16] Lemuria's place-makers are drawn from widely disparate life-worlds, ostensibly modern though they might all be. Among them are metropolitan paleo-scientists participating in the global projects of natural history, geology, ethnology, and prehistory; Euro-American esoterists and occultists writing back against the unsettling revelations of Darwinian evolution; and, most of all, fervent devotees of the Tamil language, coping with the indignities of imperial rule and with the ambiguities of national belonging in colonial and postcolonial India.

As such, much more so than Basso, Casey, and others who write in the rich disciplinary traditions of humanist geography and phenomenologically informed ethnography, I am concerned with the fate of place-worlds (like Lemuria) which travel from their sites of origin in the metropole to the latter's putative subaltern (and subalternized) margins under the pressure of the spread of colonial rule, the attendant consolidation of global capitalism, and the enabling conquest of our earth by the natural and human sciences. Place-making in the modern world, I insist, cannot ignore the colonization of imagination itself, and the numerous contradictions and conflicted intimacies of power, contestation, and resistance in an age dominated by capitalism, imperialism, and the postcolonial condition.[17] To amend Basso, we *are* the place-worlds we are compelled to imagine with languages and conceptual tools that may not be of our own making and are frequently alien to our being. Which is why I have found it necessary to leaven the useful insights of Basso's ethnography and Casey's phenomenology with the revealing truths of postcolonial theory and its searing critique of the global politics of disciplinary knowledges, the celebration of Reason, and the subordination of difference and singularity to Enlightenment universals.

As I recount in what follows, Lemuria assumes many different shapes in the labors of its place-makers as they imagine, remember, yearn for, and map this vanished place. But even as it travels far and away from its first appearance in the by-lanes of Victorian science, there are three fundamental regards in which it has a shared identity. First, in all the place-making around it, Lemuria belongs to a time before our own. This means various things to its different place-makers, but at the very least, they are all certain that this is a place that existed long before recorded history, or to put a finer point on it, before any *surviving* recorded history. As a land before our time, it reaches back to an age when the continents looked different, when giant forests loomed large, and when dinosaurs roamed the surface of the earth. Few traces remain as metonyms of its presence. Lemuria is therefore a *paleo* place-world. Its making was enabled by natural history, geology, and zoogeography, which came into their own as paleo-disciplines over the course of the nineteenth century. Lemuria's allure for all its place-makers emerges from the fact that it was a world before our own, as revealed by some of the most prestigious sciences of their times. In an age mesmerized by the global revelations of the paleo-sciences, Lemuria was a creation of those very sciences.

As a paleo place-world from a time before our own, Lemuria is by definition not empirically available to its makers for seeing, surveying, and occupying, as places routinely are in the clear and present light of everyday reality. In other words, it is a place-world that can only be summoned into existence through imagination and has no existence beyond it. It lies outside our realm of empirical experience and appearances. This is one of the many senses in which I characterize this place-world as *fabulous,* to underscore both the primary location of Lemuria in the imagination of its place-makers, and its virtual absence outside their labors. This is not to say that the imagination which generates this fabulous place-world is singular or similar. On the contrary, the paleo-scientist's fabulous place-making takes him down a very different track than those followed by the occultist or the Tamil devotee. Nor is it to suggest—as the word "fabulous" in one of its everyday senses might lead one to quickly conclude, or as some of its detractors too readily insist—that Lemuria is necessarily false or inauthentic. But it is certainly to insist that as a fabulous place-world, Lemuria is utterly unavailable outside imagination. There is simply no there *there.*

Not least, and perhaps most importantly, Lemuria is a *lost* place-world.[18] Part of the burden of this book is to demonstrate that there is no simple or singular way in which Lemuria is lost, and to show that its loss itself is multiply configured. But having said this, we can also recognize that because Lemuria is a paleo place-world belonging to a time before our own, it is no longer part of our time and no longer at hand, and hence is lost to our present. Further, as a fabulous place-world that is not available (any longer) for seeing, possessing, or dwelling, it is lost to us as well. But most of all, Lemuria is a place that is lost until its place-makers summon it into existence. If not for them, it would remain unknown, vanished, even nonexistent. Because of them, it reappears as lost.

So, to identify and follow the fabulous travels of a lost paleo place-world like Lemuria across the globe means writing a spatial history which, following Paul Carter, I understand to be "a history of the spatial forms and fantasies through which a culture declares its presence."[19] As Carter's *The Road to Botany Bay* suggests, spatial history is history "in motion," as it is about spaces and places whose pasts are contested, forgotten, disavowed, lost.[20] It is also about "place-talk," about words with which whole worlds are summoned into existence.[21] Spatial history reminds us as well that "the physical world is also a site where unrequited desires, bizarre ideologies, and hidden productivities are encrypted, so that any narration of space must confront the dilemma of geographic enigmas head on, including the enigma of what gets forgotten, or hidden, or lost in the comforts of ordinary space."[22] A confrontation with such geographic enigmas demonstrates both the premises and promises of modern place-making, but also its limits. One such geographic enigma produced out of spatial fantasies and place-talk as these swirl around in the cauldron of the new sciences of the earth's paleo past, Lemuria is a fabulously lost place-world that is ever available for making as long as it holds the interest of the human imagination and the contingent politics that underwrite this. But it is also this very imagination that allows us to see it for what it is—a place made to cope with modernity's preoccupation with loss.

LABORING OVER LOSS IN A DISENCHANTED WORLD

Lemuria as a place-world is established through varied *labors of loss,* an idea I borrow (but use quite differently) from the French philosopher Georges Bataille, who tells us that "sacred things are constituted through an operation of loss."[23] For me, labors of loss are those disciplinary practices, interpretive acts, and narrative moves which declare something as lost, only to "find" them through modernity's knowledge protocols, the very act of discovery and naming constituting the originary loss. Lost places do not exist as such in our life-worlds. They are summoned into existence and their being shaped through our labors of loss. A place declared as lost does not precede our labors of loss. It is instead their product and outcome. This is not to say that a lost place cannot (or does not) exceed the labors of loss that generate it; this happens time and again. Nonetheless, a lost place is fundamentally constituted *as such* through labors of loss that are performed around it. This is the founding premise of this book.

A standard dictionary definition of *lost* includes "wasted, no longer possessed, having wandered from the path, ruined, beyond reach, preoccupied." While the range of meanings suggested by this polysemic word is wide, the emphasis is on the negative: destruction, lack, absence, death. Outside of psychoanalysis, for which loss is a central concern as well key category of knowledge,[24] most other scholarship generally sees the preoccupation with loss (which is invariably reduced to only one of its manifestations—nostalgia) as a pathology, as a conservative fixation with primal origins and unity, as reactionary escapism, as even a social disease.[25] For instance, Ajay Skaria observes, "As post-colonial historians, we have learnt to be suspicious of narratives of loss. And rightly so. In their colonial, nationalist, or liberal forms, narratives of loss are about a transition from homogeneity to differentiation, from the originary fullness of autonomy to a degraded condition of subalternity. These narratives suffer from a pervasive nostalgia, a yearning for unity with the homogeneous past, a desire for the closure of difference, and any politics associated with them would be profoundly conservative and restorative."[26]

The tack I take here is to the contrary, for I suspend my postcolonial suspicion, or at least keep it at bay, while I explore the many ways in which the preoccupation with loss is socially and materially productive (and in ways that are not necessarily only conservative or restorative). I do not mean this merely in the well-known psychoanalytic formulation that a sense of lack and absence initiates fantasy activity, but more to the point, I am interested in how apprehensions of loss mobilize the imagination, provoke political action, and interpellate whole fields of knowledge. This interest is also fueled by the fact that even the discipline in which I am professionally located as a historian is predicated on the premise of loss. As Pierre Nora notes perceptively, "History . . . is the reconstruction, always problematic and incomplete, *of what is no longer*."[27] The very writing of history is only possible when the present and the living are fundamentally disassociated from the past and from that which is absent or dead. "A sense of loss is advanced, but its void is immediately filled with the knowledge the historian reaps from his division of past and present."[28] This sense of loss is the very ground of my disciplinary practice, for as Michel de Certeau demonstrates, history's own beginning "is one which presupposes a *lost* object."[29] "Discourse about the past has the status of being the discourse of the dead. The object circulating in it is only the absent. . . . The dead are the objective figure of an exchange among the living."[30] Loss is thus the enabling fiction that energizes the production of history, a collective practice through which we cope in modernity with being in exile from our pasts, and with the consequent displacement and estrangement such a rupture produces. Historians, be they professional or amateur, always already write in the shadow of loss. They, too, are engaged in labors of loss

My deployment of labors of loss as a conceptual category is therefore meant to foreground the productive potentiality of the rich structure of sentiment of loss without too hastily reducing it to a pathology. This focus on labors of loss allows me to consider circumstances and instances in which absence or disappearance leads to an affirmation of the present/presence, "a making present of the absent."[31] It enables me to consider how apprehensions of loss may be a source of pleasure and hope, and not just poignancy and pain. And it provokes me to examine the full range of ambiguities and ambivalences engendered by discursive operations conducted under the sign of loss. Commenting on Bataille's statement that sacred things are established through an operation of loss, de Certeau notes that "just as the disappearance of Moses allowed for the appearance of the Mosaic saga, so, then, in its many different forms, a lacuna of history makes the production of a *culture* both possible and necessary: a collective epos, a legend, a tradition."[32] Labors of loss are thus dialectically enabling because "a presence, as it vanishes, founds the obligation of writing."[33] It is in this sense of labors of loss as productive, enabling, empowering, even obligatory, that I write back, against both a master narrative of modernity which has primarily identified it with progress and gain, but also against a dominant diagnosis of the modern preoccupation with loss as inevitably and unequivocally regressive or reactionary.

As I demonstrate in what follows, the labors of loss around the paleo place-world of Lemuria are not vestiges of the archaic (although in many cases, they may appear so, and strategically), but are outcomes of and responses to various projects of scientific and colonial modernity as these come to be conducted across the inhabited world. This in itself is not surprising, for the preoccupation with loss is itself symptomatic of what Celeste Olalquiaga refers to as "the intoxication of modernity."[34] In her imaginative study of the Victorian kitsch experience, Olalquiaga connects the nineteenth-century metropolitan world's "fetishization of loss" to the excesses of a postindustrial age and to the contradictory pulls of advanced modernity.[35] "Modernity is a displaced time: it wants nothing to do with the past and looks only toward a future constantly receding on the horizon. Yet the past denied by industrialization continually reappears like a littered landscape next to an indifferent highway."[36] She reminds us that Benjamin's Angel of History has his face turned toward the past, even as the storm of progress "irresistibly propels him into the future to which his back is turned, while the pile of debris before him grows skyward."[37] The very operation of modernity's contradictory attraction to both novelty and to all that is left behind results in labors of loss around drowned continents, golden ages, fake ruins, fossil collectibles, enigmatic mermaids, and so on. So much so that loss itself is commodified "in a futile and overabundant production . . . [as] kitsch."[38]

Missing in Olalquiaga's provocative analysis of the mutually constitutive relationship between modernity and what I call its labors of loss is a discussion of disenchantment, arguably one of the most diagnostic signs of the metropolitan modern. Peter Homans has recently insisted that Weber's theory of disenchantment is "an as yet unexplored theory of collective loss."[39] If we were to follow Weber and his commentators, not only has the modern world become a place where it is no longer possible to live in union with the divine, but as importantly, as a consequence of the intellectualizations and rationalizations of its natural and human sciences, the universe, once perceived as alive and as cognizant of its own goals and purposes, is now an inert entity, hurtling about neverendingly, an immense machine of matter and motion blindly obeying mathematical laws. There is no mystery remaining in being. Reality has been rendered dreary, flat, and utilitarian, "leaving a great void in the souls of men which they seek to fill by furious activity and through various devices and substitutions."[40] Our lives stand impoverished, reduced to an endless pragmatic and instrumental pursuit of meaningless activities. "The disenchantment of the world is thus marked . . . by a loss of the archaic power of fictions to command belief."[41]

Years ago, as Alex Owen reminds us, the historian Owen Chadwick warned his colleagues to "beware of the word *Entzauberung*."[42] Heeding this warning, and also recognizing that disenchantment manifests itself very differently in the metropole than it does in the (post)colony, I suggest that a good many of the labors of loss around Lemuria, albeit not all, are attempts to re-enchant a life-world from which magic and mystery are being steadily leached out. I interpret Euro-American occult labors of loss as attempts to invite spirit back into a world from which it has been dismissed by the material sciences. In striking contrast to this religious retort to the perceived disenchantment of the world, we encounter an essentially secular response across the globe in colonial India, where Tamil place-makers brought back into play beliefs and notions that the British had attempted to purge in their drive to rationalize Indian society and liberate it from its thralldom to tradition. Even the paleo-scientist's labors are not entirely devoid of a sense of wonder and mystery as he contemplates the newly plumbed depths of the ocean, studies a surprising prehistoric rock formation, or unearths a beguiling fossil here, a missing link there. Disenchantment thus produces many counter-responses beyond a simple re-divinization of the world, and one of the burdens of this book is to demonstrate how the variegated labors of loss around Lemuria amply demonstrate this.

In fact, that these labors of loss around Lemuria have been performed with the connivance of "disenchanted" science, and even enabled by it, lends credence to Alex Owen's argument that what we see in modernity is not so much a totalizing disenchantment of the world at the expense of enchantment, but "an uneasy co-existence" of the two.[43] As I demonstrate later, even as the steady march of the sciences of the earth's deep past demystifies timeless mysteries, others rush in to fill the void opened up by the resulting disenchantment. Labors of loss around Lemuria occupy the vortex of the dialectic constituted by the opposing pulls of the will to disenchant and the rush to re-enchant. Modernity's discontents are thus both disabling and enabling for the preoccupations of loss around this vanished land.

MODALITIES OF LOSS

The lostness of Lemuria is predominantly established in these labors of loss through a place-making modality I characterize as catastrophic. In its literal sense, traceable back to its origin in the Greek *katastrophe,* catastrophe means overturning, ruin, and conclusion. In modern geology, it signifies "a sudden and violent change in the physical order of things, such as sudden upheaval, depression, or convulsion, affecting the earth's surface and the living beings upon it by which some have supposed that successive geological periods were suddenly brought to an end." More generally, as the *Oxford English Dictionary* goes on to note, a catastrophe is also "an event producing a subversion of the order or system of things." Not surprisingly, in the labors of loss around Lemuria, which intend to demonstrate its tragic, even fatal, disappearance forever, this is the most ubiquitous term deployed to designate the fate of this hapless, lost place-world, its very semiotic range enabling the multiple agendas of its different place-makers. Even paleo-scientists whose disenchanted disciplines were increasingly uncomfortable with it frequently resort to the vocabulary of catastrophism—"sudden earth movements," "convulsions," "ocean upheavals"—in laboring over the loss of Lemuria.

Ever since Charles Lyell's well-placed attacks on catastrophism in the 1830s, many in the scientific community have hoped that at long last one more irrationality in the premodern paraphernalia of miracle-mongering has died a quiet death. Yet recent scholarship shows that the revamped catastrophism of the nineteenth century drew its strength from the empirical findings of the new earth sciences. Modern catastrophism and uniformitarianism were not the pitched, polar opposites they were made out to be by their respective advocates, and neo-catastrophism is finding scattered support in the metropolitan scientific community as an explanation for dramatic changes in the earth's long and dynamic history.[44] Stephen Gould also notes the inherent conservatism that underwrote catastrophism's putative Other in the nineteenth century—Lyell's gradualism, "the quintessential doctrine of liberalism as it faced a world increasingly engulfed by demands for revolutionary change." Thus, Gould observes, "When scientific theories of a static world order collapsed toward the end of the eighteenth century, a new ideology rose to justify social stability within a world now dominated by ceaseless change. If change is intrinsic and fundamental, what could be better than a notion that it must proceed with excruciating slowness, move from one system to another through countless intermediary stages, and always be weighted down by an inheritance from the past?"[45] In contrast, some of the more radical ideologies of the nineteenth century, like Marxism, are closer to a catastrophic rather than a gradualist style of thought. More recently, for Foucault, as Clifford Geertz affirms, "History is not a continuity, one thing growing organically out of the last into the next, like the chapters in some nineteenth-century romance. It is a series of radical discontinuities, ruptures, breaks, each of which involves a wholly novel mutation in the possibilities for human observation, thought, and action."[46]

So, in spite of the bad press that catastrophism has received in the centuries during which a disenchanted modernity has consolidated itself, a catastrophic style of thought has quietly persisted in the interstices and margins, and occasionally even at the very center, of metropolitan science. As some historians of science observe, this is not surprising because notwithstanding the outright hostility of the Anglo-American paleo-scientific establishment to catastrophism, everyday events in the natural world themselves appeared to affirm that sudden, dramatic, and violent change resulting in convulsions, ruptures, dismemberment, and disappearance was the fate of the earth, not Lyell's "slow and insensible" transformations. So, events such as the great earthquakes that suddenly rocked Lisbon in 1755 or Assam in 1897, violent volcanic eruptions such as those in Indonesia at Sumba in 1815 or Mount Pelee in Martinique in 1902, destructive tidal waves like the tsunami unleashed by the Krakatoa explosion in 1883 which killed thousands, not to mention the fierce cyclonic storms that periodically hit the east coast of India, destroying life and property, kept alive both a scientific conviction in catastrophism as well as an educated (and not-so-educated) public interest in it.

As others have noted, post-Enlightenment catastrophists in the paleosciences of the nineteenth century were a different breed altogether from their miracle-mongering predecessors who "gave free play to fantasy, who rashly invoked supernatural causes, and allowed their geological researches to be dictated by a priori metaphysical beliefs."[47] Instead, they were empirical realists, painstakingly documenting abrupt breaks in sediments and fossils, scrutinizing the sudden extinction of whole species, and subjecting the geological evidence gathered from all over the world to a literal read. They spurned explanations that smacked of the miraculous, the supernatural, or the occult. They turned their backs on diluvialism, one of the chief preoccupations of premodern catastrophism, which after the 1830s rarely rears its head. Even the radical distinction that was once drawn between the catastrophist and his uniformitarian Other breaks down when we realize that modern catastrophism acknowledges that the earth's past is dominated by long periods of gradual, quiet change, occasionally punctuated by sudden movement and transformation. And the focus increasingly has shifted to the singular and rare event that might have caused sudden disappearances and dramatic extinctions. It is out of this context of what we might call disenchanted catastrophism that the earliest labors of loss around Lemuria first emerge among metropolitan paleo-scientists.

But outside the rarefied circles of professional paleo-science there also flourished a flamboyant version of catastrophism that found a large popular audience in "the world of autodidacts, penny newspapers, weekly encyclopedias, evening classes, public lectures, worker's educational institutes, debating unions, libraries of popular classics, socialist societies, and art clubs."[48] It is this world that bought and consumed the catastrophic writings of the likes of Antonio Snider-Pellegrini, an American living in Paris who in 1858 published a book entitled *La création et ses mystères dévoilés* (The Creation and Its Mysteries Revealed). Bringing together current geological theories of a rapidly shrinking earth and catastrophic notions of ruptures and dismemberment with the Genesis account of the creation of life in seven days ("epochs"), Snider-Pellegrini suggested that each epoch was marked by a cataclysm, "until on the fifth day all the lands of the earth were concentrated in one large, unstable mass along which ran a giant fissure oriented approximately north-south. The Deluge occurred on the sixth day, as volcanic gases poured through the fissure, forcing the New and Old World continents apart and causing a sudden contraction of the earth. Thus oceanic waters were forced over the continents and the Atlantic was born."[49] As can be expected, Snider-Pelligrini's book made hardly a dent in paleo-scientific circles, but years later, occult labors of loss around Lemuria picked up on it, and on others like it that publishers regularly churned out, filled with gloom-and-doom predictions of giant meteors and runaway planets crashing into our globe, of the impending tumble of the earth off its axis, and of worldwide cataclysms. Immanuel Velikovsky's *Worlds in Collision* (originally published in 1950 by no less than Macmillan) is an example closer to our time of this kind of popular—and enchanted—catastrophism. The disenchanted skeptic and the professional scholar dismiss books like *Worlds in Collision* as sensationalist, as patent nonsense, and above all, as misappropriation of good science.[50] In a disenchanted time when "the miraculous has been excluded from geologic explanation,"[51] popular catastrophic thinking smacks of a miracle-mongering that is far more dangerous precisely because it resorts so openly to the contemporary findings of the earth sciences in pursuing its fabulous agenda.

Given the turbulent, even catastrophic, fate of catastrophism in our times that I have hastily sketched here, I run the risk of being accused of miracle-mongering myself when I propose that we pay attention to the catastrophic style of thought that underwrites labors of loss around place-worlds like Lemuria. I argue we attend to it because it has allowed many in the colonies, as well as in the metropole, to salvage the subjugated truths that a scientific modernity has shrugged aside, to recuperate the wondrous and the fabulous that its disenchanted knowledge-practices have disallowed, and to pursue enchanted visions of the earth's past outside the professional disciplines. Most of all, catastrophism has allowed those who resort to it to think loss. It appears to enable those who labor over irrevocable and total loss to explain it, account for it, even rationalize it. But it is clear that it is catastrophism which constitutes the original loss with which they are all so preoccupied, ensuring that it remains irrecoverable forever. The logic of irrevocable loss is critically and crucially underwritten by catastrophism. So much so that I would suggest that it would be almost impossible to think a fabulously lost paleo place-world like Lemuria outside of catastrophism.

Apprehensions of catastrophic loss facilitate, and are in turn fed by, another place-making imagination that I characterize as fabulous. I have already noted that as a paleo place-world that is not present, Lemuria is beyond reach or attainment, and hence, definitionally fabulous. The category of the fabulous is intriguing, incorporating within its semantic range meanings that are mutually contrary. Thus, the fabulous is that which is absurd, mythical, and legendary, but also astonishing, marvelous, incredible, breathtaking. The fabulous, like the fantastic, another similarly semantically charged category on which much has been written, is that which approaches the impossible, which lies beyond the usual range of facts, which makes "leaps into other realms."[52] When I characterize Lemuria as a fabulous place-world, I mean to keep in play all these meanings in order to suggest that fabulous place-making is a process of thinking imaginatively—and enchantingly—about places not actually present or existing, about elsewheres beyond the usual range of facts, "beyond the known, beyond the accepted, beyond belief."[53] I am interested in exploring the limits of the impossible as this manifests itself in Lemuria's place-making, as I am concerned with flagging the sentiments of marvel and wonder that frequently accompany the labors of loss around it.[54] In so doing I follow Rosemary Jackson in her suggestion that "in a secularized culture, desire for otherness is not displaced into alternative regions of heaven or hell, but is directed towards the absent areas of this world, transforming it into something 'other' than the familiar, comfortable one. Instead of an alternative order, it creates 'alterity,' this world re-placed and dis-located."[55] What also makes Lemuria fabulous is that it is seemingly of this world even while it is not. It is a place-world that is "neither entirely 'real,' nor entirely 'unreal,' but is located somewhere indeterminately between the two."[56] I attempt to capture Lemuria's tantalizing indeterminacy through the category of the fabulous.

But I also deploy the category oppositionally and subversively[57] in order to keep reminding my reader—and myself—that in the disenchanted disciplines within which we function, lost continents like Lemuria or Atlantis have been typically dismissed as fabulous and fantastic, hence fictitious and false, and not worthy of our critical attention. There are important precedents to such a dismissal of those imaginations which cannot be readily accommodated within a disenchanted disciplinary view of Earth as an empirically knowable, measurable, and traversable place that fits into our usual range of observable and available facts. In colonial India, for example, the category of the fabulous was ubiquitously deployed by the British administrator-scholar to disavow all those "native" spatial imaginations which were not "relentlessly matter-of-fact and empirical."[58] Located in a geographical tradition influenced by Locke and Hume that was "deliberately dry and untheoretical, an antidote to romance and imagination,"[59] the colonial archive abounds with statements such as the following, made by the missionary-educationist John Murdoch:

> The different modes adopted by Hindus and Europeans in framing systems of geography are well worthy of notice. A Hindu, without any investigation, sat down and wrote that the centre of our universe consisted of an immense rock, surrounded by concentric oceans of ghee, milk, and other fluids. To induce men to believe his account, he then pretended that it was inspired. Europeans, on the other hand, visited countries, measured distances, and after very

careful investigation, wrote descriptions of the earth. Which is the more worthy of credit?[60]

That a similar situation had come to pass in the metropole is also apparent from a remarkable essay by Joseph Conrad in which he insists that the "dull imaginary wonders of the dark ages" that constituted what he characterizes as "Geography Fabulous" came to be eventually replaced by the unquenchable search for certitude and truth in the age of "Geography Militant." "Geography is a science of facts, and [it] devoted [itself] to the discovery of facts in the configuration and features of the main continents." Yet geography's prehistory in Europe had been tainted, and it had to fight its way to "truth through a long series of errors. It has suffered from the love of the marvelous, from our credulity, from rash and unwarrantable assumptions, from the play of unbridled fancy."[61] For Conrad, as indeed for most professional geographers of his generation, geography eventually triumphed as a science when it was able to break the enthrallment to fanciful and fabulous speculations about the earth, in order to reveal the real truth about the world in which we live.

Given this hegemony of the real and the visible, how may we attend to those place-making imaginations that are not necessarily rooted in disciplinary geography's normative planetary consciousness that transformed the globe into a disenchanted place over the course of the imperial nineteenth century, that disavowed imagination in favor of empirical reason, and that consolidated (the metropolitan) man as the all-knowing subject and master of all he surveyed?[62] How do we handle those forms of spatial labors in modernity—such as those around Lemuria—which carry traces of the precolonial, the non-modern, and the non-metropolitan? This is also why I deploy the category of the fabulous. I do so to bring back into theoretical play those imaginations about space and place that have been rejected by the sciences of the metropolitan modern as fantastic. I do so to expose the potential of such fantastic geographies—Conrad's Geography Fabulous—to destabilize the knowledge-empires of these sciences, even as I explore the inherent limitations of their transgressions and disruptions. The fabulous geographies of Lemuria lead me to consider a range of sentiments about our earth that have been banished to the margins with the rise of the disenchanted sciences, with their positivist preoccupation with empirical observation, causal reasoning, and objectivist certitude. They compel me to foreground the poetics of creativity and the politics of imagination in place-making, and to come to terms with the undertow of the marvelous, the magical, and the wondrous in attitudes toward space. Not least, they remind me, as I hope they do my reader, that our earth can continue to be an enchanted realm even after being colonized by modern science.

HISTORY AGAINST LOSS

In the chapters that follow, I document the dominant labors of loss around Lemuria as performed by metropolitan paleo-scientists (chapter 2), Euro-American occultists (chapter 3), and devotees of Tamil in colonial and postcolonial India (chapter 4). Chapter 5 explores how the catastrophic place-making of Lemuria results in the production of fabulous geographies of loss, and chapter 6 focuses on what I characterize as cartographic labors of loss, as its place-makers attempt to map Lemuria. I conclude with some reflections on the preoccupation with loss and dispossession in a century that has also been obsessed with presence and possession.

The labors of loss around Lemuria I document here are *off-modern.* Svetlana Boym, from whom I borrow this efficacious notion, writes that off-modern practices are those which take a detour away from deterministic narratives and explore "sideshows and back alleys rather than the straight road of progress."[63] The European off-modern writers and artists she examines concern themselves with the "hybrids of past and present," offering a critique of "both the modern fascination with newness and no less modern reinvention of tradition."[64] As Boym notes, some of the meanings of the adverb *off* include "aside, extending and branching out from, and absent or away from work or duty." Building on her work, I suggest that the off-modern is ostensibly modern, but not wholly in it, or even of it. It starts from the modern, and may even eventually return to it, but by and large it meanders off, moving away from the mainstream to reside in its margins. For the off-modern, the modern is indispensable but inadequate.[65]

Lemuria is an off-modern place-world. A creation of modernity, and of high scientific modernity at that, it has been virtually ignored for much of its existence by the modern, on whose edges it has lurked since it first surfaced in mid-Victorian England. As we shall see, the labors of loss that generate Lemuria deploy all manner of thought and practice with which the modern, especially the scientific modern, has been uncomfortable or even shunned, such as catastrophism and clairvoyance. This was especially true in colonial India, itself on the margins of modernity yet ostensibly of it, where Tamil's devotees consciously and strategically recuperated the archaic in the very name of modern science and the modern nation. As for the men (and the few women) who have labored over Lemuria as its place-makers, even though several of them are prominently, and even quintessentially, modern for their own times and in their own place, their labors around this lost place-world are frequently (although not always) sideshows and off-stage activities, footnotes and even whispers along the way as they go about their everyday pursuits, be they paleo-science, occultism, or Tamil devotion.

This is also what makes their labors of loss *eccentric.* As we shall see from the brief biographies I provide along the way, some of Lemuria's place-makers have been considered eccentrics, and a few even proudly declare themselves as such. While a textbook definition of eccentricity may declare it to be "predominantly inadequate or passive psychopathy,"[66] the only sense in which I use the term is its literal one. The eccentric is that which is not-of-the-center: not centrally placed, remote from or removed from the center, out of the way. As lowly subjects of a vast colonial empire, Tamil's devotees are by definition eccentric, and their labors of loss even more so, as they struggled to make a place for themselves against an imperial center that barely recognized their existence. Even as citizens of independent India, their labors of loss remain eccentric, barely known outside an intimate realm of Tamil speakers (if even there), on the margins of the putative nation. In the metropole's labors of loss as well, Lemuria maintains its eccentricity, whether it be in the place-making of Euro-American occultists who pitched themselves against an institutionalized Christianity, or in the paleo-sciences, where the lost continent is generally confined to a tentative conjecture or a speculative footnote. Rarely, if ever, does it come to occupy the center of any of the sciences that gave rise to it in the first place, or in the occult imaginary where it continues to thrive to this day.

Off-modern and eccentric, place-making around Lemuria is also by implication *oppositional*, although the extent of the subversion varies in intensity across the labors of loss that I focus upon. Surprisingly and paradoxically, it is when it first appeared in the sciences that it was least subversive, and where, if only for a brief moment, Lemuria became an exemplar of established truths about the earth that some really prominent figures promoted, albeit hesitantly and speculatively. But outside the rarified circles of the metropolitan paleo-sciences, Lemuria's off-modern and eccentric presence has always been contradictory, embattled, subaltern, even scandalous. Occultism's labors of loss are pitched against the gathering dominance of Darwinian evolution, while Tamil devotion takes on, if only unsuccessfully, one of the primary weapons with which modern colonialism sought to conquer "the native mind," historicism itself. Across the board, these labors of loss have remained oppositional, even in Tamil India, where they have managed to secure some amount of official blessing, after a fashion.

So, this is what makes Lemuria and the labors of loss that summon it into existence provocative for me, as a historian located in a professional discipline that has hitherto given them scant regard, even spurned the likes of them. These off-modern, eccentric, and oppositional labors compel me to explore the poetics and politics of enchantment, and examine as well the limits of writing about Lemuria from within the confines of a discipline that has been vigorously dedicated to disenchanting itself over the past two centuries in order to establish its credentials as a credible and reputable modern knowledge-form.[67] There are those who will undoubtedly argue that the very consideration I give to such off-modern and eccentric attempts to counter disenchantment is troubling, even dangerous, in an intellectual and political climate where all manner of archaic irrationalisms, benighted religiosities, and accursed traditions, not to mention just plain old-fashioned racism and bigotry, have been barely kept at bay, in spite of scientific and enlightened modernity. Our disenchanted disciplines, they will insist, have been essential to the Enlightenment project of modernity through which knowledges have been demystified and democratized, rational forms of social organization developed, human emancipation furthered, and everyday life enriched. I find this hard to dispute, especially as a beneficiary myself of such blessings. But it is equally impossible to ignore the fact that in this process, as recent scholarship demonstrates persuasively, this very Enlightenment project has also cast out local truths, subjugated knowledges, discarded sensibilities, and destroyed much in the non-European world but also on the margins in the metropole itself, that have not suited its own rationalizing, imperializing, and globalizing agendas.[68]

The spirit in which I query the disenchantment of history is very much in resonance thus with those who ask practitioners of my discipline to take heed of the heterogeneities, pluralities, and subalternities that have resisted or survived the onslaught of universalizing Enlightenment. In order to understand the preoccupation with lost worlds and lost times, the historical sensibility that I perforce adopt is best described as counter-disenchanted. I see such a sensibility as inherently off-modern in its focus on the sideshows, the crevices, and the back alleys rather than on the main highway of historical development. Off-modern, it seeks to go beyond a linear, deterministic, and progressive narrative that is always on stage—history-on-tap, in a manner of speaking—where there are few surprises, nasty or otherwise, where there is little left to wonder about, much less to bring us up short. I also conceive of such a sensibility as eccentric in that it consciously situates itself in the peripheries and provinces, away from power centers, conspicuously undermining the status quo, the normative, and the normalizing. Such a sensibility is as well oppositional, in that it deliberately locates itself among the scandalous presences and subjugated knowledges that reveal the operations of domination and mastery. Not least, in querying the disenchantment of history in order to explain the fascination with the likes of Lemuria, I take cognizance of loss as a category of knowledge, as a discursive formation, and as a condition of being in whose shadow we have always already practiced this disenchanted discipline.

Chapter 2

Science in the Service of Loss

It is not too much to say that every spot which is now dry land has been sea at some former period, and every part of the space now covered by the deepest ocean has been land.[1]

THE LOST WORLDS OF DEEP TIME

Lemuria first surfaced to visibility in the by-lanes of Victorian science, but the foundations for the metropolitan fascination with Earth's lost worlds and vanished pasts were laid in the closing decades of the eighteenth century with two important developments. The first of these was the discovery of "deep time" in the 1780s.[2] Up until then, most scientists and educated opinion considered the earth to be about 6,000 years old. Yet this reckoning, based on Biblical chronology, was soon at odds with the nascent science of geology, which was fast revealing that the earth's surface had undergone vast transformations at a rate that could not be accommodated within such a short time span. Beginning with the Comte de Buffon, who estimated the age of the world to be around 75,000 years in 1774, many scientists progressively jettisoned the Christian calendar in favor of a new secular chronology in which the birth of Earth as a functioning planet was pushed further and further back in time. In Robert Wood's estimation, "to join battle with the 'prejudice of human time' (i.e., to accommodate all past times to the scale provided by human memory) was to prove the great crusade of the heroic age of Geology."[3] By the opening years of the nineteenth century, the limits of humanly remembered time had been blasted. The bottom had dropped out of a hitherto finite earth history, opening up a deep (and to some, a dark) abyss, waiting to be filled by human imagination.

As important was the realization that in its journey through deep time, the earth had undergone massive changes that had radically altered its surface. Not only was it not created at one stroke nor was it heading toward an inevitable doomsday, as orthodox Christian theology would have it, but it had also not remained static. Instead, "from the top of the mountain to the shore of the sea . . . everything is in a state of change," wrote James Hutton in his revolutionary *Theory of the Earth* (1788).[4] In phrases he made famous, Hutton wrote evocatively of "a succession of worlds" that had followed one after another, so that "we find no vestige of a beginning, no prospect of an end."[5] As a consequence of the ceaseless rhythm of erosion and sedimentation, continents were slowly worn down and, sinking to the bottom of the ocean, were raised again as new continents. The earth is therefore a self-renewing creation of successive worlds, surfacing, disappearing, and re-emerging. As Loren Eiseley observes, "In this eternal hurrying of particles across the surface of the land, in the dissolution of previous continents with all their varied life, there emerges once more into Western thought the long shadow of illimitable time."[6] From the early years of the new nineteenth century, William "Strata" Smith's geological maps, soon to be followed by others, visually captured these former worlds as the science of cartography helped reveal, layer by layer, Earth's passage through illimitable time.[7]

It was not just continents that appeared and disappeared over the *longue durée* of deep time, but all manner of living beings as well. As Eiseley writes, from the closing years of the eighteenth century there was growing awareness that "the past life of the earth . . . might offer marvels no living eye had beheld. . . . An anonymous contemporary writer spoke [in 1812] in an awed tone of perished species and the mystery of how new species originated."[8] The term "lost species," which had been originally formulated in the sixteenth century, found a renewed life in the latter half of the eighteenth and the opening decades of the nineteenth centuries, as natural historians, paleontologists, and fossil hunters became preoccupied with "vanished beasts" who had roamed the earth in "former ages."[9] In French paleontologist Georges Cuvier's influential theory of catastrophism, "life in those times was often disturbed" as a consequence of periodic oceanic floods and the subsidence of lands which punctuated earth's passage through deep time: "Numberless living things were victims of such catastrophes: some, inhabitants of the dry land, were engulfed in deluges; others, living in the heart of the seas, were left stranded when the ocean floor was suddenly raised up again; and whole races were destroyed forever, leaving only a few relics which the naturalist can scarcely recognize."[10]

As Cuvier's catastrophism gained a following, especially in continental Europe, the creation and extinction of whole species in former ages came to be diligently documented, so that, as Eiseley writes, "between the achievements of Smith and Cuvier, the public had finally become excited and convinced that a past world existed."[11] By the middle decades of the nineteenth century, therefore, the ground had been cleared for Charles Darwin and his "fevered search"—and that of the sciences of paleontology, prehistory, and ethnology—for "missing links" and "extinct forms."[12]

Deep time, a succession of worlds in former ages, extinct species, missing links: these, then, were the principal ingredients of the labors of loss in which paleo-scientists participated over the course of the nineteenth century. Their disciplines—natural history, geology, paleogeography, prehistory, and ethnology—as these were then practiced, were premised on a sense of loss, acutely concerned as these were with recovering and restoring, if only to human knowledge, that which had vanished into deep time. These sciences were also underwritten by the hubris of a positivist paradigm which assumed that global and complete knowledge is possible and attainable. As Thomas Richards suggests, the imperial nineteenth century witnessed, at least in Britain, the merger of the Romantic project of comprehensive knowledge with the Victorian project of positive knowledge.[13] Even as European colonial rule claimed the entire planet, numerous exotic species—floral, faunal, and human—became available to the metropolitan scientist, enriching his laboratories and databases. The whole became progressively attainable to the scientist and the scholar, "collectively imagining a not-too-distant future when all species would be identified, all languages translated, all books catalogued."[14] That which had been deemed disappeared and lost could—and would—be recovered through modernity's all-reaching knowledge practices.

Not surprisingly, it is among the proud practitioners of these metropolitan sciences that place-making around lost worlds like Lemuria first began, moved as they were by the conviction that it was indeed possible to ascertain what the earth had been like in its former ages. Geology, paleogeography, archaeology, paleontology, prehistory, ethnology, and the nascent field of oceanography added an immense, even unfathomable, depth to humanly imaginable time, even as they speculated about a succession of former worlds hitherto unknown to moderns: worlds that might have been catastrophically destroyed by earth movements and ocean surges; worlds populated by such mysterious creatures as the Megalosaurus and the Pterodactylus, Neanderthal Man and *Pithecanthropus erectus*, now long dead and gone; worlds that clearly demonstrated that the earth as we know it today, with its current configuration of continents, oceans, and life-forms is but one episode in the vast panorama of planetary history.

THE ECCENTRIC BIRTH OF LEMURIA

In contrast to Atlantis, whose lofty origins lie in the influential reflections of Plato in the mid-fourth century B.C.E., Lemuria had a relatively obscure and humble birth in 1864 in a short essay called "The Mammals of Madagascar" that appeared in a brand-new periodical published out of London called *The Quarterly Journal of Science.* Its progenitor was the reputable English zoologist Philip Lutley Sclater (1829–1913), who just a few years before, in 1858, had published what would become an influential template for the worldwide geographical distribution of fauna.[15] In the course of his biogeographical work, Sclater had become intrigued by the zoological makeup of the island of Madagascar, which appeared to lack in most of the terrestrial life-forms that characterize the neighboring continent of Africa.[16] In particular, he was taken with the anomaly posed by the only primate to be found in the island, the lemur.[17] He observed that "while 30 different species of Lemurs are found in Madagascar alone, all of Africa contains some 11 or 12, while the Indian region has only 3." He sought to explain this distribution pattern by insisting that "some land-connection must have existed in former ages between Madagascar and India, whereon the original stock, whence the present Lemuridae of Africa, Madagascar, and India, are descended, flourished."[18] Because the Lemuridae are closer to American monkeys than to the simians of the Old World, there might also have been a terrestrial connection between Madagascar and the lands that today make up the New World. On the basis of these zoogeographical propositions, he concluded his brief essay thus:

> The anomalies of the Mammal fauna of Madagascar can best be explained by supposing that anterior to the existence of Africa in its present shape, a large continent occupied parts of the Atlantic and Indian Oceans stretching out towards (what is now) America to the west, and to India and its islands on the east; that this continent was broken up into islands, of which some have became amalgamated with the present continent of Africa, and some, possibly, with what is now Asia; and that in Madagascar and the Mascarene Islands we have existing relics of this great continent, for which as the original focus of the "*Stirps Lemurum,*" I should propose the name Lemuria![19]

By the mid-1860s, when "The Mammals of Madagascar" first appeared, there had been other scientists who had already invoked a submerged continent in the Indian Ocean to account for the geographical distribution of specific fauna (although Sclater himself did not specifically refer to these prior studies). Thus, in a paper that was read before the prestigious Linnean Society in London in November 1859, Alfred R. Wallace (1823–1913), who later became an acerbic critic of sunken continents as an explanatory device in zoogeography, had offered a similar explanation of the anomalous fauna of the Celebes, for "facts such as these can only be explained by a bold acceptance of vast changes in the surface of the earth."[20] Similarly, another of Wallace's contemporaries, the English geologist Searles V. Wood (1830–84) had also hypothesized, on the strength of growing geological and biological evidence, the existence of a giant southern continent during what we today call the Mesozoic era.[21] Well before these English scientists, as the German ornithologist Gustav Hartlaub reminded his colleagues in 1877, the French natural historian Etienne Geoffrey Saint-Hilaire had speculated in the 1840s that Madagascar was part of a fourth continent in the Indian Ocean which future researches would confirm was, "as regards its fauna, much more different from Africa, which lies so near to it, than from India, which is so far away."[22]

Thus, Sclater's originality did not lie in summoning into existence a Mesozoic southern continent now drowned in the waters of the Indian Ocean, but in christening it "Lemuria." Naming, Ian Barrow insists, "is a parental prerogative, conveying both a sense of ownership and an acknowledgement of responsibility. . . . The namer asserts control . . . and privileges. . . . More than any other activity, naming gives voice to vision; a name encapsulates our understanding of what we see."[23] Similarly, Paul Carter writes, "It [is] through the act of naming that a space [is] delineated as having a character, as something that could be referred to."[24] Indeed, by the very act of conferring a name on it, Sclater's essay may have assured the putative land connection between Africa and India a future life and biography that other paleo-territories, left unnamed, have lost out on. Naming it was thus crucial to the spatial constitution of Lemuria as an object of knowledge, for to name a space is to turn it into a negotiable place.[25]

Yet, having named Lemuria with such casual elan, Sclater himself did not do anything more with this now-vanished place-world to which his labors of loss had given birth, referring to it merely in passing in a couple more of his scholarly essays.[26] But others in the metropolitan scientific community soon stepped into the arena and, as we will see, Sclater's drowned place-world surfaces every now and then well into the 1960s, sometimes in passing, at other times more prominently, to exemplify the truths of Earth as it once was—an ever-shrinking planet scoured by submerged landbridges, sunken continents, and vanished territorial connections.

Indeed, it is this that accounts for the very presence of Lemuria, however precariously and fleetingly, in Euro-American paleo-scientific place-making in the latter nineteenth and early decades of the twentieth centuries, for it had a role to play, albeit eccentrically, in the two major intellectual formations that dominated the study of the earth in these times. These were natural history, or more specifically, biogeography, and geology, especially paleogeography, both of which really came into their own in these years as imperial and global knowledge-systems dedicated to the new science of the earth. Both biogeography and paleogeography thrived with the increasing flow of geological, paleontological, and biological data into metropolitan scholarly societies from far corners of the world as European scientists seized the economic and institutional opportunities provided by the expansion of colonial power. Both sciences were interested in the ancient distribution of land and water, the former in order to explain the geographical dispersal of life forms, past and present, and the latter, to uncover the buried history of the earth. And both sought connections between parts of the world that may today be separated and distinctive but perhaps not so in the remote past, for the distribution of life-forms as revealed by fossil remains and living species was uncovering a geography of Earth quite at odds with that which appeared on modern maps.[27]

Through the middle of the twentieth century, most paleo-scientists were steadfastly committed to the idea that the earth's continents and oceans had always occupied the place that they currently do on its surface. This "fixist" idea was to be eventually jettisoned after the 1960s, with the acceptance of a "mobilist" theory of the earth as a planet whose surface was constituted by moving plates and drifting continents. But until that happened, scientists on either side of the Atlantic subscribed to one of two doctrines: that the continents had not fundamentally changed in their contours (the view of the so-called "permanentists"); or that they—or parts thereof—periodically sank to the bottom of the ocean only to re-emerge again as a consequence of the contraction process to which the earth, formerly a fiery hot sphere, was subjected as it cooled over time (the position of the "contractionists"). In turn, contractionists fell into two camps: "gradualists," generally belonging to the English-speaking world of Charles Lyell and his theory of uniformitarianism, who envisaged the process of submergence and reappearance of land as gradual, ceaseless, and routine; and "catastrophists," largely European, who subscribed to the notion that the history of the cooling earth was marked by periodic cataclysmic disturbances followed by quiet readjustments.[28] Submarine continents and drowned landbridges, therefore, were staples of geological, and particularly biogeographical, thought among contractionists, and especially of the catastrophists among them. Through the early decades of the twentieth century, these included a large cross-section of the European paleo-scientific community, who rejected the permanentist position first articulated in 1846 by the American geologist James Dana that "once a continent, always a continent; once an ocean, always an ocean."[29] They instead believed, with Charles Lyell, that "every spot which is now dry land has been sea at some former period, and every part of the space now covered by the deepest ocean has been land."[30]

In fact, well before Sclater's labors of loss had produced Lemuria, other landbridges and continental connections had been summoned into existence in zoogeography, primarily to account for the terrestrial dispersal of animals.[31] As one advocate of such former connections wrote in 1925, "It is *impossible* in zoogeography to arrive at an acceptable explanation of the distribution of animals if no connections between today's separate continents are assumed to have existed."[32] Thus, an ancient Atlantic continent was invoked by the English biologist Edward Forbes in 1846 to explain the faunal similarities between the Azores, Madeira, and the Canary Islands and those of North America and Europe. In 1853 Dalton Hooker similarly posited a land connection between New Zealand, Tasmania, and South America.[33] And the Scottish naturalist Andrew Murray wrote eloquently in 1866 of a "Miocene Atlantis" linking Europe and America "where the North

Atlantic now rolls."[34] So much so that Charles Darwin—who himself in his earlier days had not been averse to submerged land connections—complained in 1856 to his colleague and friend Charles Lyell that the latter's followers were with great abandon raising *and* sinking whole continents:

> Here poor Forbes made a continent to North America and another (or the same) to the Gulf weed; Hooker makes one from New Zealand to South America and round the world to Kerguelen Land. Here is Wollaston speaking of Madeira and Porto Santo "as the sure and certain witnesses of a former continent." . . . And all this within the existence of recent species! If you do not stop this, if there be a lower region for the punishment of geologists, I believe, my great master, you will go there.[35]

It is in such an intellectual environment, preoccupied with the mysterious submarine and the catastrophically drowned, that the birth of Lemuria in 1864 has to be placed. Recently, Peter Tyson has remarked, in commenting on Sclater's founding proposition, that "In the absence of a better idea, Lemuria took hold in the scientific imagination. Coming sixty years before Alfred Wegener proposed the theory of continental drift, and perhaps a century before the scientific community began to accept it, the idea of a conveniently placed continent that then conveniently vanished without a trace seemed a plausible if hardly defensible explanation."[36] I would not agree with this hasty assessment which seems to suggest that Sclater's vanished place-world was a gross, albeit convenient, aberration in metropolitan thought. Although it carries a singular name, Lemuria belongs to a family of other similar entities that had been summoned into existence by the various paleo-sciences and hence is not unique from this perspective. What makes its fortunes unusual is what happens to it *outside* the realms of metropolitan science, from where it quietly disappeared once the fixist, contractionist, and catastrophic theories that sustained its birth went out of fashion. With their demise, as we will see, it too vanishes from the labors of loss of the paleo-scientist.

LEMURIA AS A FAUNAL HIGHWAY

From the mid-1860s, with the continued expansion of geology and natural history in both British India and southern Africa, a growing number of scientists based in the metropole, as well as reporting in from the periphery, wrote in favor of a former landmass in the Indian Ocean in the Mesozoic era. Few explicitly called it Lemuria though, preferring alternative names such as "Africano-Indian continent,"[37] "Indo-Oceania,"[38] "Indo-Madagascar peninsula,"[39] or "Indo-African Continent."[40] It also surfaced in discussions in metropolitan learned societies like the British Association of the Advancement of Science, and the Royal Geographical Society.[41] Its Mesozoic existence was reaffirmed in at least two presidential addresses of the influential Geological

Society of London, in 1870 and in 1890, the former by a biologist of no less a stature than Thomas Huxley (1825–95),[42] the latter by the colonial geologist William T. Blanford (1832–1905), who offered a spirited defense of submerged continental connections in light of mounting criticism by permanentists.[43] In 1917 it also came under extensive scrutiny by the German paleogeographer Theodor Arldt (1878–19??), who after discussing the various theories in support of and against its existence, lent it his learned assent.[44] As I discuss later, many metropolitan scientists even cartographically visualized its contours well into the 1960s, even as the paleogeographical tide turned against drowned continents and submerged landbridges with the growing acceptance of plate tectonics theory in the earth sciences.

Of course, there was considerable disagreement over Lemuria's extent, the duration of its existence, and the role it had played in the paleo-world. There was also little consensus on whether it had been a "landbridge," a whole "continent," or merely a territorial extension of other continents. For most geologists, it was a Mesozoic land connection having little to do with the dispersal of birds and especially of higher mammals.[45] For zoologists like Edward Blyth and Gustav Hartlaub, following in Sclater's footsteps, it was a distinctive faunal region, which they named "Lemurian," populated by its own eccentric set of mammals and birds common to Madagascar and India. As such, it had survived at least into the early Tertiary period.[46] In 1870, as we will shortly see, the biologist Ernst Haeckel even declared it "the probable cradle of the human race, which in all likelihood here first developed out of anthropoid apes."[47] Such differences notwithstanding, the accumulating floral, faunal, geological, and paleontological evidence was interpreted as pointing to some kind of former land connection between Africa and India. And there was growing consensus about its progressive fragmentation and eventual submergence in the Indian Ocean sometime in the early Tertiary.[48]

All the same, almost from the start, these paleogeographical labors of loss around Lemuria were undermined on many fronts. First, adding to the lack of consensus over its name—which took away from its integrity as a singular entity—was a growing sense that Lemuria might have only been a part of a larger continent that had spread over the southern hemisphere. Sclater's original formulation itself hinted at this, and in a later report to the British Association for the Advancement of Science, in 1875, he limited the term "Lemurian" to Madagascar and its adjacent islands, instead of using it—as some of its more enthusiastic advocates were wont to do—for a continental bridge that extended all the way to India and perhaps beyond.[49] Similarly, in 1887, the German paleontologist Melchior Neumayr (1845–90) used fossil correlation to envision a giant Brazilian–Ethiopian continent of which the "Indo-Madagascar Peninsula" was a mere appendage (Fig. 1).[50] But the real nominal as well as conceptual subordination of Lemuria had already happened a couple years earlier, in 1885, in the Austrian Eduard Suess's

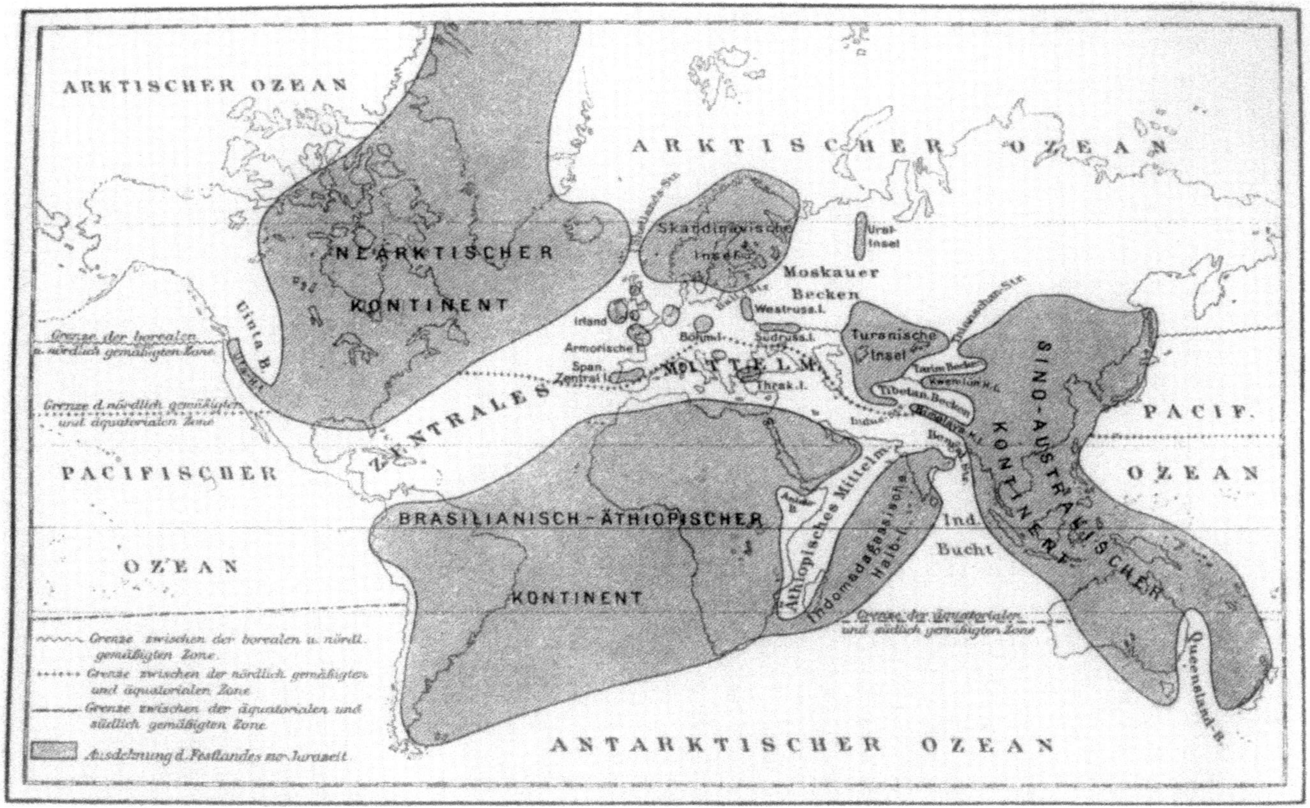

Figure 1. Melchior Neumayr, Map of the geographical distribution of the Jurassic seas. From Neumayr, *Erdgeschichte*, vol. 2.

enormously influential *Das Antlitz der Erde (The Face of the Earth),* which introduced "Gondwanaland" into the lexicon of paleogeography.[51]

A mega-continent which had once occupied almost the entirety of the southern hemisphere, Gondwanaland was christened as such by Suess "after the ancient Gondwana flora which is common to all its parts*; it corresponds to a large extent with the Lemuria of zoologists.*" [52] At one stroke, Suess's rechristening overshadowed Sclater's earlier act of naming, since Gondwanaland soon displaced Lemuria as the object of paleogeographical labors of loss in the metropole.[53] Although Lemuria continues to crop up every now and again over the next half century in paleogeographical place-making, both the name and the land that it designated were increasingly subordinated to the super-continent Gondwanaland, which was relegitimized in the twentieth century by first, continental displacement theory, and then later, by plate tectonics, both of which have had no place for Sclater's vanished place-world. The concept of Gondwanaland was increasingly attractive to scientists, who sought to account for the similarity in floral and geological remains not just between India and Africa but also in what is now Australia and South America, and perhaps even Antarctica. In the growing tendency from the late 1850s toward imagining into existence large supercontinents,[54] a relatively minor landbridge like Lemuria was lost in the shuffle, limited as it was both by name and association with just the lemur and its dispersal.

But even prior to Suess's rechristening, Lemuria's status as a faunal bridge had been challenged in the influential writings of the naturalist Alfred Russel Wallace.[55] As I noted earlier, in the 1850s Wallace had been among the first to propose a former Indian Ocean continental mass that had subsequently subsided, and he had been an advocate of other such landbridges which had been invoked in zoogeography. By the mid-1860s, however, with the ascendancy of permanentism among English paleoscientists, he began changing his stance, proposing instead that "land connections could be inferred only in special instances where the geological evidence, as well as distributional data, was overwhelming."[56] Increasingly for Wallace, Lemuria was not one of these instances. In his *The Geographical Distribution of Animals* (1876), which soon became the standard textbook of zoogeography in the metropole, he cautiously described Sclater's Lemuria as "a supposed submerged continent extending from Madagascar to Ceylon and Sumatra, in which the Lemuroid type of animals was developed. This is undoubtedly a legitimate and highly probable supposition, and it is an example of the way in which a study of the geographical distribution of animals may enable us to reconstruct the geography of a bygone age."[57]

At the same time, he was reluctant to go along with colleagues like Edward Blyth who wanted to transform Sclater's Lemuria into a distinctive zoological region, nor was he entirely sure of either its extent or of its function as the terrestrial means through which the lemur had spread.[58] Soon after, in a much-quoted essay published in the *Proceedings of the Royal Geographical Society,* he insisted that the lemur had had a much wider range in the Eocene than its contemporary distribution showed and that it could have migrated to its present habitats from Europe. There was no reason to invoke a Lemuria to account for the geography of the lemur. "Lemuria, therefore, may be discarded as one of those temporary hypotheses which are useful for drawing attention to a group of anomalous facts, but which fuller knowledge shows to be unnecessary."[59]

His growing skepticism bloomed into outright hostility by 1880, when he published his magnum opus *Island Life,* which denounced the ongoing labors of loss around former landbridges as paleogeographically unsound, for the "testimony of geology . . . upholds the . . . theory of the stability of our continents and the permanence of our oceans. . . . Yet so easy and pleasant is it to speculate on former changes of land and sea with which to cut the gordian knot offered by anomalies of distribution, that we still continually meet with suggestions of former continents, stretching in every direction across the deepest oceans."[60]

Given this *volte-face,* it is not surprising that Wallace now had this to say about Lemuria: "The supposed Lemurian continent is constantly referred to by quasi-scientific writers, as well as by naturalists and geologists, as if its existence had been demonstrated by facts, or as if it were absolutely necessary to postulate such a land in order to account for the entire species of phenomena connected with the Madagascar fauna, and especially with the distribution of the Lemuridae."[61]

If "the supposed Lemuria" had existed at all, this was in so remote a period that higher animals had not appeared on earth, by which time it had disappeared. It certainly could not be invoked for solving the problem of geographical distribution "any more than the hypothesis of an Atlantis solved the problems presented by the Atlantic Islands and the relations of the European and North American flora and fauna." Yet, "the Atlantis [*sic*] is now rarely introduced seriously except by the absolutely unscientific. . . . But 'Lemuria' still keeps its place—a good example of the survival of a provisional hypothesis which offers what seems an easy solution of a difficult problem, and has received an appropriate and easily remembered name, long after it has been proved to be untenable."[62]

Wallace's is perhaps the most well-articulated rejection of paleogeographic labors of loss around a submerged Lemuria from the permanentist's position of "once a continent, always a continent; once an ocean, always an ocean," to which he had been gradually converted through the 1870s. Like other permanentists, all that Wallace was now willing to concede was that while minor elevations and subsidence happened along their coastal margins, it was impossible that whole continents would drown. Continents were always already stable, permanently occupying the place that they currently do on the surface of Earth in their current form. Although they could grow by accretion along their edges, the old cores were stable and permanent. Flooding might happen, but this did not lead to submergence, as the contractionist would have it. The permanentist's cause had received a boost by the 1870s with the discovery of the principle of isostasy, which showed that it was impossible for the lighter continent to sink into the denser sea floor. Subscribers to this tectonic principle insisted that because continental platforms were in hydrostatic equilibrium with ocean basins, this implied continental permanency. It was as impossible for the lighter continents to vanish into a denser sea floor as it would be for a piece of wood or block of ice to sink to the bottom of a pool of water.[63]

Although isostasy was recognized by the early years of the twentieth century as "the most fundamental of all the laws to apply to the earth,"[64] the theory of continental permanence ran into trouble because it could not solve the pressing problem of the geographical distribution of animals the way landbridges so conveniently did. Because permanentists could not effectively account for similarity in life-forms across widely separated lands, many biologists and other contractionists therefore continued to invoke drowned territorial connections as the best explanation possible for the earth's zoological evidence, isostasy notwithstanding.[65] So much so that as late as 1925—three-quarters of a century after the permanentist challenge had been launched—the German geologist Carl Diener insisted that "a dry-land connection between the Indian peninsula and southern Africa via Madagascar is an inescapable feature of the Permian and Triassic periods on zoogeographical grounds."[66]

Amid these ongoing debates over submerged continents and sunken landbridges between biologists and geophysicists, contractionists and permanentists, the German meteorologist Alfred L. Wegener (1880–1930) proposed his theory of continental displacement, which also threatened the truth of a drowned Lemuria.[67] Articulated first in 1912–15, elaborated in 1919 and again in 1922, and finessed in 1929, the theory was premised on the assumption that in the late Paleozoic era, all landmasses had been fused together into the super-continent of Pangaea, which began to fragment about two hundred million years ago. At that time, in the southern hemisphere, "Antarctica, Australia and India lay adjoining South Africa, and with the latter and South America formed, until the beginning of the Jurassic period, a single large—even if partly submerged at times by shallow water—continental area, which in the course of Jurassic, Cretaceous and Tertiary time split and crumbled into small blocks which drifted away from each other in all directions" (see Fig. 2).[68] While South America and Africa drifted apart, the fate of India was different:

> It was originally connected by a long continental tract, mostly, it is true, covered by shallow seas, to the Asiatic continent. After the separation of India from Australia on one side (in the Lower Jurassic) and from Madagascar on

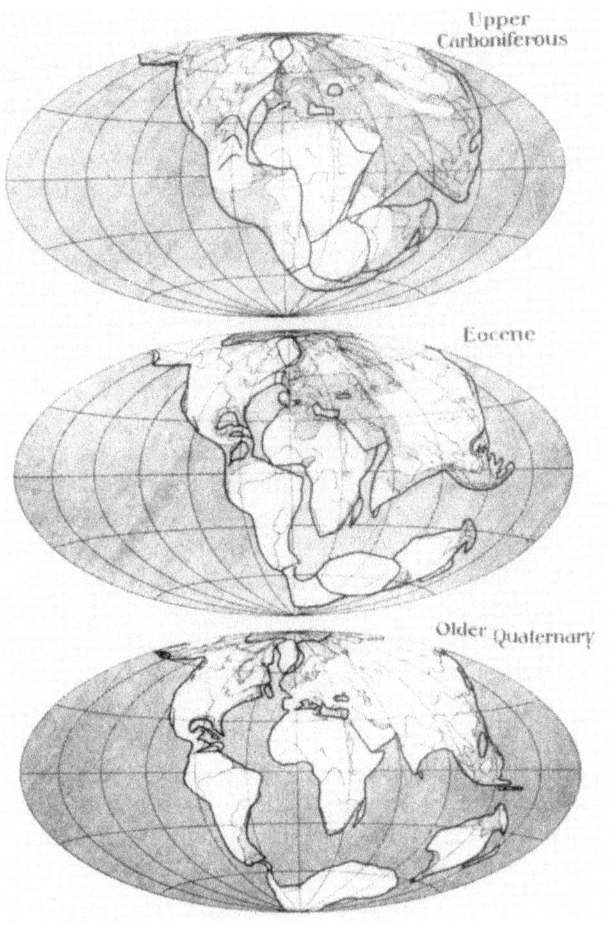

Figure 2. Alfred Wegener, Reconstruction of the map of the world for three periods according to the displacement theory. From Wegener, *The Origin of Continents and Oceans.*

> the other (during the transition from the Chalk [Cretaceous] to Tertiary), this long connecting portion was more and more folded together through the continuous gradual approach of India to Asia and constitutes today the mightiest mountain folds of the earth, the Himalayas and the numerous folded ranges of the high lands of Asia.[69]

Wegener called this entire process "The Lemurian Compression" and included an accompanying illustration.[70] As a result of this compression, "we obtain a displacement of India of about 3000 km. India must therefore have lain near Madagascar before the thrusting began. *No room remains for a sub-*

merged Lemuria, in the older sense."[71] As Ursula Marvin observes, "The compression cleared the Indian Ocean of land without recourse to a sunken continent of Lemuria."[72]

When Wegener invokes Lemuria in his place-making (which is not often), there is considerable ambiguity regarding its status. In one instance, it appears to be Sclater's Lemuria, a continental connection linking India and Madagascar; in another, it is a "long connecting piece of land," occasionally covered with shallow seas that linked India with Asia, that eventually gets compressed to form the Himalayas. In yet another instance, "Lemuria" is basically synonymous with "India," whose configuration itself changed through the various editions of his book.[73] These ambiguities notwithstanding, it is important that Wegener's displacement theory uses Lemuria, but puts it to very different purposes than that intended for it by Sclater and other zoogeographers. Rather than positing that Lemuria was flooded out of existence through submergence, Wegener compresses it into oblivion, as the super-continent of which it formed a part was dismembered and the resulting fragments drifted apart. "Instead of the subsidence to more than 4 km depth . . . which is impossible isostatically over such an area, we believe that this bridge was compressed to form upland Asia."[74] In Wegener's theory of drifting continents, there was simply no place for drowning ones.

For Wegener's supporters, the advantages of continental displacement theory were many, not least of which was the fact that it could account for the geographical dispersal of fauna without having to invent, willy-nilly, continental connections stretching across thousands of miles of ocean that subsequently drowned, defying the principle of isostasy. In the words of the South African geologist Alexander du Toit (1878–1948), an ardent advocate of drift theory at a time when it had few:

> Biologists have . . . been demanding an inordinate number of such links; indeed, were all the deduced connections to be plotted on a globe, they would recall the criss-crossing system of deep-sea cables and bear as little relation to one another in time and place; few parts of the oceans would not at some period have been thus spanned. Such a haphazard network, being palpably absurd, indicates that current ideas of continental connection are *fundamentally unsound* and so induces us to accept the principle of Drift. This in turn wipes out straightaway all those mythical lands, such as Atlantis, Archhelenis, Archipacificis, Archinotis, Lemuria, Tasmantis, Flabellites Land, etc., that have been conjured up to explain the life-distribution of the earth.[75]

Wegener's continental displacement theory was for many decades ferociously resisted on both sides of the Atlantic by a paleo-scientific community that was steadfastly committed to a fixist conception of our planet which assumed that the continents are essentially static and cannot move laterally on the surface of the earth. So much so that, faced with the "fantasy" of moving continents, even permanentists were willing to turn to landbridges to account for the geographical distribution of animals. As Robert Wood wryly observes, "For most geologists it was patently simpler to sink a landbridge 4 kilometers than to move a continent 4000 kilometers."[76] Consequently, in the middle decades of the twentieth century, in the course of opposing Wegener's theory of continental displacement, there was a renewed revival of interest in landbridges like Lemuria. In Wood's words, "The most economical solution to the problem of reconciling past and modern zoogeographies came from the construction of landbridges. They were substantial enough to explain all the biological enigmas, and as slim and ephemeral as was required to avoid too overt a contradiction of any supposed physical law. . . . Landbridges were to provide the geologist's chief alternative to drifting continents."[77]

Contrary to du Toit's hope, therefore, Lemuria did not altogether disappear with Wegener's compression and instead was reincarnated as an "isthmian link" in an essay published by the American geologist Bailey Willis (1857–1949) in 1932.[78] In this influential scheme, which held its own, at least in American geology, until the 1960s, Willis did not name the landmass linking India, Madagascar, and Africa specifically as "Lemuria," but nevertheless Sclater's old place-world did reappear, if only as a temporary "Africa–India" ridge, to account for the otherwise inexplicable features of Madagascarian biogeography.[79] As Ursula Marvin notes, "To many scientists Willis' isthmian links were the perfect compromise between the doctrine of permanence and the Suessian idea of submerged continents. While conforming to the requirements of isostasy, the isthmian links allowed for the migrations of land and shallow marine flora and fauna . . . but they were too small to upset the water balance in the oceans. Above all, their existence was reversible—they were easily built and destroyed."[80]

Wegener and Willis were the last big theorists in the metropole in whose place-making Lemuria warrants discussion as an Indo-African landmass now submerged in the Indian Ocean.[81] From the 1970s, as new theories of seafloor spreading and plate tectonics came to dominate paleogeographic place-making, it virtually disappears in the growing preoccupation with moving crustal plates and spreading ocean floors, not even meriting an occasional footnote.[82] To paraphrase Wegener, there is little room in these influential place-making activities for Lemuria in *any* sense of the term.

Scholarship today on nineteenth-century geology, zoogeography, or natural history only occasionally invokes Sclater's lost continent in passing, sometimes as a curiosity item or as yet another illustration of the quixotic Victorian mind.[83] Thus, in an analysis of the famous Wallace Line and its author, published a few years ago, Penny van Oosterzee writes:

> According to Lyell and everyone before him, the continents moved—but only up and down. Using this worldview to explain the many faunal similarities be-

> tween distant lands, whole continents had to rise from the ocean depths, forming faunal highways. In this way the romantically named "Lemuria" rose, streaming from deep below the Indian Ocean, like some ancient Atlantis, to levitate lemurs from Madagascar to India and hence to the Malay Archipelago.[84]

Or here is Gabriel Gohau in a recent history of geology:

> Early twentieth-century authors believed the existence of former landbridges between present-day continents explained fauna and flora from one continent to another. Numerous imaginary bridges were proposed: Africa and Brazil were connected by the bridge called Archhelenis, and Europe and North America by Archatlantide. Smaller landbridges across the Indian Ocean joined Madagascar to India.[85]

Such scattered comments aside, it would be difficult to find Lemuria tucked away in even a footnote, its virtual absence in today's scholarship at odds with the tantalizing presence it did have in paleogeographical place-making, albeit eccentrically, for close to a century from the 1860s. And its amazing off-modern adventures in contexts and in places far outside the hallowed pages of metropolitan science have scarcely been noted.

Yet, the foundations for these adventures were surely laid by the labors of loss of the late-nineteenth- and early-twentieth-century paleogeography. First and very importantly for its future career, especially outside its metropolitan habitat, Lemuria was imagined by paleobiologists and geologists as a "continent," even though there were many who also refer to it as a "connection," a "bridge," or an "isthmus." True, there was no consensus, as we have seen, on its size or reach. For some, it was limited to the islands around Madagascar; for a majority, it extended across the Indian Ocean to India; and for still others, it spread out into both the Atlantic and Pacific. Notwithstanding this, the fact that it was granted a continental status from the start is crucial, for continents have been the basic territorial units of knowledge-production for everything from earth history to the evolution of life and the spread of "civilization," as Martin Lewis and Karen Wigen have persuasively demonstrated.[86]

Also, paleogeographical place-making stands out among others in that it presents Lemuria as a primeval land that flourished and disappeared far prior to the appearance of humans on earth. To be sure, there are differences of opinion regarding when Lemuria finally subsided. For most geologists, it was a Mesozoic landmass, submerging in the Indian Ocean sometime at the dawn of the Tertiary period.[87] For the biologically inclined, concerned as they were with accounting for the dispersal of the lemur and other mammals, it lasted well into the Tertiary, disappearing sometime in the Eocene epoch, or perhaps even as late as the Miocene.[88] But absolutely and clearly, this was not a continent inhabited by man, even in the latter's most primeval form. Not surprisingly, these labors of loss are free of any regret, grief, or nostalgia over the disappearance of this place-world, although a fascination with its very existence, and its subsequent vanishing, is clearly evident.

Perhaps most saliently, of all the vanished place-worlds generated by Euro-American paleogeography, Lemuria has had—in spite of its fleeting, episodic, and marginal presence—the most interesting and varied career *outside* the rarified realm of metropolitan science. As I have noted, there were other, similar faunal highways that were summoned into existence, such as Archhelenis linking Brazil and Africa, Archigalenis across the Pacific linking Asia and the Americas, or Hipparion connecting Florida to Europe. Yet it is telling that of all these former place-worlds, Lemuria alone has spawned a wide array of labors of loss, spanning a whole spectrum of imaginative universes ranging from the mundane to the magical. In this regard, it rivals Atlantis, that paradigmatic lost continent of them all. But unlike Atlantis, which rose in the prescientific and premodern imagination, Lemuria first surfaced in modern science, as a possible solution to one of the leading biogeographical problems of the nineteenth and early twentieth centuries. Why it alone of similar place-worlds should have had the kind of career that it has had is, of course, one of the main concerns of this study, and I begin to address this question by turning to the ethnological imaginary of the metropole, which, too, performed its own labors of loss around Sclater's Lemuria.

LEMURIA AS THE CRADLE OF MANKIND

> The Indian Ocean formed a continent which extended from the Sunda Islands along the southern coast of Asia to the east coast of Africa. This large continent of former times Sclater, an Englishman, has called *Lemuria*, from the monkey-like animals which inhabited it, and it is at the same time of great importance from [*sic*] being the probable cradle of the human race, which in all likelihood here first developed out of anthropoid apes.[89]

So wrote the German biologist Ernst Haeckel (1834–1919) in 1870 in the second edition of his best-selling *Natürliche Schöpfungsgeschichte,* translated into English in 1876 as *The History of Creation.*[90] Haeckel's identification of Lemuria as "the probable cradle of the human race" distinguishes his labors of loss from the place-making of natural historians, biologists, and geologists. In so identifying it, Haeckel wrests Lemuria from the world of zoogeography, where it had circulated as a faunal highway and paleocontinental connection, and inserts it instead into the all-important grand narrative of the primeval history of man that so many were attempting in the second half of the nineteenth century. At the very least, this means that in his evolutionary and ethnological labors of loss, Lemuria lingers on into the Pliocene instead of disappearing in the early Tertiary period, or earlier, as it does for the zoogeographer or the geologist. As a Pliocene place-world, it enables both the appearance of humans and their dispersal across the globe as so many races. In other words, for the first time, Sclater's lost continent comes to rest within the horizon of human reckoning, leading some even to boldly suggest that it was its submergence and loss that might be remembered in legends of an antediluvial world prior to the Noachian Deluge.[91]

Provocative though this might be, the ethnological labors of loss around Lemuria are generally even more tentative and speculative than the paleogeographical. For one, their evidentiary base is not as convincing as the paleogeographer's fossil ferns or ancient rocks, not to mention the living lemur of the biologist. How, indeed, could material traces of the first man be recovered to bolster Lemuria's candidacy as the birthplace of mankind, if these now lie drowned in the waters of the Indian Ocean? Further, while some of the most prominent names in metropolitan and colonial natural history, biology, and geology participated in paleogeographical labors of loss around Lemuria, ethnology's place-makers are relatively unknown, with the important exception of Haeckel. Finally, ethnological place-making around Lemuria has a shorter life span in the metropole, petering out by the early years of the twentieth century, even though it thrived in other guises elsewhere. Nonetheless, these labors of loss are critical to the passage of Lemuria out of the rarified world of high science into wider circuits of circulation and consumption. Indeed, if Haeckel had not intervened and announced that it could have been the cradle of the human race, the "Paradise" out of which humanity fanned out to populate the world, Lemuria might have followed the same trajectory as the numerous other submerged landbridges and sunken continents of paleogeography and remained confined to the learned tomes of the natural and earth sciences. Instead, the zoogeographer's faunal bridge comes to inhabit the history of humanity, becoming both the probable birthplace of man as well as the causeway through which he dispersed throughout the globe.[92]

"No subject has lately excited more curiosity and general interest among geologists and the public than the question of the Antiquity of the Human Race." So began Charles Lyell's widely read *Geological Evidences of the Antiquity of Man,* published in 1863.[93] This was indeed "the question of questions for mankind—the problem which underlies all others and is more deeply interesting than any other," as Thomas Huxley noted soon after in his widely quoted *Man's Place in Nature.*[94] Haeckel's *Natürliche Schöpfungsgeschichte* boldly set out to tackle the problem, and answer the question. Just a few years before its publication, Charles Darwin had published his pathbreaking *Origin of Species* in 1859. From then on, finding the answers to three interrelated questions became the consuming intellectual preoccupation of scientists (and others) for at least the next century and more.

When did man first appear on earth? *Who* were his nearest ancestors? *Where* did he first appear? Although the second question far outstripped the others in the intensity of the debate around it, I am particularly concerned with the discussions surrounding the third, in which Lemuria puts in an eccentric and off-modern appearance.[95]

At the time Haeckel proposed Lemuria as a probable candidate for the cradle of the human race, Africa was the reigning favorite among most Darwinians as the most likely place one might find the first traces of human origins. But there were other contenders, especially among those who did not find Darwin's propositions attractive. Some continued to place faith in a Mosaic cradle in the Middle East; others did not want to surrender the primacy of Europe, especially in regard to this all-important question. While the Americas and even Australia found their advocates, more and more eyes were turning toward Asia, which was deemed the most likely human birthplace by the 1890s. And it remained so until the 1950s, when the discovery of an increasing number of humanoid and human fossil remains in Africa shifted attention back to that continent.[96] It is revealing of Lemuria's eccentricity in metropolitan place-making in the half century after 1870 that its candidacy is rarely noticed, and even less frequently embraced, even by those who had obviously read Haeckel and agreed with his extension of Darwin's theories. And there were quite a few dissenters. Thus, a few years after Haeckel's work was published in German and then in English, the noted French ethnologist Paul Topinard (who supported an African cradle) discussed—and dismissed—his theory of the Lemurian origins of man in the following terms:

> [Haeckel] says that this very remote ancestor is an ape of the old continent, a Pithecian, which was itself derived from a Lemur, and this in its turn from a Marsupial. He even gives it the name Lemurien—a term borrowed from Mr. Sclater; and as the focus of this series of transformations, a continent now submerged, of which Madagascar, Ceylon and the Sunda Islands are the remains. . . . This theory is painful and revolting to those who delight to surround the cradle of humanity with a brilliant aureole; and if we were to boast of our genealogy and not of our actions, we might indeed consider ourselves humiliated.[97]

Another French ethnologist, Armand de Quatrefages, who favored a Central Asian homeland, noted in 1883 that "no facts have as yet been discovered which authorize us to place the cradle of the human race elsewhere than in Asia. There are none which lead us to seek the origin of man in hot regions either of existing continents, *or of one which has disappeared. . . .*"[98] The Rev. John Mathew similarly insisted, "It can only be a last resort to account for the distribution of races by the submergence of hypothetical regions."[99] The American anthropologist Daniel Brinton, who himself favored a "Eur-African" birthplace, insisted that Lemuria had disappeared long before man appeared. "The hypothesis, therefore, advanced by Haeckel and favored by Peschel and other ethnographers, that the Indian Ocean was once filled by the continent "Lemuria" and that there man appeared on the globe, must be dismissed so far as man is concerned, as in conflict with more accurate observations."[100]

Indeed, and notably, Haeckel himself was quite tentative in his advocacy, and although till the very end of his scholarly life he continued to invoke Lemuria as the probable cradle of mankind, he also kept open the possibility that it might just as likely be in Asia or Africa that the first humans appeared. Thus, in 1870, when he first appropriated Sclater's Lemuria to propose it as the site of mankind's birth (see Fig. 3), he wrote—and note the many cautious hedges:

> The probable primeval home, or "Paradise," is here assumed to be *Lemuria,* a tropical continent at present lying below the level of the Indian Ocean, the former existence of which in the tertiary period seems very probable from numerous facts in animal and vegetable geography. But it is also very possible that the hypothetical "cradle of the human race" lay further to the east (in Hindostan or Further India), or further to the west (in eastern Africa). Future investigations, especially in comparative anthropology and paleontology, will, it is to be hoped, enable us to determine the probable position of the primeval home of man more definitely than it is possible to do at present.[101]

A few year years later, he similarly noted with equivocation in his *Anthropogenie* (1874) that man first appeared, "probably during the Diluvial period in the hotter zone of the Old World, either on the mainland in tropical Africa or Asia, or on an earlier continent (Lemuria—now sunk below the waves of the Indian Ocean), which stretched from East Africa (Madagascar, Abyssinia) to East Asia (Sunda Islands, Further India)."[102] Here, too, as in his earlier work, he bestows the enchanted label of "Paradise" on Sclater's lost continent, which, as we will see, has some far-reaching consequences for its trajectory in the labors of loss around it beyond metropolitan science.

His tentativeness notwithstanding, why was Haeckel interested in proposing Lemuria as the probable cradle of mankind when he joined the heated discussions sparked off by Darwin's work? Haeckel was an even more thoroughgoing evolutionist than Darwin or Huxley, moving beyond where they stopped on the tendentious question of man's immediate ancestor. Darwin had only dared to speculate that "a member of the anthropomorphous subgroup gave birth to man."[103] And all that Huxley was willing to suggest was that man had appeared out of the "gradual modification of a man-like ape or from the same stem."[104] But Haeckel went all out, and through his "Chain of Animal Ancestors of Man" he systematically leads us from the earliest lumps of matter, or "monera," through stage 18, when the Lemur puts

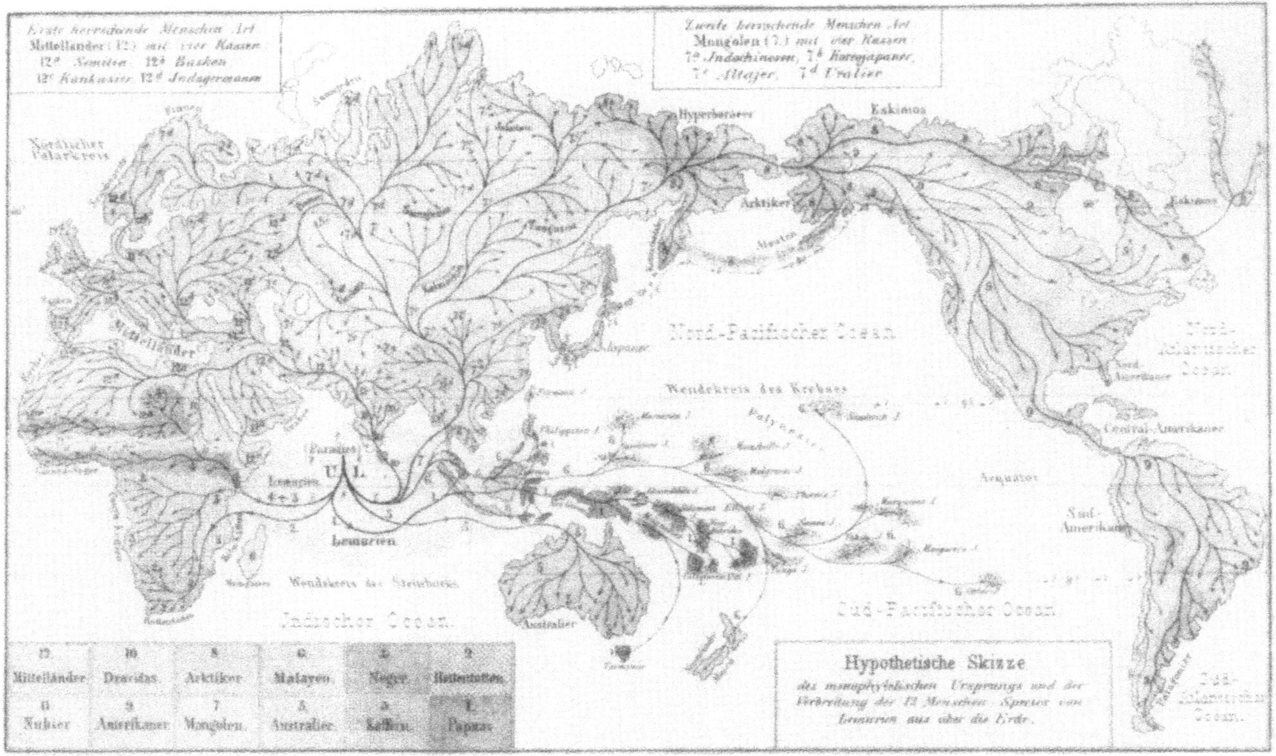

Figure 3. Ernst Haeckel, A hypothetical sketch of the monophyletic origin and extension of the twelve races of Man from Lemuria over Earth. From Haeckel, *Natürliche Schöpfungsgeschichte.* By permission of the British Library, Shelfmark 7002e18 plate XV Hypothethische Skizze.

in an appearance, and stage 20, which belongs to the "man-like apes" (anthropoides), to the final stage 22, whence appears man as we know him today.[105] The intermediate stage 21 in Haeckel's scheme is occupied by a speechless, "ape-like" creature with an upright walk, which he variously refers to as *Pithecanthropus alalus, Homo primigenius,* or simply Ape-Man. The birthplace of Ape-Man was possibly Lemuria, which is where "the gradual transmutation of catarrhine apes into pithecoid man probably took place in the tertiary period."[106] From Lemuria, the speechless Ape-Man walked out and populated the world, going east, west, and north as the continent progressively submerged, sometime in the early Pleistocene. As he did so, he learned language but also diversified into twelve different species. Thus, in Haeckel's place-making, the Lemurian was not fully human and it was only after he departed the drowning continent, whose ultimate disappearance cut him off from his brethren, that he acquired speech, ascended to full humanity, and racially disaggregated into numerous species.

At one level, spanning as it did both Africa and parts of Asia, Lemuria was a good compromise cradle of mankind, incorporating within its borders the two areas that were most favored as sites of human origin by contemporary scientists. It was also conveniently located in the Indian Ocean, whose tropical reaches had long been favored in the European imaginary as a potential site for the lost Eden.[107] A submerging continent where man first appeared but had not yet developed speech and attained full humanity was also a good way for Haeckel to reconcile the ongoing fraught debate between monogenists and polygenists on the origins of human diversity.[108] The monogenist in Haeckel wanted a single cradle—"Paradise"—where man first emerged, but the disappearance of the Lemurian landmass connecting the various branches of primeval humanity meant that the different primeval languages developed independently of one another, in turn producing racial differences between their speakers. So, in Haeckel's labors of loss, Lemuria functioned for human speciation very much like the paleobiologist's landbridge, whose appearance allowed for the dispersal of various terrestrial life-forms across the globe but whose disappearance produced the differences in formerly related species. As in zoogeography, so in anthropogeography as well, submerged continents became a convenient way to establish essential unity even while disavowing sameness. The evolutionary logic of loss—the common ancestral form cannot but be lost—lubricates the enabling fiction of affirming unity even while underscoring the difference between the races.

Not least, the proposition that a submerged continent like Lemuria was the original cradle of mankind solved the problem of producing material proof in the form of fossil remains of its inhabitants. As Peter Bowler has noted, "Its disappearance would account for the lack of fossil intermediaries."[109] In the face of missing material evidence, Haeckel offered other kinds of proof, from comparative anatomy, embryology, and the burgeoning field of (colonial) ethnology:

> We as yet know of no fossil remains of the hypothetical primeval man (Homo primigenius) who developed out of anthropoid apes during the tertiary period, either in Lemuria or in southern Asia, or possibly in Africa. But considering the extraordinary resemblance between the lowest woolly-haired men, and the highest man-like apes, which still exist at the present day, *it requires but a slight stretch of imagination* to conceive an intermediate form connecting the two, and to see in it an approximate likeness to the supposed primeval men, or ape-like men.[110]

And here Haeckel, like so many of his contemporaries, fell back on the explanatory formulations of scientific racism, which posited that the putative lower races of the present are relics of human ancestors, "living fossils" who would provide clues as to what ancestral man might have looked like, offering "a glimmer of the ape beneath the human envelope."[111] In much of ethnology and anthropology, as these sciences were then practiced in the metropole *and* in the colonies, "the lower races were made to bear the greater part of the burden of animal descent, thus sparing cultured whites some of the humiliation of being no more than higher apes."[112] For Haeckel, a man of his times after all, the Hottentots, Caffres, Negroes, and above all the Papuans, all of whom he classified as the "Ulotrichi" ("woolly-haired"), were the closest living representatives of Primeval Man:

> All Ulotrichi, or woolly-haired men, have slanting teeth and long heads, and the color of their skin, hair and eyes is always very dark. All are inhabitants of the Southern Hemisphere. . . . They are on the whole at a much lower stage of development, and more like apes, than most of the Lissotrichi or straight haired men. They are incapable of a true inner culture and of a higher mental development. . . . No woolly-haired nation has ever had an important "history."[113]

Given this understanding of such peoples, and given the fundamental operating assumption that the lower races of men are living fossils, all Haeckel had to do was to use "a slight stretch of imagination" to conjure up the following portrait of his Ape-Men, the inhabitants of Lemuria, modeled on the Ulotrichi:

> The form of their skull was probably very long, with slanting teeth; their hair woolly; the color of their skin dark, of a browning tint. The hair covering the whole body was probably thicker than in any of the still living human species; their arms comparatively longer and stronger; their legs, on the other hand, knock-kneed, shorter and thinner, with entirely undeveloped calves; their walk but half-erect. This ape-like man very probably did not as yet possess an actual human language, that is, an articulate language of ideas.[114]

The Lemurian thus conveniently helped Haeckel solve the vexing problem of the "missing link," even while his otherness allowed the biologist to demonstrate how much farther the white European had progressed beyond his less-fortunate black brethren, who continued to resemble even today the primeval inhabitants of Sclater's lost continent.

Some of the spatial and ethnological dimensions of Haeckel's evolutionary labors of loss had already been anticipated in the place-making of other paleo-scientists, who had seized upon the idea of a submerged Indian Ocean continent that accounted for the geographical distribution of mammals to explain the geographical distribution of man as well. Thus, writing in 1862, the well-known English anatomist Richard Owen (1804–92) proposed to the Geographical Society in London that the inhabitants of the Andaman Islands—who were increasingly portrayed in the contemporary ethnological literature as "Negritoes" and among the most primitive of humans—"might be the representatives of an old race belonging to a former continent that had almost disappeared."[115] Building upon Owen's tentative suggestion, the Scottish naturalist Andrew Murray (1812–78) wrote in 1866 that, along with the Andamaners, the inhabitants of sub-Saharan Africa, Madagascar, Mauritius, the Bourbon Islands, the southern part of Arabia, the remote hills of India, some of the islands of the Malay archipelago, New Guinea, Australia, and New Zealand were all members of the so-called "Black race," one of the two great divisions of mankind. And he insisted that these folks are "remnants of the inhabitants of the great submerged continent . . . alluded to by Professor Owen," and that this lost continent had provided passage for the black race as it dispersed through the southern half of the globe.[116] Haeckel's originality, therefore, lay not so much in drawing the now-submerged continent in the Indian Ocean into the ethnological place-making of his times, but in associating it with Sclater's Lemuria. The lemur's place in humanity's ancestry allows Haeckel to tentatively identify Lemuria—the putative homeland of the lemur—as "Paradise" and not just as the former habitation of the ancestors of the black races of humanity, as Owen and Murray had characterized it. Herein lies the distinctiveness of his labors of loss around Lemuria.

Haeckel's tentative identification of Lemuria as the birthplace of mankind had at best a minimal impact on metropolitan speculations over human antiquity.[117] Even his numerous followers in Germany's paleo-scientific community did not pick up on it, although as I note later, there were others on the occult fringe on whom it left its mark.[118] But the immediate effects outside Germany were quite spectacular if we follow the anthropological career of the Dutch anatomist Eugene Dubois (1858–1940) who in 1891–92 discovered the fossil remains of the so-called "Java Man" and declared him "Man's ancestor."[119] Some scholars have observed that Dubois was swayed by Haeckel's bold evolutionary theory, and especially by his speculation that the "East Indies" might have been the first of the places to which man dispersed from Lemuria.[120] He was also convinced that the gibbon (which was native to these islands) was closer to humans than were African apes. Thus, he enlisted in the Dutch colonial service in 1887 and began his quest for the material traces of the first human. In 1891 he found fossils in Trinil which resembled those from the Siwalik deposits in India.[121] When he subsequently found other fossil remains in 1893–94, he called the species they represented *Pithecanthropus,* choosing the same name that Haeckel had used a few decades before for his intermediate Ape-Man who had inhabited Lemuria.[122] A thighbone showed that, true to Haeckel's formulation, *Pithecanthropus* had an erect posture, and a chin fragment was deemed to confirm his speechlessness. Dubois thus concluded, "This was the man-like animal which clearly forms such a link between man and his nearest known mammalian relatives as the theory of development supposes . . . the transition form which in accordance with the teachings of evolution must have existed between man and the anthropoids."[123] Dubois's discovery was hailed back in Europe, but his characterization of his fossil as the "missing link" between ape and man was seriously doubted by many of his peers.

Although Haeckel himself insisted that Dubois's *Pithecanthropus* was "indeed the long-searched for 'missing link,' for which, in 1866, I myself had proposed the hypothetical genus Pithecanthropus, species Alalus,"[124] the Dutch discoverer of these fossils did not jump to the conclusion that his discovery necessarily proved the existence of Lemuria. Instead, all he appears to have soberly noted is that "the factual evidence is provided that, as some people have already suspected, the East Indies was the cradle of humankind."[125] Nevertheless, Dubois's discovery was used for a brief while to shore up the candidacy of Lemuria as the birthplace of man, most notably by Augustus H. Keane (1833–1912), gentleman ethnologist, Professor of Hindustani at the University College, London, and a former vice-president of the Anthropological Institute. Anxious like everyone else of his time to find the "missing link," Keane seized upon Dubois's *Pithecanthropus* to insist that there had been "a single cradle-land, from which the peopling of the earth was brought about by migration."[126] Insisting that the femur of Dubois's fossil was distinctly human and not just "ape-like," Keane went on to observe that "it ante-dates all other human remains hitherto discovered, and . . . of living races the nearest akin are the Australians, Andamanese, Bushmen, thereby lending support to the view that these low races spring from a common primeval stock, which originally inhabited the now vanished Indo-African continent. This pliocene inhabitant of Java may thus in a sense be taken as the long sought-for 'First Man.'"[127] Empowered by his belief that the Pliocene First Man of Java most resembled the Australians, Andamanese, and Bushmen of the present day, Keane went on to describe him—as Haeckel had done a few decades before—as of average height, with the ability to fashion "rude stone implements," and furnished with a cranial capacity of about 1000 c.c. that placed him "just midway between Gorilla and the highest present races (Europeans, 1500 c.c.)."[128]

In this regard, Keane was merely restating claims that he had already made in *Ethnology* (1896), an earlier synthesis of all contemporary writings on the subject of human antiquity. Like Haeckel, Keane, too, assumed that contrary to what the permanentists were asserting, there had been substantial changes in continental forms in earth's paleo past:

> Despite the views put forward by Wallace and others regarding the stability of the Continents, the inhabitable regions of the globe have certainly undergone considerable modifications since the appearance of the Hominidae in their several geographical areas. Doubtless Wallace is right in rejecting Sclater's "Lemuria," as unnecessary to account for the range of the Lemurs. But he cannot reject the "Indo-African Continent," which replaces Lemuria in the Indian Ocean, and which is established on a solid foundation by naturalists associated with the *Indian Geological Survey*. . . . Hence, although belonging mainly to secondary [Mesozoic] times, considerable sections of the Indo-African continent, such as are still represented by Madagascar, the Chagos, Seychelles, Mascarenhas, and other smaller groups, must have persisted far into the tertiary epoch.[129]

Keane's Indo-African continent was different from Haeckel's Lemuria not just in name, but also in its extent, including as it did parts of Australia, thus providing a passage for the movement of aboriginal Australians to their present habitat from the submerging land in the Indian Ocean. He, in fact, occasionally referred to the vast continental place-world as the "Indo-Austral region" or the "Indo-African and Austral Continents."[130] But, like Haeckel, Keane dated human origins to the Pliocene, and he insisted that man's dispersal out of his primeval cradle took place in the Pleistocene and was made possible because of the continental connections that then existed between areas now widely separated by oceans:

> Thus when the pliocene precursor, wherever evolved, began to spread abroad, he was free to move in all directions over the eastern hemisphere. Like the anthropoid allied forms, he could have wandered, say, from the Indo-African Continent, either eastwards to India and to Malaysia, where are now the gibbon and orang, or westwards to Africa, where are now the chimpanzee and gorilla, and thence northwards to Europe. . . . From the Indo-African Continent the road was also open through Australasia towards New Zealand, and from India to the shores of the flooded central Asian depression.[131]

And, like Haeckel, Keane insisted as well that "although man had but one origin, one pliocene precursor, men had several separate places of origin, several pleistocene precursors."[132] So, the conjuring up of an Indo-African and Austral continental landmass, which subsequently disappeared, enabled him, as it had Haeckel, to reconcile the monogenist and polygenist perspectives.[133] Just as it had been drawn into the dominant concern of Victorian natural history and paleogeography with the ancient distribution of Earth's territories and life-forms, Lemuria also participated, however fleetingly, ephemerally, and eccentrically, in one of the most contentious issues of late-nineteenth- and early-twentieth-century ethnology and race science—the debate between monogenesis and polygenesis.

In the metropole, ethnological labors of loss petered out by the early years of the twentieth century. By 1908, even Keane had modified his general thesis to suggest that the East Indian Archipelago, which Dubois had favored, was the "human cradle," and that the Indo-African Continent (which survived into the Pleistocene) had merely served as the pathway through which Early Man migrated to different parts of the globe from his "Javanese cradle."[134] As rival candidates who could back up their claims with fossil proofs emerged, Lemuria's status as the cradle of mankind eroded, although every now and then, we get a hint of it as, for instance, in popular works like H.G. Wells's *The Outline of History*—much cited in Tamil labors of loss—which speculated in passing that the "nursery" of mankind "may have been where now the Indian Ocean stands."[135] However much the evolutionary logic of loss might have helped reconcile monogenism and polygenism, in the long run, Lemuria's candidacy as the cradle of mankind was weakened by the growing consensus among paleo-scientists that even if land connections between continents had existed in the earth's deep past, these would have disappeared well before humans appeared. And by the 1930s, the very attractiveness of a former continent like Lemuria which might have served as a singular birthplace for all of humanity before it conveniently submerged, allowing mankind's numerous races to develop along different lines, was eroded by the increasing disapproval of theories that posited a single, hallowed Garden of Eden.[136] Instead, "evolution operates not upon one single line and one single species, but upon multiple lines, some converging and some diverging, and upon large groups of animals. Nor is this process restricted to a single continent."[137]

But Lemuria's disappearance from evolutionary and ethnological labors of loss in Europe does not mean that others did not seize upon it and make it their own. Across the Indian Ocean, in Australia, a series of essays published for a decade and more between 1896 and 1909 in the Sydney-based periodical *Science of Man: Journal of the Royal Anthropological Society of Australasia* took up the cause of Sclater's lost land.[138] With suggestive titles such as "The Lost Continent in the Indian Ocean: The First Home of Mankind," "The Original Home of the First Men," and "The Locality of the First Home of Mankind," these essays identify Lemuria as "the original home of lemurs, monkeys, apes, and primitive men," and they transformed Haeckel's tentative suggestions into confident assertions:

> The proofs are now recognised as sufficient to demonstrate that there have been great changes of elevation, and depressions in all the continents, as shown by the surveys and soundings of the river valleys of former times. So that the submergence of "Lemuria" and "Atlantis" is now recognized by all

competent students who have most fully investigated this subject. The lands now submerged were the dry land connection over which ancient men passed from their earliest homes. It is therefore shown by modern soundings, surveys, and explorations, as well as by ancient traditions, that the first home of mankind was in lands now submerged beneath the Indian Ocean, and there mankind had developed and evolved side by side with the Anthropoid.[139]

As with Haeckel, the submergence of Lemuria provides a convenient resolution here as well to the problem of the absence of material remains of the first humans. "It is not to be expected that the men of the primitive or earliest type will ever be discovered, for they are buried beneath the Indian Ocean."[140] In contrast to Haeckel, however, who had suggested that the inhabitant of Lemuria had been an "Ape-Man," these Australian labors of loss distance themselves from this formulation, insisting instead on the enormous gap between humans and apes. "It is erroneously thought that men had developed from the apes, but it is far more likely that men originated from some animal on a land now sunk beneath the Indian Ocean. And that none of the intermediate links of the evolutionary chain of beings have been preserved, but all have been drowned in the submergence."[141] Nonetheless, one could imagine what the intermediate link between animal and human might have been like, for "in the countries surrounding the Indian Ocean are still to be found several kinds of the primary men who wandered from their original home, some of these are yet to be found in the Andaman Islands, others are the Negritos of Equatorial Africa and the Bushmen of the Cape Colony, also the Negritos of India, the Phillipines [*sic*], and Melanesia, also the peculiar man found in Java by Dr. Duboise [*sic*], and the Papuans of India, Indonesia, and Melanesia."[142] The essays even offered a timetable for the peopling of the world from Lemuria, as first the Negritos, followed by the Negroes, the Dravidians, the Mon, the "red race," and finally "the white race" wandered off the drowning continent from the Miocene age.[143]

Europe's ethnological labors of loss around Lemuria also found an audience in colonial India. As I discuss at length in chapter 4, Haeckel's tentative contention that Lemuria had been home to the Promalay, primeval ancestor to the Dravidians of southern India, was seized upon by British administrator-scholars and colonial ethnologists, as well as Tamil-speaking intellectuals, in their labors of loss around Sclater's drowned continent.[144] At the same time, Keane's alternate hypothesis for the peopling of India by pre-Dravidian aborigines, "dark peoples, probably of aberrant Negrito type" who had migrated eastward from their primeval Indo-African homeland,[145] is also picked up for elaboration, mostly outside the Madras Presidency, by Indian ethnologists and amateur-scholars increasingly preoccupied with the complex race sociology of the subcontinent and the tantalizing business of deducing originary homelands.[146] Thus, writing in the pages of the Calcutta-based *Modern Review*, B. C. Mazumdar engaged the issues that were at the heart of the European ethnological enterprise, drawing heavily upon Keane's work. "Now that the doctrine of evolution is regarded by all scientific men of eminence as an elementary truth like the Copernican and Newtonian doctrines, it will quite do, if I make the bare statement that the mighty ancestor or precursor of man was a furry animal of arborial habits."[147] He then insisted that when Darwin had proposed southern Africa as the birthplace of man in the 1850s, he had been unaware of the existence of Lemuria, "as it was not then definitely established that there existed once a vast continent stretching from the south of Africa and extending to Malaysia, joining the Indian peninsular region with it."[148] Now that science had confirmed its existence, Keane must certainly be right in proposing that "all the conditions point to these Indo-African and Austral lands as the most probable centre of evolution":

> "There rolls the deep" where our *pitris* [ancestors] first assumed the form of men and had a happy existence of many thousand years. Who knows if our scientific men will not obtain as a result of their devotion to the *pitris,* some consecrated bones of theirs to identify conclusively the holy shrine of the earliest *pitriloka* [ancestral world].[149]

Following Keane, Mazumdar, too, maintained that peninsular India, which was "then not connected with the other parts of Asia to the north-west," offered a home to the migrating Negritos from the south during the later Paleolithic, "when the disastrous submergence of the Indo-African continent continued in an appalling manner."[150] Later, when the Dravidians entered India from the northwest once the subcontinent was connected with the Asian mainland, "the small remainder of the Negrito people disappeared very likely in the general body of the Dravidians, with the result that where the Negrito element asserted prepotency [*sic*] on the borderland of the Aryans and the Dravidians . . . , separate Kolarian tribes originated."[151] The fact that the Kolarian languages are similar to the languages of some Australian tribes is the one remainder—and reminder—in Mazumdar's place-making of the original Indo-African-Austral connection.[152]

This contention, that the Kolarians were the original inhabitants of India who had settled in the subcontinent after their Indian Ocean homeland drowned, also came to the consideration of Sarat Chandra Roy (1871–1942), a pioneering Bengali anthropologist and founder of the quarterly journal *Man in India.*[153] In 1912, in his monograph entitled *The Mundas and their Country,* Roy joined the ongoing debate among many in colonial India in the early decades of the twentieth century about the original inhabitants of the subcontinent, and came down on the side of "the great Kol race" and its "typical representative," the Munda.[154] But he was less sure about the original homeland of the Mundas. Invoking as possibilities both "the now-submerged hypothetical continent of Lemuria which has been supposed to have once connected India with Madagascar and Africa," as well as regions to the northwest of India, he cautiously concluded that this question would "perhaps ever remain hidden from view in the midst of ages."[155] Years later, when he gave the Presidential Address to the Seventh Session of the All-India Oriental Conference in Baroda in 1935, he remained tentative as he offered an overview of the ethnic prehistory of the subcontinent in which he identified a pre-Dravidian populace which "might have entered India from either the north-east or as appears more probable, from the north-west, or perhaps from submerged 'Lemuria' on the south-west." Wherever he came from, the Proto-Australoid was "the true aborigine of India," eventually making way for the Dravidian from the Mediterranean.[156] Inspired by Roy's speculations, C.S. Srinivasachari, a professor at the newly established Annamalai University in the Madras Presidency, also invoked Lemuria as the former homeland of the pre-Dravidian peoples of southern India, such as the Kurumbas, the Sholagans, the Irulans, the Chenchus, and the Yenadis, as well as the Veddahs of Ceylon. These pre-Dravidians "were the representatives of a submerged Negrito element" whose former habitat might have been the submerged continent of Lemuria, Srinivasachari postulated. These are, therefore, the original aboriginals of India, although they might have entered the subcontinent from the now-submerged Lemuria.[157]

So it is that Europe's ethnological labors of loss around Lemuria undergo a subtle but important shift as they travel to the periphery. The "grand question" of human origins, in which Sclater's submerged continent had gotten entangled through Haeckel's place-making intervention, is here appropriated to answer the more pressing issue of who was first in the subcontinent. In colonial India, this meant, above all, a determination of whether the hallowed Aryan was an outsider or was indigenous to the subcontinent. While the dominant view espoused by European ethnologists and colonial administrator-scholars maintained that the Aryans had migrated into India, from the closing years of the nineteenth century a growing number of Indian intellectuals insisted to the contrary, bolstered by the rising tide of a militant Aryanist–Hindu nationalism.[158] For one among these intellectuals, Abinas Chandra Das, a lecturer in history at Calcutta University, the metropolitan labors of loss around a drowned Lemuria came to be particularly useful for his theory that "the Aryans were autochthonous in the Punjab (or Sapta-Sindhu, as it used to be called in Vedic times), or at any rate, had been living in the country from time immemorial."[159] Quoting liberally from Wallace, Blanford, and Haeckel, whose paleo-scientific findings only corroborated the much more ancient ritual evidence of the sacred *Rig Veda,* Das insisted that up to the Miocene, and perhaps even into the early Pliocene, Sapta-Sindhu was completely cut off by an intervening sea from the southern peninsula which was part of the vast continent of Lemuria which had extended from Burma and Southern China on the east, to East and South Africa on the west, and from the Vindhya Hills in the north to Australia in the south. At this time, Lemuria was inhabited, among others, by Kolarians and Dravidians who had absolutely no ties to the Aryans who lived to their north, separated as they were by the Rajputana Sea:

> Whether this continent was the original cradle of mankind or not, there can be no doubt that man existed here from very early times, and that his creation in this continent was made possible only after the creation of anthropoid apes which were his nearest approach [*sic*]. We have already got evidence of the existence of Pliocene man in the valley gravels of the Narmada and of Miocene man in Upper Burma. It can, therefore, be safely surmised that man had existed in this continent long before the time when the greater portion of it was submerged in consequence of a violent cataclysm. *Though Sapta-Sindhu was not directly connected with it, conditions similar to those of the lost continent must have prevailed there, which favored the creation of a family of human beings entirely different from that of the Southern Continent; and these were the progenitors of the Aryan race who, having been endowed with higher mental faculties, developed a civilisation which was destined to dominate the whole world, and uplift the entire human race.*[160]

A map that Das appended to his book visually and cartographically underscored the spatial distance and distinctiveness between the civilized Aryan, safe and secure in his pure Sapta-Sindhu homeland in the Punjab, and the savage Kolarians and Dravidians, a whole continent away on Lemuria.[161] The progressive fragmentation and submergence of Lemuria, and the drying up of the Rajputana Sea, led to a territorial connection opening up between the two homelands and the dissemination of Aryan civilization into peninsular India. But not everyone benefited from these catastrophic transformations, and Das goes on, in the best tradition of Victorian and colonial ethnology, to discuss the bestiality and savagery of the "primitive races" of southern India who even today have not really progressed beyond their ancestors who had once inhabited Lemuria, "little removed from the condition of anthropoid apes or brutes."[162]

As I noted earlier, by the 1920s, metropolitan labors of loss around Lemuria as the birthplace of mankind, or as the former homeland of all the black races of the world, had died a quiet death among paleo-scientists. It is a measure, however, of the global and colonizing reach of European thought that ideas wrought in the far-off capitals of the metropole in another century continued to have the kind of life that they did in the periphery, far after they had expended themselves in their originary sites of production and far beyond the original projects for which they were intended. Almost a century after Europe's natural historians and ethnologists had first proposed a kinship between all of Earth's black races based on their putative common ancestry on a land that now lay submerged beneath the waters of the Indian Ocean, Leopold Senghor, the then-President of Senegal, gave a talk in Madras at the International Institute of Tamil Studies in May 1974. He referred to the former land connection between southern India and Africa, and reiterated—as others had before him, in Europe and elsewhere—that this ought to be considered "the cradle of Mankind." But he did not stop there. He noted that this land connection was perhaps "engulfed by the ocean during the Neolithic revolution, that is to say, the period of prehistory when *Homo Sapiens* achieved his 'first revolution,' by laying the foundations of the recorded civilizations through new techniques he had invented." Not surprisingly, "the early civilizations which arose in the valleys of the Nile, the Tigris and Euphrates, and lastly of the Indus, bore the marks of black men." All that was needed was "for archaeologists and prehistorians to have a chance to explore the depths of the seas, to discover old lithic industries or human skeleton fossils, in the area stretching from East Africa to southern India"[163]

For those living in the black periphery, Europe's ethnological labors of loss around Lemuria came with their share of burdens as well as possibilities. These identified the southern hemisphere as the location of "Paradise" and identified First Man as having as his nearest kin the black races of the world. Yet any pride that black races could take in this putative achievement, the chronological preeminence of having been the first humans to show up on earth, was clearly offset by the fact that in these same acts of evolutionary place-making, the Lemurian was as primitive as they come, barely human for Haeckel, and midway between a gorilla and the European in intellectual capacity in Keane's work. Yes, he may have been First Man, but in the evolutionary logic which generated the Lemurian, this only underscored his primitiveness and his bestiality—and his distance, geographically, temporally, and culturally, from the civilized white European on his own distant continent. Not surprisingly, the metropolitan paleo-scientist, be he a biologist like Haeckel or an ethnologist like Keane, expressed little regret over the fatal drowning of Lemuria, for its inhabitants were, after all, closer to the "lower" races of the black diaspora, the ones who had barely progressed, even in the present, and who had little hope of either culture or history. They were literally the mirror opposite of the civilized European. This is perhaps not least of the reasons that, for those outside the metropole who wanted to define themselves in opposition to the white European, the putative black inhabitants of Lemuria become potentially useful to claim—but only after they had been transformed from barely human to the earliest civilized representatives of mankind, as we will see.

LEMURIA AND PALEO-SCIENTIFIC LABORS OF LOSS

Are there any lessons to be learned here from what is clearly an off-modern moment in the grand progress of modern European science? As I have noted, few scholars who have studied the development over the past century and a half of the various paleo-sciences—natural history, geology, paleogeography, ethnology, prehistory—have bothered to even mention Lemuria, let alone examine its eccentric eruption in the place-making activities of some of the most important practitioners of these disciplines, scientists of the stature and influence of Ernst Haeckel, Thomas Huxley, Philip Sclater, Eduard Suess, and Alfred Wallace. Yet, Lemuria was a child of modern science, which gave birth to it and nurtured it for close to a century when it participated, albeit fleetingly and ephemerally, in some of the most pressing problems that preoccupied these men, questions such as the former distribution of land in relation to oceans on the earth's surface, the geography and evolution of terrestrial life forms, the birth of man, and the sheer diversity of humanity. The fact that these prominent men of science even resorted to Lemuria, if only eccentrically and sometimes oppositionally, to answer these questions may seem absurd in retrospect, but I have documented that lost continents, submerged worlds, and vanished territorial connections were quite critical in those years to contemporary theories of the earth and to the paleo-sciences that sustained them. Yet, the very march of science led to a repudiation of these theories as fantastic, and along with this repudiation, the place-making labors of the men who fostered them also disappear into oblivion, rarely to be quoted again by metropolitan scientists who pride themselves on having transcended their ancestors' fanciful foibles. However, away from the rarified corridors of Euro-America's paleo-scientific establishment—in contexts as different as occultism and in lands as distant as India—the labors of loss of Sclater and Haeckel continue to have a purchase that far outlived their use in the metropole, as we will shortly see.

Indeed, it is the very fact that Lemuria was a creation of science that makes it so attractive to all those place-makers, outside science's hallowed circle, who invariably invoke a Wallace or a Keane to bolster their own labors of loss. This is more than a case of science bestowing legitimacy, credibility, and respectability on projects whose agendas and goals did not coincide with those of the Euro-American natural historian, paleogeographer, or ethnologist. As importantly, the paleo-sciences also generated a new, exciting vocabulary (Gondwanaland, Tethys, and so on), and a new, alluring terminology (landbridges, ocean floors, subsided continents, and the like), which were mobilized to imagine the lost place-worlds of Earth's deep past, as well as new technologies of place-making such as the geological map (which, as I note in a later chapter, becomes ubiquitous in all labors of loss around Sclater's disappeared continent).

But, most importantly, these paleo-sciences make loss itself—in the form of lost worlds, lost times, and lost species—into an object of scientific knowledge production. Certainly, the premise that Earth as we know it today is only the latest in a succession of former worlds that stretch back into deep time is critical to the operating logic of these sciences. But, as importantly, there is the conviction that through scientific modernity's various knowledge practices, the disciplined scientist could and would apprehend what had vanished in former ages. It is possible, in other words, to speak back to loss with the instruments of modern science. The evidentiary base with which the paleo-scientist combats loss is mundane: rocks and fossils, earth movements, and ocean floor configurations. His labors of loss are contingent, based on empirical evidence that is always incomplete and limited, adhering as he does to the protocols of science. They are also tentative, speculative, and multiply hedged, lacking the spectacular reach of some other, transparently enchanted labors that we will encounter later. And they are dispassionate and disinterested as they methodically go where their science inevitably takes them, even to the point of repudiating earlier labors. Nonetheless, these labors of loss are not entirely disenchanted, as they reach beyond that which is immediately apparent, available, and attainable in order to reveal former worlds and extinct pasts that once existed but do no more. They do so with the conviction that ultimately the whole is knowable and ascertainable. In the final analysis, Lemuria's significance emerges from this, from the part it plays, albeit eccentrically, in the scientist's effort to battle loss by endeavoring to complete the human stock of knowledge. A disenchanted globe stands poised on the road to reenchantment.

Chapter 3

Occult Losses

Redemption from the rationalism and intellectualism of science is the fundamental presupposition of living in union with the divine.[1]

OCCULT VANISHINGS

From the late 1870s Sclater's Lemuria embarks on its most enduring journey in the metropole as it is drafted into the proliferating labors of loss of Euro-American occult.[2] Consequently, it joins the ranks of other vanished, hidden, or secret lands—the ubiquitous Atlantis, of course, but also places with wondrous names such as Hyperborea, Mu, Pan, or Shamballa—that dot the modern occultscapes of Euro-America. Occultism's place-making has scarcely been scrutinized by scholars, and yet this is enormously revealing of the labors of loss that distinguish its modernity.[3] Thus, my primary goal here is to examine Lemuria's place in occult labors of loss around disappeared worlds and vanished pasts. In so doing, I also consider why lost, hidden, or secret places are important to modern occultism. The esotericist preoccupation with lost continents has been widely disparaged by both professional and freelance scholars,[4] yet they fail or refuse to consider why occultism fetishizes lost places. In focusing on occult labors of loss around Lemuria, therefore, I suggest that instead of dismissing outright esotericist place-making as dangerous flights from reason or utter nonsense, we may instead learn something from them as well of modernity's preoccupation with loss.

Lemuria circulates in occult circles at a time when much of the known world's geography was radically reconfigured, in the closing decades of the nineteenth century, by the consolidation of Euro-American imperialism, and then, as the twentieth century wears on, by the forces of global capitalism. These, as we well know, have virtually left no part of the known world untouched. Modern occultism undoubtedly benefits from empire, especially its explorations of the Orient, and its wondrous discoveries of ancient wisdom far older than Judeo-Christianity. All the same, the so-called occult revival of the later nineteenth century was also a reaction to empire's materialist excesses, its scientization of the globe, and its participation in the disenchantment of our world. As traditional esoteric favorites like Egypt, India, and even Tibet come under metropolitan influence, ever new sites—submerged, subterranean, extraterrestrial, astral—are conjured up for occult colonization. These are transformed from terrae incognitae into esoteric place-worlds that are drawn into the cosmic drama of the human spirit. The imperative to find spaces and places in and around our earth that are available for occult colonization in the age of imperial and global capital is one important context for the fascination with lost continents in esoteric circles over the past century.

Just as important, with the rise and consolidation of the various paleo-sciences, our earth ceased to be "shaped by the benevolent hand of God . . . populated by the plants, animals and men that He created."[5] These sciences excluded as unscientific, and hence illegitimate, "almost all that had previously made [it] rich in cosmological meaning." They were instead replaced by "an astonishing drama of vanished worlds," but emptied of human presence and agency, disembedded from the sacred history of man.[6] "Man, convinced at first, in his naive innocence, that the world was made for him, has now been told by the time voyagers that, at a period not very remote, geologically speaking, the human form is no longer to be found."[7] In Paolo Rossi's account of the long revolution in European thought that preceded the disciplining of these paleo-sciences in the nineteenth century, "the difference lies not only between living at the center or at the margins of the universe, but also between living in a present relatively close to the origins (and having at hand, what is more, a text that narrates the entire history of the world), or living instead in a present behind which stretches the 'dark abyss' . . . of an almost infinite time."[8] In contrast, and for many centuries prior to this revolution, "the history of man was conceived as coextensive with the history of the earth. An earth not populated by men seemed meaningless, like a reality that was somehow 'incomplete.'"[9]

Occult labors of loss around place-worlds like Lemuria seek to complete the history of Earth rendered incomplete, unmanned, and a-theized by the physical sciences. As one occultist put it, the *paleontological* history of mankind might well begin "with the fossils embedded in the diluvial deposits of the Quaternary period." But a *spiritual* history reaches back to Creation itself.[10] In distinction, therefore, to the paleo-scientist's preoccupation with Lemuria, occultism's labors of loss are transparently enchanted. Their quest for lost wholeness and lost unity, for the lost Word and lost wisdom, reintroduces into the contemplation of Earth's deep past all the mystery and magic that had been banished by the material sciences, at the same time that the latter's findings are used to fabulate new geographies of Spirit and new histories of Man.

I therefore locate the emergence of occult labors of loss around Lemuria in the late nineteenth century within what Alex Owen identifies as a "new dialectic of modernity" marked by an "uneasy co-existence of a distinctively modern Weberian disenchantment and a converse aspiration to noumenal experience."[11] Not surprisingly, these labors bear numerous marks of this dialectic in their conflicted intimacy with modern science. On the one hand, they strategically clothe themselves in the vocabulary and findings of the paleo-sciences. On the other, they seek to go beyond science's partial and inadequate truths in their efforts to create a complete "science of the spirit," also variously referred to as the "science of the invisible" and the "science of the hidden." It is Lemuria's place in these labors on behalf of Spirit that this chapter explores.

OCCULT LEMURIA

Lemuria has been variously configured in the century and more since its first appearance in the late 1870s in the modern occult imaginary. From this dense archive, I have chosen three place-making moments for what they tell us about occultism's different labors of loss, enchanted though they might all be. The first is the Theosophical moment of the 1880s, in which Lemuria is transformed from the paleogeographer's lost continent, uninhabited by humanity, into the submerged home of the "Third Root-Race," progenitors to Man. I then turn to its recasting as a drowned Pacific Paradise from the 1920s. Finally, I explore its New Age reincarnation from the 1950s as a lost utopian world of wisdom and well-being. Although the concerns of each of these moments is different, and the social, political, and cultural milieu in which they operate are varied, they are all united by the imperative to reclaim an Earth a-theized by the material sciences. It is this imperative that motivates the varied occult labors around Sclater's lost place-world.

Theosophy's Lemuria: Home of the Third Root-Race

> The Story of Man, as discovered by occult investigation, can be briefly told. He is developed in seven clearly marked stages called Root-Races. The first three were occupied in the work of building a serviceable physical body and developing the senses of hearing, touch, and sight. No physical traces will ever be found of the first two, for their bodies were made of such fine matter that no fossils could be left, and they did not build cities or temples. The third race has more in common with our own. It inhabited the continent of Lemuria in the Secondary Period, and it was therefore a contemporary of the gigantic saurians.[12]

The occult future of Sclater's Lemuria is more or less assured by the fact that it put in its first appearance, albeit briefly, in a work that has become a classic in esoteric circles. In 1877 Helena P. Blavatsky (1837–91), the charismatic but also much-maligned Russian cofounder of the Theosophical Society, published her first synthesis of occult thought, *Isis Unveiled,* while she was in residence in New York.[13] In the course of rehabilitating discarded cosmogonical legends from all over the world, "based on a knowledge by the ancients of those sciences . . . [which] were far better acquainted with the fact of evolution itself . . . than we are now," Blavatsky noted that "the garden of Eden as a locality is no myth at all."[14] Instead, it could be located in the "great lost continent" situated south of Asia, extending from India to Tasmania. This lost continent, named Lemuria, "a dream of scientists," is perhaps the same as Atlantis.[15] "If the hypothesis now so much doubted, and positively denied by some learned authors who regard it as a joke of Plato's, is ever verified, then, perhaps, will the scientists believe that the description of the god-inhabited continent was not altogether fable."[16] As we will see, this recuperation of Atlantis by resorting to the paleo-scientist's Lemuria is crucial to Theosophical place-making over the next decade, as a result of which Plato's lost land becomes the birthplace of humanity, the home of the Fourth Root-Race from which the majority of Earth's more advanced peoples today are descended.

Three years later, after she had moved to India, "the cradle of the [human] race,"[17] Blavatsky began expanding her labors of loss around submerged continents by suggesting that they had served as the home of great civilizations far more ancient than those discovered by the material sciences. These vanished place-worlds also accounted for the fact that "nations so antipodal to each other as India, Egypt, and America" had "nearly identical architecture and arts."[18] So, while paleo-scientists conveniently used drowned continents to account for connections between geographically separated flora, fauna, and human races, Blavatsky resorted to them to solve the puzzle of avowed similarities in civilizations in different parts of the worlds that her contemporaries typically explained by turning to theories of diffusion. She also incorporated lost places like Lemuria and Atlantis into a cyclical scheme of cultural evolution that explicitly critiqued the reigning theories of linear progress in which modern civilization occupies the pinnacle of cultural development. As a leading exponent of the "ancient wisdom family,"[19] Blavatsky rehabilitated knowledges dismissed as archaic by the sciences but really more useful than anything scientific modernity had to offer. These knowledges had been produced in former place-worlds like Lemuria now lost to the ocean, but they had left their mark in legends that circulate today in parts of the world that are outside the influence of the West and its materialist sciences. Lemuria might be the "dream" of scientists, but ancient peoples had known about it all along—a claim that was to prove enormously productive in twentieth-century occultism. The Theosophical moment has thus been critical to bringing Lemuria in from nature into culture.

Most critically for its occult future, Blavatsky relocates Lemuria to the

Pacific Ocean, and starting with the infamous Mahatma letters, also begins to distinguish it from Atlantis.[20] So, the (apocryphal) Master Koot Hoomi intones, "Lemuria can no more be confounded with the Atlantic Continent [Atlantis] than Europe with America." Consequentially for its future, it is also identified for the first time as the home of the "Third Root-Race," whose remnants may even today be found in "the flat headed aborigines" of Australia, but who once upon a time had authored great civilizations. "Greek and Roman and even Egyptian civilization are nothing compared to the civilizations that began with the 3rd race."[21]

These scattered labors of loss around Lemuria from the late 1870s and early 1880s are finally systematized in Blavatsky's 1888 magnum opus, *The Secret Doctrine,* the foundational narrative for all Theosophical thought (and for modern occultism, more generally).[22] Subsequent Theosophical labors of loss around Lemuria are more or less footnotes to this master text,[23] which itself claimed to be a commentary on the "Archaic Records" contained in the *Stanzas of Dzyan,* written in the Atlantean language Senzar that was revealed to Blavatsky in clairvoyant communication. Regardless of Blavatsky's insistence on the Atlantean authorship of her visions, a wide range of contemporary writers clearly inspired her ideas on Lemuria. These included paleo-scientists like Haeckel and Wallace, both of whom, as we have seen, wrote about it.[24] Lemuria also fleetingly appears in Ignatius Donnelly's blockbuster *Atlantis: The Antediluvian World* (1882), which in one assessment was the principal source of Blavatsky's own labors of loss around vanished lands.[25] Blavatsky herself quotes at length from the French writings of Louis Jacolliot, whose own labors of loss around a submerged Pacific continent (that he insisted he learned of from Sanskrit and Polynesian legends) came to her attention just as *Isis Unveiled* was going to press.[26] Neither Donnelly nor Jacolliot, however, linked Lemuria to Atlantis, as it is in Blavatskian Theosophy, although they might have been responsible for the shift in its location in her place-making from the Indian to the Pacific Ocean. American spiritualists whom Blavatsky read, like Thomas L. Harris[27] and John Newbrough,[28] had also written about vanished place-worlds in their own occult histories of mankind.

Thus, lost continents were clearly in the spiritualist and occult air when Blavatsky arrived in the United States in 1873 after her already varied travels across Europe and Egypt. But it is with the publication of *The Secret Doctrine* in the late 1880s that they become staples of modern occultism and remain so to this day. And Theosophy has played no small role in assuring them of this. For a religious movement concerned with the recovery of lost ancient wisdom—a *philosophia perennis*—in order to cope with what was widely perceived as the loss of faith and belief in an age of advancing materialism, the scientific speculations about lost continents in the earth's past came in handy. In these vanished paleo-worlds, ancient beings (who, rather than the ignominious simians of the paleo-scientists, were the real ancestors of humanity) lived out their lives, the memory of which has been left as traces in lost and secret records that only the Theosophist can access through clairvoyance, telepathy, and other occult means. In Blavatsky's labors of loss, as well as those of her followers in the Theosophical Society over the next half century and more, Lemuria retains its old persona as a former continent, now vanished, but attains a new function. It resurfaces as the home of the Third Root-Race in Man's evolution through seven epochs, each of which gets staged on a specifically named continent that covers the surface of our Earth during the long period of its awakening after its cosmic slumber.[29] While the Theosophists and their writings have come under considerable scholarly scrutiny, much of it not particularly complimentary to their cause or its epistemology, their place-making, although noted *en passant,* has scarcely been scrutinized. And yet few other aspects of the Theosophical cosmology so clearly illustrate the productive eclecticism of "eastern" spirituality, "western" occult, and contemporary science that sets it apart from other late-nineteenth-century religious movements in Euro-America. For, in the name of "occult science," or "science of the spirit," the Lemuria of paleo-scientific place-making is appropriated and put to the diametrically opposed purpose of vigorously challenging the history of Earth and humankind that was then being reconstructed by Lyellian geologists and Darwinian evolutionists, especially the latter's avowed insistence that man had evolved from ape.[30] In its place, the Theosophist offered an alternate genealogy for Man, through the medium of clairvoyant communication with (apocryphal) Mahatmas far outside the reach of the material(istic) world of the geologist, the natural historian, or the ethnologist: "Every unprejudiced person would prefer to believe that Primeval Humanity had at first an Ethereal—or, if so preferred, a huge filamentoid, jelly-like Form, evolved by Gods or natural 'Forces,' which grew, condensed throughout millions of ages, and became gigantic in its physical impulse and tendency, until it settled into the huge, physical form of the Fourth Race Man—rather than believe him created of the dust of the Earth (literally), or from some unknown anthropoid ancestor."[31]

In this process, Blavatsky also challenged the Creation myth of orthodox Judeo-Christianity whose hallowed Adamic ancestor is replaced by the giant Lemurians of the Third Root-Race and their successors, the Atlanteans. So she boasts: "The modern Anthropologist is quite welcome to laugh at our Titans, as he laughs at the Biblical Adam, and as the Theologian laughs at the former's pithecoid ancestor. . . . Occult Sciences claim less and give more, at all events, than either Darwinian Anthropology or Biblical Theology."[32]

In this, as in other regards, there is remarkable consensus among the Theosophists on the place of Lemuria in the history of Earth and of mankind, in contrast to the differences and dissonance that plague the paleo-sciences. For one, they are all agreed in consistently naming the continent "Lemuria" in their place-making, insisting that science itself had attested to its existence.[33] As Blavatsky wrote, "Lemuria is half the creation of Modern Science, and has therefore to be believed in."[34] Theosophists are also agreed that Lemuria was a Mesozoic (or "Secondary") continent which flourished from the end of the Permian until its eventual disappearance, after a series of cataclysmic events, sometime in the Eocene.[35] In contrast to the paleo-scientist who shies away from giving his fabulous imagination full rein in this matter, the Theosophist uses occult technologies of "reading" and "seeing" to fantasize about the ecoscape of Lemuria, as in the following example from W. Scott-Elliot's much-quoted *The Lost Lemuria* (1904):[36]

> The Lemurian man lived in the age of Reptiles and Pine Forests. The amphibious monsters and the gigantic tree-ferns of the Permian age still flourished in the warm damp climates. Plesiosauri and Icthyosauri swarmed in the tepid marshes of the Mesolithic epoch, but, with the drying up of many of the inland seas, the Dinosauria—the monstrous land reptiles—gradually became the dominant type, while the Pterodactyls—the Saurians which developed bat-like wings—not only crawled on the earth, but flew through the air. The smallest of these latter were about the size of a sparrow; the largest, however, with a breadth of wing of more than sixteen feet, exceeding the largest of our living birds of today. . . . [So] it is written in the stanzas of the archaic Book of Dzyan.[37]

As occult place-making gathers momentum over the course of the century, other such fantasies of Lemuria's landscape follow. Although they lack the intimacy of the ancestral homeland of the Tamil devotee, they nonetheless begin to translate Sclater's lost place-world from the remote terra incognita of the paleo-sciences into a more recognizable place, albeit one populated by monstrous animals and fantastic beings.

In Theosophy's place-making, Lemuria was peopled by the Third Root-Race, itself a successor to two others, an unnamed First Root-Race which had flourished on a continent called "The Imperishable Sacred Land," and the Second Root-Race of Hyperborea, a "*bona fide*" continent of the North Pole.[38] In contrast to the First Root-Race and to the Hyperboreans, the Lemurians were not form-less, speech-less, sight-less, or sex-less. Indeed, in the course of his evolution through seven stages (or "sub-races"), the Lemurian progressively developed a material body, began to walk erect, started to use his vision (with the help of a Third Eye), learned to speak (albeit in monosyllables), and, most importantly, took to sexual reproduction after millions of years of asexual procreation. This was a defining moment—about halfway through the Lemurian cycle, 18 million years ago precisely, when the fourth of the Lemurian sub-races evolved—for this is when Man attains humanity, by receiving "the gift of the mind."[39] Although Blavatsky herself was surprisingly reticent about the average Lemurian's physical appearance, her followers were quick to step into the breach and offer fabulous pen-portraits. Thus, Scott-Elliot's Lemurian of the fifth sub-race

> was gigantic, somewhere between twelve and fifteen feet. His skin was very dark, being of a yellowish brown colour. He had a long lower jaw, a strangely flattened face, eyes small but piercing and set curiously far apart, so that he could see sideways as well as in front, while the eye at the back of the head—on which part of the head no hair, of course, grew—enabled him to see in that direction also. He had no forehead, but there seemed to be a roll of flesh where it should have been. The head sloped backwards and upwards in a rather curious way. The arms and legs (especially the former) were longer in proportion than ours, and could not be perfectly straightened either at elbows or knees; the hands and feet were enormous, and the heels projected backwards in an ungainly way. The figure was draped in a loose robe of skin, something like the rhinoceros hide, but more scaly, probably the skin of some animals of which we now know only through its fossil remains. Round his head, on which the hair was quite short, was twisted another piece of skin to which were attached tassels of bright red, blue and other colours. In his left hand he held a sharpened staff, which was doubtless used for defence or attack. It was about the height of his own body, viz., twelve to fifteen feet. In his right hand was twisted the end of a long rope made of some sort of creeping plant, by which he led a huge and hideous reptile, somewhat resembling the Plesiosaurus. The Lemurians actually domesticated these creatures, and trained them to employ their strength in hunting other animals. The appearance of this man gave an unpleasant sensation, but he was not entirely uncivilised.[40]

This profile of the Lemurian was partly modeled, like Haeckel's and Keane's, on the Theosophist's fantasies of the black man who was declared his "degenerate" descendant. But it was also based on Blavatsky's contrary insistence that the Archaic Records speak of "towering giants of godly strength and beauty."[41] Even at his most developed, the Lemurian of the seventh sub-race retained "the projecting lower jaw, the thick heavy lips, the flattened face, and the uncanny looking eyes," although he "had by this time developed something which might be called a forehead, while the curious projection of the heel had been considerably reduced."[42]

It is Lemurians of the seventh sub-race who (along with their immediate predecessors of the sixth sub-race) developed "an important and long-lasting civilisation," under the benevolent guidance of divine elders from Venus called the Lhas, "the highly evolved humanity of some system of evolution which had run its course at a period in the infinitely far-off past."[43] These Lemurians were still barely human, "on the verge of attaining true manhood," but nonetheless, with divine supervision, they learned to use fire as well as the art of spinning and weaving. While they had no dogma or institutionalized religion, they followed simple codes of conduct and moral precepts, and worshipped "a Supreme Being whose symbol was represented as the Sun."[44]

They also built large cities and "Cyclopean" buildings, the remnants of some of which might still be seen in places like Easter Island.[45] Here, as in so many other regards, Theosophists insisted that their access to "lost" records enabled them to offer solutions to many a mystery that puzzled the archaeologist and the historian dependent on a limited material archive.

Thus, in its labors of loss around Lemurian society, of which it offered the earliest descriptions, Theosophy associated for the very first time Sclater's faunal bridge and Haeckel's savage "cradle" with that semantically charged Victorian category, "civilization." In ethnological place-making in the metropole, as we have seen, the inhabitants of Lemuria were barely human, one step above the ape on the ladder of evolution. Theosophists, however, intent as they were in rescuing man from the ignominy of simian descent, bestow a "civilized" status on the Lemurian, savage though he might appear in comparison to his more "evolved" successors on Atlantis: "Our modern Geologists are now being driven into admitting the demonstrable existence of submerged continents. But to confess the existence of the continents is quite a different thing from admitting that there were men on them during the early geological periods—ay, men and civilized nations, not Paleolithic savages only; who, under the guidance of their *divine* Rulers, built large cities, cultivated Arts and Sciences, and knew Astronomy, Architecture, and Mathematics to perfection."[46]

Indeed, in the Theosophical vision of human history, the ancient civilizations of Mesopotamia, Egypt, China, India, and Mesoamerica, so painstakingly being uncovered by classicists, archaeologists, and historians through their study of material remains, were but survivals of far earlier "higher lost cultures" which flourished not "on any of our present continents, but on an earth with a different division of land and water—when the Sahara was fertile soil, India an island, Central Asia a tremendous sea, when landbridges the size of continents connected Asia with Australia and Europe with America."[47] The submerged continents and landbridges of paleogeography thus provide the setting for Theosophy's lost civilizations of humanity, enabling it to write *longue durée* histories of mankind that extended far back into Earth's deep time which had been emptied of human achievement by the disenchanted material sciences. All the same, although Theosophy presents Lemurians of the seventh sub-race as "civilized," they are still not credited with the origins of all of human civilization; that privilege is reserved for the Atlanteans who followed in their wake. Nonetheless, this original Theosophical innovation of putting Lemurians on a road out of bestial savagery is useful for succeeding occult labors in which Lemuria is recast not just as the cradle of man, but also the birthplace of all human civilization and high culture, as we will shortly see.

About 700, 000 years before the Eocene, Lemuria began to break apart. Its catastrophic destruction was precipitated by an outburst of volcanic fire, not unlike that caused by the eruption of the Krakatoa in 1883 or of Mount Pelee in 1902 that impressed contemporary Theosophists, as it clearly did so many others.[48] Earthquakes soon followed, and Lemuria progressively fragmented and subsided, "leaving only such fragments as Australia and Madagascar behind, as traces of its story, with Easter Island, submerged and re-uplifted."[49] Even while this was coming to pass, an elect group of Lemurians had been led away to a spur of the submerging continent located at "latitude 7 degrees north and longitude 5 degrees west, which a reference to any modern atlas will show to lie on the Ashanti coast of today."[50] It was there, out of the ashes of the Lemurian continent and the seed of its dying inhabitants, that the Fourth Root-Race was born on the new continent of Atlantis.

The yoking of Lemuria's fate to Atlantis's fortune is another key Theosophical innovation that sets its apart from the place-making of the paleo-scientists of the previous chapter—and is yet another reason for the continuing popularity of Sclater's vanished continent in the occult imaginary. For the Theosophist, Atlantis had long existed, from the very origins of Lemuria, as a province of that continent. And yet, unlike Lemuria, "the dream of the scientists," Atlantis did not have science's blessing, and not least of the Theosophist's missions is to rehabilitate this mystical land which was responsible for nurturing the "veritable and complete human races—the Fourth and the Fifth."[51] Accordingly, Blavatsky complained in 1888, "Atlantis is denied, when not confused with Lemuria and other departed Continents, because, perhaps, Lemuria is half the creation of Modern Science, and has, therefore, to be believed in; while Plato's Atlantis is regarded by most of the Scientists as a dream."[52] Similarly, Scott-Elliot noted in 1904, "Although the lost continent of Atlantis has so far received scant recognition from the world of science, the general consensus of opinion has for long pointed to the existence, at some prehistoric time, of a vast southern continent to which the name of Lemuria has been assigned."[53]

This, too, accounts for Lemuria's significance in Theosophical place-making, for it enables the conversion of the "dream" of Atlantis into a hard "scientific" fact:

> The *Atlantic portion of Lemuria* was the geological basis of what is generally known as Atlantis, but which must be regarded rather as a development of the Atlantic prolongation of Lemuria than as an entirely new mass of land upheaved to meet the special requirements of the Fourth Root-Race. Just as in the case of Race evolution, so in that of the shifting and re-shifting of Continental masses, no hard and fast line can be drawn as to where a new order ends and another begins. Continuity in natural processes is never broken. Thus the Fourth-Race Atlanteans were developed from a nucleus of Northern Lemurian Third-Race Men, centered, roughly speaking, toward a point of land in what is now the mid-Atlantic Ocean.[54]

So, Sclater's Lemuria enables the Theosophist to rescue Atlantis from the fuzzy world of myth and legend and to relocate it in the more precise world of paleogeography, where it now comes to rest firmly and securely in the geologist's Eocene.

Lemuria and Atlantis have more than an umbilical connection; they are also envisioned as the homelands of the two Root-Races of the future, the Sixth and the Seventh. This, too, emerged in the fertile place-making of Blavatsky, whose Mahatma letters included Master Koot Hoomi's prediction that "when they reappear again, the last seventh Sub-race of the sixth Root-Race of present mankind will be flourishing on 'Lemuria' and 'Atlantis' both of which will have reappeared also (their reappearance following immediately the disappearance of the present isles and continents), and very few seas and *great waters* will be found then on our globe, waters as well as land appearing and disappearing and shifting periodically and each in turn."[55] Years later, Annie Besant (1847–1933) and Charles W. Leadbeater (1847–1934) put it a little differently when they wrote that in the twenty-eight century, America—the home of the Fifth Root-Race—would be shattered into pieces by earthquakes and volcanic outbursts (just as Lemuria had been, way back in the Eocene). At that time, a new continent would emerge in the Pacific to serve as the brand new home of the more evolved Sixth Root-Race. "Gradually will that new continent be upheaved, with many a wild outburst of volcanic energy, and the land that was once Lemuria will arise from its age-long sleep, and lie again beneath the sun rays of our earthly day."[56] As we will see, this prediction proves to be enormously productive for New Age labors of loss around Lemuria as the future utopian world of wisdom and well-being that would provide salvation for a materialist mankind that had lost its original unity with Spirit.

Notwithstanding the many innovations that Blavatsky brings to Sclater's Lemuria, and notwithstanding the fact that the "science of the spirit" that fueled her fertile fabulous place-making self-consciously poised itself against the materialist sciences of its time, there is little doubt that she was enormously indebted to the paleo-scientists' tentative speculations about the submerged Indian Ocean continent on which she, as indeed her followers, are quite clearly parasitic. In fact, in the very opening pages of volume 2 of *The Secret Doctrine,* Blavatsky marks this debt when she introduces Lemuria to her occult reader in the following terms:

> The third Continent, we propose to call Lemuria. The name is an invention, or an idea, of Mr. P.L. Sclater, who, between 1850 and 1860 [*sic*], asserted on zoological grounds the actual existence, in prehistoric times, of a Continent which he showed to have extended from Madagascar to Ceylon and Sumatra. It included some portions of what is now Africa; but otherwise this gigantic Continent, which stretched from the Indian Ocean to Australia, has now wholly disappeared beneath the waters of the Pacific [*sic*], leaving here and there only some of its highland tops which are islands.[57]

In a footnote, Blavatsky went on to observe that while Alfred Wallace had disapproved of Sclater's Lemuria and opposed the existence of a vast continent linking Africa and India, in a private letter to her "he admits that a much closer proximity of India and Australia did certainly exist, and at a time so very remote that it was 'certainly pre-tertiary.'" The renowned scientist had, however, conceded in this private letter that "'no name has been given to this supposed land.' *Yet the land did exist.*" And this land was, indeed, Lemuria, Blavatsky insisted.[58]

Similarly, when Annie Besant described the formation of Lemuria, she did so in terms that were quite explicitly geological—and that borrowed from the contemporary language of paleogeography and its fantasies of subsiding and erupting land formations:

> Meanwhile, the earth is slowly changing. . . . The huge sea to the south of Plaksha [Hyperborea] covered the desert of Gobi, Tibet, and Mongolia, and from the southern waters of this the vast Himalayan chain emerged. Southwards the land slowly appeared, stretching from the foot of the Himalayan range, southward to Ceylon, Sumatra, to far off Australia and Tasmania and Easter Island; westwards, till Madagascar and part of Africa emerge, and claiming Norway, Sweden, east and west Siberia and Kamschatka from its predecessor—a vast continent, the huge Lemuria, cradle of the Race in which human intelligence appeared. Shālmali, it is called in ancient story.[59]

As I noted in the previous chapter, notwithstanding arguments (and evidence) to the contrary, submerged continents and drowned landbridges were frequently invoked in paleo-scientific place-making in the later half of the nineteenth century, as were catastrophic explanations for their loss. The routines of planetary life—the convulsions of nature and the great earth movements—that the paleo-sciences were painstakingly discovering, are seamlessly incorporated into an esoteric vision of the cosmos, so much so that the Theosophist even boldly declared, "We have a sufficient block of geological knowledge already in our possession to fortify the cosmogony of the esoteric doctrine."[60]

All the same, occult place-making is different because of the use to which contemporary geological truths are put. So, where the paleo-scientist finds Lemuria necessary to account for the geography of terrestrial life-forms, and for the geography of early man, the Theosophist turns to the lost continent to spatialize the journey of Spirit and to narrate the geography of Being—also variously referred to as "the Pilgrim" or the "Human Monad"—as it travels through earth's history. Needing as they did both millions of years as well as a stable planetary surface to spatialize their alternative evolutionary narrative, the Theosophists found former continents like Lemuria (and Hyperborea and Atlantis) necessary for "the work of building a serviceable physical body" for Man as he journeyed from Spirit back to Spirit.[61]

But all this could happen only with some important changes. For one, for the paleo-scientist Lemuria was an Indian Ocean continent, although as Sclater had suggested in his christening act, it could have extended into the Atlantic as well. The Theosophist as well imagines Lemuria as an Atlantic landmass, for this facilitated the key mission to recuperate Atlantis which scientists might deny, but which occult records verified, as a province of the continent confirmed by paleogeography to have actually existed. But more often than not, Blavatsky and her followers were inclined to refer to Lemuria as a Pacific continent, even while acknowledging that it extended across the Indian and Atlantic oceans.[62] This geographical shift may reflect the American location of many of the early Theosophists, as well as the growing American readership of their occult writings, even as it enables the Pacific adventures of Lemuria in the twentieth century.

Second, for the paleogeographers, Lemuria was just one among the many landbridges or continental masses that spanned the Mesozoic oceans, and there was no attempt to suggest that it was the most important. Theosophists, however, constructed a spatial hierarchy of continents to parallel their racial hierarchy. For more than 200 million years, through the course of the Mesozoic, Lemuria was the only continent that existed on the face of the earth, although in its early years, remnants of Hyperborea were still around; in Lemuria's later years, even while it was subsiding, fragments of Atlantis began to congeal. And there was no disagreement (as among paleo-scientists) over Lemuria's continental status. As I document later, the English Theosophist Scott-Elliott even drew maps to graphically illustrate its continental proportions.

Finally, in Theosophical place-making, although Lemuria does eventually disappear as a result of catastrophic geological events, these are identified—drawing upon the riveting contemporary examples of the eruptions of Krakatau and Mount Pelee—as volcanic outbursts and earthquakes, rather than as subsidence or oceanic floods. And in contrast to the paleo-sciences, where it is never heard from again after its disappearance sometime in the late Mesozoic or early Tertiary—until, of course, Sclater's labors of loss rediscover it—Lemuria does resurface in the Pacific to stage the further evolution of the human spirit as it manifests itself in the Sixth Root-Race.

Like the ethnologist, the Theosophist, too, deems Lemuria to have been the birthplace of mankind, "the cradle of the Race in which human intelligence appeared."[63] Here, the figure of Ernst Haeckel looms large in Theosophy's place-making, and indeed, Blavatsky saved some of her choicest vitriol for the German biologist who had dared to go one step further than even Darwin and had identified the Lemurian as the missing link between ape and man. So, she disavows Haeckel's appropriation of Sclater's Lemuria:

By a curious coincidence, when selecting a familiar name for the continent on which the first Androgynes, the Third Root-Race, separated, the writer [referring to herself] chose, on geographical considerations, that of "Lemuria," invented by Mr. P.L. Sclater. It was only later that, on reading Haeckel's *Pedigree of Man,* it was found that the German "Animalist" had chosen the name for his late continent. He traces, properly enough, the centre of human evolution to Lemuria, but with a slight scientific variation. Speaking of it as that "cradle of mankind," he pictures the gradual transformation of the anthropoid mammal into the primeval savage!![64]

For Blavatsky and her Theosophical followers, although Haeckel had been right in identifying Lemuria as the cradle of mankind—in fact, beating them to the punch in this regard, their disavowals notwithstanding—he had clearly gotten the story of human evolution completely wrong, as all materialistic science was wont to do, by tracing it all the way back to a speck of protoplasm. "As to the idea that Haeckel's Moneron—a pinch of salt!—has solved the problem of the origin of life; it is simply absurd."[65] Further, they noted that while the German materialist was "correct enough in his surmise that Lemuria was the cradle of the human race as it now exists, but it was not out of Anthropoid apes that mankind developed."[66] This was the part of Haeckel's evolutionary labors of loss that Theosophists found most abhorrent. Like so many of their Victorian contemporaries, they were victims of what the paleontologist William King Gregory calls "pithecophobia, or the dread of apes—especially the dread of apes as relatives or ancestors."[67] In the Theosophical evolutionary narrative with which Blavatsky and her followers boldly coped with Darwinianism, apes only appear millions of years *after* the speaking human being. Their anatomical resemblance with man, of which material science made so much, was a consequence of the "sin of the mindless," namely, intercourse between beasts and the Lemurian in his fifth sub-race phase (after he had discovered sex).[68] So anxious was Blavatsky in establishing a distance between the ape and the human that her Lemuria did not even include Africa within its borders: "Eastern Africa, by the bye, was not even in existence when the Third Race flourished."[69]

This in itself is also revealing of another reason why Lemuria assumed significance for Theosophy in the late nineteenth century. Because it had been identified by the paleo-scientist, albeit tentatively and speculatively, as the possible cradle of mankind, it is yet another site on which to wage the battle against Darwinian evolutionary science, especially as this was being currently propagated by enthusiasts like Haeckel. Hence the Theosophical insistence that Lemuria had indeed been the cradle of man, but that man had descended from Spirit rather than from a mere animal. The logic of Theosophy's elaborate cosmic evolutionary history of seven successive continental formations on and through which Spirit worked its biography is one that was shaped by the need to counter Darwinian evolution with an alternate geography of Being that needed many worlds, indeed, many planets. As Alfred P. Sinnett insisted as early as 1883, "The evolution of man is not a process carried out on this planet alone. It is a result to which many worlds in different conditions of material and spiritual development have contributed."[70] Thus, the Darwinian theory of evolution had only discovered a portion, "unhappily but a small portion—of the vast natural truth. But occultists know how to explain evolution without degrading the highest principles of man."[71] Evolutionary science could look in vain for the missing link between ape and man because "that link which unites man with his real ancestry is searched for on the objective plane and in the material world of forms, whereas it is safely hidden from the microscope and the dissecting knife." All things, especially Man, "had their origin in Spirit—evolution having originally begun from above and proceeded downward, instead of the reverse, as taught in the Darwinian theory."[72] Indeed, as Peter Washington astutely observes, Blavatsky "transform[ed] evolution from a limited sociobiological theory into an explanation of everything from atoms to angels. Instead of opposing religion with the facts as presented by Victorian science, she attempt[ed] to subsume those facts into a grand synthesis that makes religious wisdom not the enemy of scientific knowledge but its final goal."[73] In other words, Blavatsky and her fellow Theosophists sought to end the monstrosity of a world without god that evolutionary science had ushered in. Lemuria, in turn, played a small but telling role in this consequential struggle.

While Haeckel was roundly condemned for his fatal materialist mistake of singling out a simian ancestor for man and for degrading humanity by tracing the sacrality of life to a speck of protoplasm, there was one important area of agreement. As I earlier noted, the German biologist had suggested that the Hottentots, the Caffres, the Negroes, and the Papuans were the closest living representatives of his Ape-Man, and he had even fantasized about what his Lemurian would have looked like on this basis. This meets with Blavatsky's rare approval: "Professor Haeckel must also have *dreamt* a dream and seen for once a *true* vision!"[74] Speaking through the voice of Master Koot Homi, Blavatsky had earlier intoned, in the Mahatma letters, "Behold, the relics of that once great nation [Lemuria] in some of the flat headed aborigines of your Australia."[75] But it was not just Australians who had a Lemurian connection. In Annie Besant's place-making from a few years later, "the aboriginal Australians and Tasmanians, now well-nigh extinct, belong to the seventh Lemurian sub-race; the Malays and Papuans have descended from a cross between this sub-race and the Atlanteans; and the Hottentots form another remnant. The Dravidians of southern India are a mixture of the seventh sub-race with the second Atlantean sub-race. *Where a real black race is found, such as the negro, Lemurian descent is strongly marked.*"[76]

Indeed, Theosophy generates a complex geography of human races in which all the black peoples of the world are either Lemurians or their degenerate descendants, while the most advanced peoples of today—white Caucasians—are members of the fifth Root-Race, far removed from them.[77] In the Theosophical evolutionism, as Spirit—or Monad or Pilgrim—works its way through the history of the earth, it "is compelled to incarnate in, or rather contact, every race"[78] As it marches across the history of the earth, Spirit manifests itself in the form of the various Root-Races and sub-races which it successively sheds as it surges upward toward our present Fifth Race, the most perfect so far. Those who get left behind—referred to variously as "sluggards" and "failures"—are destined to stagnate. Arguably, this enchanted evolutionary vision is much more racist and hierarchical than that espoused by many a contemporary disenchanted materialist, for millions of years separate the white Anglo-Saxon from the black aborigine whose origins are ascribed to "the racial decay" that besets the seventh subrace in the closing years of Lemuria's life on earth.[79] Further, rather than emerging from the more perfected forms of the Fourth Root-Race on Atlantis, as the majority of northern humanity do, the blacks of the world—"fallen, degraded semblances of humanity"—are deemed to be descendants of a "root-race" that was ultimately transcended by other, superior forms.[80] Lemuria is handy in this regard as well, allowing the Theosophist to not only place the lower, degraded specimens of humanity in a different time, but also to isolate them further from the more evolved races by tracing their origins to a totally different continental configuration.[81]

Even a century after Blavatsky first published her occult reading of the lost *Book of Dzyan,* her Theosophical followers continued to repeat her labors of loss with virtually no change, even though the paleo-sciences that had enabled them in the first place went on to repudiate the idea of drowned continents and moved on to other theories of the history of the earth's deep past.[82] But what is remarkable is the reach of Blavatskian place-making outside formal Theosophy. For instance, fantasy fiction writers like E. Charles Vivian and Lin Carter have made use of the Theosophists' Lemuria as the setting for their adventure stories.[83] Less benignly, as Nicholas Goodrick-Clarke's important study of the occult roots of German Fascism shows, Theosophical labors of loss also left their mark on the "Ariosophy" of Guido von List (1848–1919) and Jorg Lanz von Liebenfels (1874–1954), whose writings described a prehistoric Aryan golden age, racially pure and governed by the principles of a lost esoteric knowledge that the Ariosophists sought to revive. Such was the popularity of Theosophy in Germany that both Blavatsky's *The Secret Doctrine* and Scott-Elliot's *The Lost Lemuria* were available in German by the early years of the twentieth century. List and von Liebenfels put Theosophical evolutionism to work in their own place-making around a racially pristine prehistoric Aryan golden age that would serve as a template for the present. In this process, Lemurians were Germanized as the descendants of the mythical Teutonic giant Thrudgelmir, and they were eventually overcome—much as our own world's non-Aryans—by the racially superior Aryans who succeeded them.[84] As Goodrick-Clarke notes, these German occultists found Blavatsky's racial hierarchization convenient for their own project of exalting the Teutonic Aryan as the pinnacle of human development. At the same time, the Theosophical recasting of lost continents as former homelands peopled by distinct races was a useful strategy to adopt for these German Aryanists. Indeed, how much more effectively could one separate the Aryans from the more degenerate races of mankind than by isolating the latter on a continent that had belonged to a different geological time, a time when man had barely been human?[85]

Perhaps the most famous of Blavatsky's European followers was Rudolf Steiner (1861–1925) who, prior to his formal espousal of Anthroposophy, wrote extensively of Lemuria.[86] Steiner's labors of loss around Sclater's vanished continent were based on the so-called "Akashic Records," "Records of the Ether," which are "as much an undeniable reality as mountains and rivers are for the physical eye."[87] Given the occult premise of the essential unity of lost ancient wisdom, the Akashic Records reveal Lemuria's story to be remarkably similar to that found in Blavatsky's Archaic Records. All the same, given his Anthroposophical inclinations, Steiner also traced back to the Lemurian period the development of will power, as well as the powers of clairvoyant imagination which later either atrophied or utterly disappeared in man. Lemurian women, in particular, played a key role in the development of man's spiritual and psychic powers.[88] Anthroposophy, therefore, sought to rekindle those powers which once mankind had possessed—way back in Lemurian and Atlantean times—but which are now lost. Even after Steiner turned his back on Theosophy, preferring to systematize his own brand of occultism that was more firmly rooted in Christian esotericism rather than Oriental mysticism, the place-making labors of his "science of the invisible" continue to be faithful to Blavatsky's vision. His *Cosmic History: Prehistory of Earth and Man* (1923) repeats most of the occult assertions of his earlier labors around the lost continents of Lemuria and Atlantis.

Another offshoot of Theosophy which continued to be indebted to Blavatskian labors of loss was the Rosicrucian Fellowship, founded in 1907 in southern California by Max Heindel (1865–1919).[89] In Heindel's *The Rosicrucian Cosmo-conception* (1911), Theosophy is leavened with mystical Christianity, and Lemurians are cast in an Adamic role, although Lemuria itself does not appear to have been an idyllic Garden of Eden, its landscape being hot, fierce, and cataclysmic.[90] The Lemurian inhabitant of this primeval planet was a mystical creature, a "born magician," who felt himself a "descendant of the Gods, a spiritual being." He used his spiritual and clairvoyant powers to shape the world around him, utilizing his inner eye. There were Temples of Initiation, which were "High Schools for the cultivation of Will-power and Imagination, with 'post-graduate courses' in Art and Science."[91] This innocent Lemurian world underwent "the Fall" after its inhabitants' "eyes were opened." This laid the foundations for Man to become a fully conscious, thinking, reasoning being, making his way through the Atlantean and Aryan epochs. Like Blavatskian Theosophists, Heindel, too, claimed that "the greater part of the Lemurians were animal-like and the forms inhabited by them have degenerated into the savages and anthropoids of the present day."[92] When Lemuria had played its part in the cosmic evolution of mankind, it was destroyed by volcanic cataclysms, and "in its stead rose the Atlantean continent, where the Atlantic Ocean now is."[93] Heindel's narrative ends with a telling prediction, appropriate for the American context in which it was published: "From the last of the seven races of the Aryan Epoch, and from the people of the United States will descend the last of all the Races in this scheme of evolution, which will run its course in the beginning of the Sixth Epoch."[94]

Even Alice Bailey (1880–1949), an early pioneer of the New Age Movement who started her occult career around 1915 in the Theosophical Society before peeling off to form her own Arcane Society in 1923, embraces the Blavatskian place-making vision of successive lost continents which serve as the homelands of successive races. While some aspects of her occultism depart from standard Theosophy, her Lemuria retains its status as the home of the Third Race, predecessor to Atlantis, and as the former continent where Spirit on its journey through Earth first assumed a material and physical form.[95]

Thus, Blavatsky's appropriation of Lemuria from paleo-science's labors of loss in which it had at first circulated, and its insertion into an occult vision of earth's history and mankind's past, has been enormously productive and enduring. Thanks to Theosophy, Lemuria becomes a player, albeit in an off-modern and eccentric fashion, in the complex resistance mounted in post-Darwinian Euro-America by those who sought to reconcile the findings of the modern paleo-sciences with an older, religious worldview that placed man at the center of the universe. In other words, Lemuria's significance for the project of re-theizing the world lies in the fact that the material sciences themselves had revealed submerged continents and were attempting to reconstruct the life-forms that had inhabited these mysterious, now vanished, worlds. In insisting that Man, too, was one of these life-forms, and that he could not be banished from the deep time of Earth's past, Theosophy asserted that there were occult dimensions to human evolution of which material sciences were barely cognizant. As Besant and Leadbeater triumphantly recalled in 1922: "The most modern knowledge has vindicated the most ancient records in ascribing to our earth and its inhabitants a period of existence of vast extant and of marvelous complexity; hundreds of millions of years are tossed together to give time for the slow and laborious processes of nature; further and further back 'primeval man' is pushed; Lemuria is seen where now the Pacific ripples, and Australia, but lately rediscovered, is regarded as one of the oldest of lands; Atlantis is posited, where now the Atlantic rolls, and Africa is linked to America by a solid bridge of land."[96]

At the same time, unlike the physical scientists, who are constrained by the paucity of material testimony for their lost continents, the occultist can resort to clairvoyance and spirit communication to fill in the blanks. Indeed, as far as the occultist was concerned, the further back in time material science sought to push the antiquity of Man, the less capable it was of answering the all-important question of human origins. Scientists were searching in vain for clues to man's past in fossil remains and on ocean floors:

> When we resolutely turn the Soul away from earth and concentrate his attention on the Spirit . . . the Soul may reach the "Memory of Nature," the embodiment in the material world of the Thoughts of the LOGOS, the reflection, as it were, of His Mind. There dwells the Past in ever-living records; there also dwells the Future, more difficult for the half-developed Soul to reach, because not yet manifested, nor yet embodied, though quite as "real." The Soul, reading these records, may transmit them to the body, impress them on the brain, and then record them in words and writings.[97]

Theosophy's innovation lay in bringing occult technologies of clairvoyance and communion with other-worldly beings—abilities that were lost with the onset of disenchanted modernity—to bear upon the cosmic history of the vanished worlds of the earth's past, a history that had hitherto been told only from incomplete and mute material remains. That this innovation was to prove very productive is clear when we turn to the other twentieth-century adventures of Lemuria in the occult imaginary.

Lemuria in the United States: A Drowned Pacific Paradise

Theosophy, I have suggested, brings Lemuria into the realm of culture from the paleogeographer's nature by incorporating Sclater's submerged continent into the cosmic history of Man, where it is the birthplace of the Third Root-Race in the course of whose evolution humanity first attained physical form. All the same, Lemuria's status as the Garden of Eden is quite ambiguous, given the Theosophists' imperative to reserve that hallowed status for Atlantis. This is not the case, however, when stories of Sclater's lost Indian Ocean continent begin to circulate in, of all places, the United States, beginning in the 1890s. This was perhaps bolstered by a passing suggestion in Ignatius Donnelly's 1882 blockbuster on Atlantis, which mentioned "a drowned Pacific continent" that had once reached from India to South America. "Science has gone so far as to even give it a name; it is called 'Lemuria,' and here, it is claimed, the human race originated."[98]

The closing decades of the American nineteenth century saw the beginnings of attempts to fashion for a nation on the verge of global preeminence a history that would rival the Old World's in age and scale, as well as reaffirm the foundational premise of Americans as a chosen people living out their lives in their promised land that extended from sea to shining sea. The kingdom of God would be realized in and through American history.[99] These decades were also witness to a flourishing occult revival, a virtual "esoteric boom."[100] So much so that, as one contemporary commentator observed in 1928, the country was "overrun with messiahs."[101] In this climate, Blavatsky's Theosophical works, as well as Donnelly's *Atlantis,* both of which accorded a special place to America, did very well, going into multiple editions. Nonetheless, the idea of a drowned Pacific continent called Lemuria did not catch on immediately, struggling as it did to capture the imagination of a people mesmerized by the mighty Atlantis. Indeed, in 1890, one William Churchward lamented that while everyone knew Atlantis and that even children could chart its location, "who can offhand draw the lines of Lemuria? It is just as much a continent as the famed Atlantis, but who knows its former place upon the globe?" His letter to the *Brooklyn Times,* in which this 3,800-word lament appeared, included the following catchy headers: "Relics of an Extinct Race!" "The Drowned Continent of the Pacific Islands. . . . Only the Spires and Points of Land Now Peep Above the Waves But These Are Full of Interest to Travelers and Spectacled Men of Science." However, because Lemuria lacked "the two sages, Plato of Athens and Donnelly of Minneapolis," who have "made Atlantis into a household name," it was destined to linger on in the shadows of the lost Atlantic continent for quite some time.[102]

But by the 1930s, things changed, as a series of labors of loss inserted Lemuria securely, albeit eccentrically into the American "cult milieu,"[103] from where it has not been dislodged till this day. In 1908 a San Francisco occultist named Adelia Taffinder, who had obviously read her Blavatsky, Besant, and Scott-Elliot, published an essay in which she wrote about "a vast southern continent" called Lemuria whose evolution, "from the occult stand-point . . . constitutes one of the most interesting chapters of man's development."[104] Using the Theosophists' esoteric chronology, she went on, however, to give it an American twist when she wrote that "it is calculated that this submerged Southern Continent flourished 18,000,000 years ago, and that California was the center of a civilization that antedates the Continent of Atlantis, by thousands of years."[105] Where the Theosophists had suggested that Easter Island, and perhaps Madagascar, were surviving fragments of Lemuria after its catastrophic submergence, for Taffinder "the glorious land of California remained as the only monument in this part of the world to testify to the ancient grandeur of the land and the high civilization to which it gave birth."[106]

In turn, Taffinder might have been inspired to claim Lemuria as "the ancient mother of California"[107] by the 1894 publication of Fredrick Spencer Oliver's *A Dweller on Two Planets,* which established a lasting connection between Lemuria and California. The son of northern California miners, Oliver was a young Eureka teenager who insisted that he was merely serving as amanuensis to "Phylos the Tibetan," who had, once upon a time, lived on Lemuria (and, once it had disappeared, on Atlantis), but who currently inhabited the depths of Mount Shasta, which is home to a great brotherhood of masters who had preserved the perennial wisdom produced on the lost continents.[108] Oliver's biography is remarkable at several levels, not least because it points to the penetration into the remote reaches of late-nineteenth-century rural California of Theosophical labors of loss around vanished continents, of the identification of Tibet as a secret repository of the perennial wisdom of forgotten place-worlds, and of occult technologies such as automatic writing and telepathic communication with dead spirits.

Both Taffinder's identification of Lemuria as "the ancient mother of California" and Oliver's channeling of an ancient Lemurian's thoughts would have remained murmurs, lost in the cacophony of other occult and spiritualist voices that were gaining large audiences across the United States, if it had not been for the publication of a work that drew upon their claims as well as publicized them. In 1931, Wishar S. Cervé, a.k.a. H. Spencer Lewis (1883–1939), founder and Grand Master of the San Jose-based Ancient and Mystical Order of the Rosae Crucis (AMORC),[109] published his *Lemuria: The Lost Continent of the Pacific,* an occult classic that remains in print to this day.[110] The popularity of the work may lie in its attempt to link Lemuria with the early history of the Americas, as well as in its assertion that the "New World" may in fact be the oldest territory on Earth.[111] It is also possible that Cervé's appealing labors of loss around a great ancient civilization, probably American in origin, whose remnants might have survived into the present, gained an audience at a time when so many were combating the anxieties and uncertainties of the Depression.[112] In Martin Marty's assessment of religious life in the United States in the 1930s, "not content with the sober and undramatic attempts of the churches to meet something of the attainable, or to offer transcendent meaning where immediate goals seemed unattainable, a large public followed leaders who channeled their fanaticisms and furies into the realms of the impossible and the dangerous."[113] Cervé's labors of loss around a vanished Pacific paradise may be seen as a benign expression of Marty's "Depression extremism."

Cervé begins boldly by proposing that "the Garden of Eden" did not lie in "some Oriental country," as it was habitually thought, but right at home, in the United States.[114] As others have noted, from the early sixteenth century the Americas were imagined as the remnants of Plato's lost Atlantis.[115] So Cervé's originality lay not in suggesting their hoary antiquity, or even in supporting their candidacy as the birthplace of man, but in identifying these continents as remnants, not of Atlantis, but of Lemuria. In particular, the Pacific coast was singled out as a Lemurian landmass, where "we have the oldest of living things, the oldest of cultivated soil, and the more numerous relics of the human race which has reached a higher state of cultural development and civilization than any other races of man."[116] Others, as we have seen, had proposed Lemuria as the birthplace of man, but this was on the basis of its location in the Indian Ocean and its proximity to Africa and India. Cervé was among the first to draw upon Lemuria's relocation to the Pacific by Donnelly and the Theosophists in the 1880s to put in a claim on behalf of the Americas, and, in particular, the United States, as "the Garden of Eden." Herein lies his primary contribution to the proliferating labors of loss around Lemuria in the American occult imaginary.

But this was not his only innovation. Cervé also proposed that Lemuria had been "arbitrarily" named by scientists for a continent that was really referred to as "Mu" in "ancient writings."[117] In making this connection, if only in passing, Cervé enabled the ready identification of Lemuria as Mu that continues to this day, both among occultists as well as among scholarly commentators on lost continents.[118] Mu first surfaced in the pantheon of lost continents in the 1860s when the Flemish antiquarian Brasseur de Bourbourg (followed in the 1890s by the French archaeologist Augustus LePlongeon) announced that it was the ancient Mayan name for Plato's Atlantis.[119] In 1926, however, it was forcefully disassociated from Atlantis and appropriated as the name of a submerged Pacific continent by "Colonel" James Churchward (1850–1936), an English occultist living in Vermont. Churchward insisted that this renaming was authorized by ancient tablets written more than 15, 000 years ago in the long-lost Naacal language "written either in Burma or in Mu, the lost continent," which he had recovered when he had served in India, and deciphered with the help of a local "high priest."[120] The Land of Mu had once extended six thousand miles across the Pacific before it was shattered by cataclysmic earthquakes about 12,000 years ago.[121] Most importantly, "All centers of civilization had drawn their culture from a common source—Mu," and the many maps in Churchward's trilogy graphically illustrated this.[122] For the British colonel-turned-occultist, however, surviving remnants of Mu are to be found not in the United States, as for Cervé, but in the South Sea islands, some of whose populace today are descendents of the inhabitants of the lost Eden.[123] Churchward also made no attempt to link Mu to Lemuria, although the former's location in the Pacific undoubtedly facilitated this identification later on by Cervé. In so doing, Cervé brought about the convergence of the labors of loss around two vanished continents, contributing further to the density of Lemuria's presence in the occult imaginary.

Of all of Cervé's assertions, the one that ensured Lemuria would remain visible among American occultists is the suggestion that descendants of the former Garden of Eden might still be periodically seen as they reappear from their subterranean homes in northern California's Mount Shasta, which had once been a part of the now-vanished continent.[124] As do all occultists, Cervé as well narrates the destruction of Lemuria by "the greatest of all floods and catastrophes," which he dates to as recently as 50,000 to 25,000 years ago. When this happened, its inhabitants migrated with their ancient achievements to different parts of the world, including what was left of the Americas. Thus, Amerindians are descendants of Lemurians, and Mayans, in particular, are deemed of mixed Lemurian and Atlantean parentage.[125] If one were to look for "pure-blooded Lemurians existing today, however, we would have to seek for them among those who had descended from the first Lemurians and had remained on a land and in a climate as nearly like the early continent of Lemuria as possible."[126] This land was California, where Lemuria's purest established "perpetual memories to their highly advanced civilization."[127] Such a suggestion found immediate resonance at a time when California was fast consolidating its status in the American imagination as "the Golden State," as well as the haven for every new religion in town.[128] Every utopian site has been connected at some time to California: Atlantis, Arcadia, Avalon, the Garden of Eden, El Dorado, the Elysian Field, the Isle of the Blest, the Land of Prester John, and so on.[129] With Cervé's labors of loss, Lemuria seekers, mystics, and clairvoyants, as well, gravitated toward the Golden State. In particular, Mount Shasta became the new beacon for the occult, its hallowed status reaffirmed by the circulation of the notion that it was a remnant of the long lost continent of Lemuria.[130]

As I noted earlier, the first connection between Lemuria and Mount Shasta was made in 1894 in Oliver's *A Dweller on Two Planets,* but it was Cervé's book which consolidated this connection by singling out Mount Shasta as the spot where one encounters evidence for the continuing presence of ancient Lemurians among us:

> Many years ago it was quite common to hear stories whispered in Northern California about the occasional strange-looking persons seen to emerge from the dense growth of trees in that region. They would run back into hiding when discovered or seen by anyone. Occasionally one of these oddly dressed individuals would come to one of the smaller towns and trade nuggets and gold dust for some modern commodities. . . . They were tall, graceful, and agile, having the appearance of being quite old and yet exceedingly virile. They gave every indication of being what one would term *foreigners,* but with larger heads, much larger foreheads, head-dresses that had a special decora-

> tion which came down over the center of the forehead to the bridge of the nose, and thus hid or covered a part of the forehead that many have attempted to see and study.[131]

Going on to recount other mysterious occurrences that had been reported to him—cattle that responded to invisible signals, boat-shaped airships that floated through the air, chanting and singing emanating from the mountain top, mysterious energy blasts—Cervé concluded: "Thus we have had, until recently, one of the present-day groups of Lemurians hidden in isolation in modern California." This also showed that

> America has the honor and prestige of having on its soil the last survivors in a direct line of the first human race on the face of Earth. Here was the beginning of that race, and here will be the end of that race. This makes America, in fact, the oldest country of the world and yet the newest. Perhaps it is this happy combination, this rare association of the old and the new, this unique blending of the spirit of ancient culture with modern progressiveness, that makes the North American continent and its peoples from Canada to the Isthmus a great world of opportunity and golden fortunes.[132]

With these words, the foundational notions of American millennialism and nationalism—of the United States as the promised land extending from sea to shining sea—found a novel foothold in the most unlikeliest of places, Sclater's drowned continent.

Barely a year after Cervé's book was published, the *Los Angeles Times* carried a story entitled "A People of Mystery," written by one Edward Lanser, a California businessman who had recently visited Mount Shasta. Invoking some of the same sources as Cervé—including a "scientist" called Professor Edgar Larkin who had claimed to have spotted the Lemurians "in the heart of the[ir] mystic village . . . engaged peacefully in the manufacture of articles necessary for their consumption"—Lanser, too, concluded that on the slopes of Mount Shasta "live the last descendants of the first inhabitants of this earth, the Lemurians. . . . It is not, therefore, incredible that the last sons of lost Lemuria are nestled at the foot of Mount Shasta's volcano. The really incredible thing is that these staunch descendants of that vanished race have succeeded in secluding themselves in the midst of our teeming State and that they have managed through some marvelous sorcery to keep highways, hot-dog establishments, filling stations and the other ugly counterparts of our tourist system out of their sacred precincts."[133]

Long the subject of Amerindian collective memory, Mount Shasta was drawn into Anglo-American spiritual activities from the 1850s, when it began to attract all manner of mystics and mediums, clairvoyants and seers. With the publication of Cervé's book and Lanser's essay—and the re-publication in their wake of an 1899 sequel to Oliver's *A Dweller on Two Planets* called *An Earth Dweller's Return* (1940)—the Forest Service located at the base of Mount Shasta began to complain that the mountain was besieged by Lemuria seekers.[134] Foremost among these were followers of the so-called "Mighty I AM" movement that converged around the writings of Guy Ballard (1878–1939), who insisted that he has seen visions—"on the cosmic screen"—of the glorious life and civilization on Mu and Atlantis when he was taken deep into bowels of Mount Shasta by the "Ascended Master," Saint Germain. A member of the Great White Brotherhood (made famous by Blavatsky), Saint Germain's project was that of initiating the Seventh Golden Age, the permanent "I AM" Age of Eternal Perfection on Earth, for which, after searching in vain in Europe for centuries, he chose America, "the Carrier of the Cup of 'Light' to the nations of the earth in the Golden Age that is opening before us."[135] And the American he chose to convey this message to earth-dwellers was Guy Ballard. *Unveiled Mysteries* (1934), Ballard's narrative of his vision of past lives and lost civilizations, is another example of a Depression-era fantasy that offered a volatile mixture of spiritual hope and heady patriotism for a nation awash in the anxiety of the economic downturn.[136] At its height, the movement counted close to 350,000 members in California alone, and perhaps close to a million nationwide.[137]

Closer to our own time, Mount Shasta has also attracted the spiritualist-healer David Junglaus, who mounts there his "Lemurian–Atlantean Vision Wheel" to travel to other planes of existence. For Jungclaus, the mountain's sacrality is constituted by the fact that "the Akashic records," which contain details of every moment of every single individual's life story over countless eons, are housed in "secret storage areas" in the bowels of Mount Shasta.[138] Today, it is almost impossible to come across an occult publication which does not identify the mountain with Sclater's vanished place-world, so much so that a recent publication on Mount Shasta which identifies it as "home of the ancients," also describes it as "last refuge of the survivors of the Lost Continent of Lemuria."[139] And lest we think that Mount Shasta only figures in occultism's preoccupation with Lemuria, a secular reminder may be found in the "History Corner: California Legends," printed in the *San Francisco Chronicle* as recently as February 18, 2000. This narrated the story of one J.C. Brown, who insisted he had met a Lemurian deep inside Mount Shasta while prospecting for gold in 1904. The Lemurian told him, "We are . . . an ancient race of people. When the earth was young, we lived in a tropical land called Mu in the Pacific Ocean. Mu was paradise until it sank, leaving only the tops of our highest mountains above the waves, forming the islands you call Hawaii. After Mu sank, our people came to this mountain and established our village deep inside it." The raconteur, Jim Silverman, ends the column with the statement, "Real or not, Lemurian legends make good stories."[140]

It is not only occultists who were drawn to Lemuria as a drowned Pacific paradise. In 1933 Lewis Spence (1874–1955), a Scottish journalist and amateur folklorist best known for his numerous books on Atlantis, turned his attention to Sclater's lost continent.[141] "The tradition of a vast continent once occupying a large area of the Pacific basin has long exercised a charm of attraction compelling in its fascination. In some respects its geological history is more definite than that of sunken Atlantis."[142] At the same time, Spence suggested that the geological evidence for the existence of Lemuria, while substantial, was limited because it had caused Sclater, Haeckel, and other scientists to err in their conclusion that it had been an Indian Ocean continent, when it actually belonged in the Pacific.[143] But more importantly, the paleogeographer's contention that Lemuria had been a Mesozoic continent ruled out the possibility of human presence, whereas legends and traditions from the Polynesian islands—the remnants of Lemuria left today—pointed to the contrary. Spence therefore resorts to the Blavatskian strategy of subjecting ancient myths to a literal reading in his contention that Lemuria had once been inhabited by humans before its catastrophic submergence.

All the same, Spence is at his most critical in dealing with occult labors of loss around Lemuria, dismissing them as "the weakly [*sic*] effort of third-rate imaginations, wretched inventions which fall immeasurably beneath the avowed fictions of a Swift, an H. G. Wells, or an M. P. Shiel. The genuine mystic should shun this description of 'science' as coming from intelligences which, if not mischievous, are certainly equivocal and dubious."[144] So, his own labors of loss around Lemuria rescued it from the vise of the occultists and reestablished the truth of its former presence through a "logical," instead of a "pseudo-arcane," analysis of the legends that circulate in the islands which are remnants of the lost continent. These point to the existence of a former Pacific paradise which, before it was catastrophically destroyed, was home to "a white, fair-haired race which owes nothing to European admixture." This white race of rulers "instituted an extensive body of law and custom and a religion" which were subsequently bequeathed "in some measure" to the incoming Polynesians.[145] In Spence, we see how far the Lemuria of both Darwinian ethnology as well as Blavatskian Theosophy has been transformed, as the lost continent is detached from the pre-history of a black diaspora, and inserted into the antediluvian past of the white race. As Richard Ellis has noted, Spence was preoccupied with what he saw as "the moral decay" of Europe, especially Fascist Germany, a preoccupation which led him to look for new origins for white peoples.[146] Lost continents like Atlantis and Lemuria, on both of which he wrote extensively, enabled him to reclaim a new past and pristine beginnings. By locating the origins of the white race on the remote islands of the Pacific—far away from its current degenerate home in Europe, and in a region of the globe that recent imagination suggested was the last remaining Paradise left on Earth—Spence was able to put Lemuria to yet another use in its complex and much-variegated involvement in the cultural productions of the twentieth century. And he is not alone in this regard; others who labored around Lemuria as a lost Pacific paradise also claimed it for the white man. For Churchward as well, "the dominant race in the land of Mu was a *white race,* exceedingly handsome people, with clear white or olive skins, large, soft, dark eyes and straight black hair."[147] In doing so, both Churchward and Spence were not only bestowing an antediluvian past on the white race, but also claiming for it the credit for authoring the most ancient civilizations of the world, an achievement associated since the European Enlightenment with the Sumerians, Egyptians, Indians, and Chinese. Such labors of loss, as indeed those of Theosophists and the Ariosophists, are a telling reminder that even fringe cultural productions—and Lemuria is certainly no more than that in Euro-America—can play a part, albeit eccentrically, in movements of global significance such as the consolidation of modern white racism.

That by the 1930s Lemuria as a lost Pacific continent had become a part of the off-modern cultural landscape of Euro-America is also clear from its appearance in an increasing numbers of works of the "lost race" adventure-story genre. Such works have been seen as a reaction to the hegemony of realism that marked the literary productions of the industrializing nineteenth century, as well as products of new opportunities for demonstrating white male heroism in the age of empire.[148] The typical plotline of these stories takes the form of a band of intrepid adventurers, generally male and always white, and more often than not archaeologists, ethnologists, or anthropologists of some sort. They set out in search of a rumored lost civilization or a lost race (or chance upon these accidentally). Inevitably, when they reach this lost world—usually situated on a remote island or in the nether reaches of Earth—they encounter strange peoples or beings. Our white hero typically rescues a woman or two in the process of this encounter, which turns out to be frequently conflictual, and returns home triumphant with the girl and with a renewed understanding of Self.[149] G. Firth Scott's *The Last Lemurian* (1898) may well be the first fictional work of this sort to feature Lemuria.[150] Set in Western Australia, Lemuria appears as a primeval land, somewhere in the Pacific, in which our heroes have to battle the forces of evil (headed by "Tor Ymoothe, the evil Queen of Lemuria") in order to rescue a sleeping Lemurian princess and bring her back into (the white world of) modernity.[151] Other novels followed, such as Abraham Merritt's *The Moon Pool* (1918–19) and Charles Vivian's *City of Wonder* (1922), both of which drew upon Theosophical and other occult place-making around Lemuria.[152] So, in Muriel Bruce's *Mukara* (1930), which is heavily derivative of James Churchward's labors of loss around Mu which had been published a short while before, our hero, a British archaeologist named Jack Kirby, sets out to find a lost civilization in the Pacific and discovers the "Valley of Light" and the lost city of Mukara, which had been once part of Mu the Motherland, a.k.a. Lemuria. Our archaeologist-hero surmises, in true Churchwardian fashion:

> I would conclude that Mukara was in her highest glory when the continent of Mu went down, about 16,000 years ago, and that at that time, the beginning of the great pre-Mayan civilization was in progress: that civilization spread over what we now call South America, radiating from Mukara, and in all probability being the real foundation of the Mayan, the Aztec, and the Incan empires of which there are abundant archaeological remains. . . . The ruling races in the world today are the descendants of the founders of Mu's first colonies, and . . . the aboriginal races are accidental survivals, left-overs, as it were, of something that died with the major cataclysm. Here you have a suggestive theory that might, if carefully developed, serve to clarify the mysteries of the prehistoric ages recorded in the earth strata—such as the Capital Hill at Smyrna, the "terraces" of the Nile, and Niven's buried cities in Mexico. . . . Mukara was the centre of a civilization unbelievably splendid when the Uighur Empire was stretching its powerful arms from what is now the Gobi desert across Asia into Europe, all the way to Ireland, carrying the doctrines and symbols of Mu, and building the foundations of what we are and believe today.[153]

From the 1940s, fictional works place Lemuria in Earth's subterranean depths, from where its inhabitants frequently attempt to guide the misguided moderns on the surface.[154] In the Thongor of Lemuria series authored by Lin Carter in the 1960s, Lemuria appears as a primeval land in some unspecified geographical locality, populated by dragons and other monstrous beasts which our hero ("Thongor Valkarth—mightiest warrior of the ancient continent of Lemuria, before the dawn of history") battles, albeit with the help of super-technologies like airships and ray guns.[155] In fact, from the 1960s, the adventure-story featuring Lemuria mutates into science fiction, in which Lemurians battle Atlanteans or other extraterrestrial beings for control of the universe in true Star Trek and Star Wars fashion.[156]

Although none of these works have become bestsellers, the very appearance of the lost continent as a site for fictional action for both American and European authors points to Lemuria's off-modern and eccentric visibility in Euro-American popular culture.[157] A striking example is available from the 1940s when a series of articles appeared in the popular American science fiction magazine *Amazing Stories,* edited by Roy Palmer. Their author, Richard Shaver (1910–75), a Pennsylvania steel mill worker, insisted that he was acting as "a racial memory receptacle of a man (or should I say a being) named Mutan Mion, who lived many thousands of years ago in Sub Atlan, one of the great cities of ancient Lemuria!" Shaver observed that, as a memory receptacle, he remembered Lemuria "with a faithfulness that I accept with the absolute conviction of a fanatic."[158] The Lemuria of Shaver's labors of loss is far removed from the dominant American image of the continent as a vanished Pacific paradise. Instead, it is a subterranean world inhabited by giant beings with access to all manner of technology, such as teleportation gadgets, rock-piercing long distance rays, and space rockets. When it was abandoned, about twelve thousand years ago, the descendants of the giant Lemurians—christened "abandonderos"—lingered on and are today responsible for all manner of problems that plague surface-dwellers on Earth. Such was the popularity of *I Remember Lemuria* when it first appeared in March 1945 that it created what *Life* magazine termed "the most celebrated rumpus that rocked the science fiction world," even while it added tremendously to the circulation of *Amazing Stories,* which serialized Shaver's memories until 1949.[159] It is quite possible that Shaver's "paranoid" vision of Lemuria may have been shaped by the years he might have spent in a mental hospital,[160] but what is also remarkable is that a blue-collar worker from the steel mills of Pennsylvania had heard of Sclater's lost continent at all.

Another indication of the off-modern and eccentric penetration of Lemuria into American collective memory is available in a quixotic file of the U.S. State Department's Office of the Geographer.[161] The file contains the correspondence over two decades between the U.S. government and various individuals regarding a group of islands off the east coast of Panama. Although the story begins in October 1933, I will pick it up toward the tail end, when the State Department received a letter (dated November 16, 1954) on a letterhead bearing the title "Government of Atlantis & Lemuria" from one Gertrude Norris Meeker, Governor-General. In the letter, she declared that in 1943 a group of islands (two hundred miles south and west of Florida, eight degrees north of the Equator, and three miles offshore of Panama and Costa Rica) was the "private Dynasty or Principality . . . named "Atlantis Kaj Lemuria." The letter also informed the State Department that "any trespassing in these islands or Island Empire is a prison offense." The State Department's Special Advisor on Geography, Sophia A. Saucerman, responded politely on December 7, 1954, that "in the conduct of the foreign relations of this Government, the Department of State does not recognize any so-called 'private Dynasty or Principality named Atlantis Kaj Lemuria.'" Meeker replied to this disavowal by offering a brief history of her "Principality," which she insisted had been founded in 1917 by a Danish seaman, John L. Mott, at a time when Germany was at war with Denmark. Mott and his friends had not wanted to return to a war-torn Europe, and instead they had settled on these islands and sought recognition for them as the remnant, first, of the lost continent of Atlantis, then of Mu/Lemuria.[162] Meeker concluded, acerbically, "I am not some quirk hunting a so-called 'lost continent'—these islands exist and do belong to my dynasty."[163] A few year after this exchange between Meeker and the U.S. government, the State Department received a letter, dated February 23, 1957, from one Leslie Gordon Bell, who declared that he had been retained as "legal counsel for the heirs of a Country, State, or Principality known as Atlantis & Lemuria and consisting of a few islands in the Caribbean (in the vicinity of the Virgin Islands)." Because the islands belonged to "the Mu Group in the Pacific Ocean, designated Lemuria," the name of the Empire is "Atlantis & Lemuria. . . . Believe me, this is not a figment of somebody's imagination."

Although in an internal memorandum, dated April 23, 1957, the government decided that "this matter is not to be taken too seriously," the Office of Geographer was also charged with determining "whether the islands of Flamingo, Odino and Thoro in the Caribbean Sea in the vicinity of the Virgin Islands, and 'the Mu Group in the Pacific Ocean' actually exist." The Acting Special Advisor on Geography made his determination and wrote to Bell, on May 15, 1957, that "the Department of State has in its files no credible information as to the existence of any modern State or Empire bearing the name of the legendary lands, or countries, referred to as Atlantis & Lemuria. The United States does not recognize any so-called Principality of Atlantis & Lemuria and the Department of State has no information that would indicate that the 'Principality of Atlantis & Lemuria' has been recognized by other governments."

The story of Atlantis-Kaj-Lemuria does not end here, though. In January 1958 Craig Hosmer, a member of the U.S. Congress from the 18th District in California, took an interest in Gertrude Meeker's affairs. He wrote a letter, dated January 13, 1958, to the State Department on behalf of the California-based *Southland Magazine.* Evidently, the magazine had carried a story (dated July 11, 1954) on Meeker, noting that "she had been bequeathed all rights, titles and properties of a constitutional monarch in a scattered island empire. . . . These were once known as the Danish Virgin Islands West but are now recognized internationally as the 'Atlantic and Pacific Empire of Atlantis and Lemuria.'" Hosmer went on to inform the State Department that he had learned that "Mrs. Meeker understands that by renouncing her U.S. citizenship she could become Queen of these islands, but as a citizen she can rule as governor-general." And then came the punch line: "She states that she is getting ready to do some leasing for development work on some of these islands." The Congressman wondered whether these islands were, "in the opinion of the State Department, a constitutional monarchy, an individually owned (Mrs. Meeker) principality, or the actual property of some country."

Faced with a letter from a U.S. congressman, the State Department felt compelled to reply, and quickly. On January 17, 1958, three days after receiving Hosmer's letter, an Assistant Secretary responded, "To the best of our knowledge, no scattered islands, amounting to an island empire exists. With the possible exception of a few shoals or uninhabited rocks, all islands in the area in question are under the jurisdiction of some sovereign state, or are disputed by two states. However, the Geographer of this Department is *most willing* [my emphasis] to make a geographical study of this matter if we in turn can have more information upon which to base our investigations." The file ends with this letter.

In addition to these exchanges, there are other letters from sundry individuals, writing from various parts of the United States, asking the State Department about the Principality of Atlantis, "Atlantis Kaj Lemuria," or "the empire and government of Atlantis and Lemuria." In each case, the State Department responded patiently but formulaically: "No credible information has been received with reference to the existence of any modern country (neither empire nor principality) bearing the name of the legendary continent of Atlantis, beneath the Atlantic, and Lemuria, beneath the Pacific."

I have discussed this obscure file at some length for several reasons, including the fact that it is a telling sign of the extent to which Euro-American labors of loss around Lemuria had reached a certain degree of popular circulation by the 1950s, albeit these have been largely limited to California, which had offered a hospitable home for such off-modern and eccentric fantasies from the 1890s. To the best of my knowledge, it is also the only instance in which a modern metropolitan state gets entangled in the many quixotic adventures of this vanished place-world. But most of all, it is a wonderful instance of how lost continents, by virtue of the fact that they are lost, and hence in no one's active possession, can provide a fertile opportunity for all manner of projects, including political ones such as the establishment of alternate states, in order to secure freedom from persecution (as in the case of Mott) and the renunciation of citizenship and real-estate development (as in the case of Meeker).

The labors of loss around Lemuria as a drowned Pacific paradise are marked by three key elements which distinguish them. First, Lemuria is unambiguously a lost Garden of Eden, fulfilling the yearnings and hankerings of a post-industrial spiritualist generation searching for pristine origins. No longer a remote paleo-continent inhabited by lemurs, dinosaurs, and Mesozoic monsters, nor the home of bestial "ape men," or of a "Root-Race" barely at the threshold of full humanity, it is instead the abode of the "most perfectly formed human beings who ever lived; straight as an Indian [*sic*] and as perfect a specimen of manhood as the Infinite Intelligence and the Eye of God could visualize."[164] Like Blavatsky before him, Cervé, too, insisted that "there is no evidence that warrants the belief that the Lemurians were descendants of any lower species of the animal kingdom. In fact this belief is abhorrent to every profound student of the Lemurian civilization."[165] Not surprisingly, in this moment of occult place-making, Lemuria survived well past the deep time to which it had been consigned by the paleo-scientist. Instead, it subsided barely 50,000–12,000 years ago for Cervé, and only 12,000 years ago for Churchward.[166] This is the principal reason that memories of its existence continue to linger on in ancient traditions, although the place-maker laments the fact that when the continent disappeared catastrophically, it took its records with it, which can be only recovered now through occult technologies (or, in Spence's case, through a literal reading of folk stories).

Further, not only are Lemurians of this occult moment genuinely human, they are also the most civilized of all beings, and Lemuria itself is recast as "the real cradle of the civilization of man."[167] The Theosophical attempt to civilize the Lemurians had rested on the claim that they had built "Cyclopean" monuments, and perhaps had led some kind of spiritual life, but there is very little sense that future generations could learn from this Lemurian example. This is yet another important transformation in the status of this place-world by the 1930s, as Lemurians are recast as epitomes of spiritual and technological perfection. "We are probably now treading the same road which our forefathers trod over 100,000 years ago."[168] Even while Lemurians were technologically advanced, they did not abandon, as moderns so clearly had, a life of spiritual and moral perfection, following as they did "the highest idea of God produced for the sole purpose of manifesting God."[169] As we will see, this transformation of Lemurians into paragons of both science and spirit has important consequences in the New Age moment that was to follow.

Not least, for the first time in the metropole, this occult moment produces some amount of grief over the loss of the continent, grief that is akin to the nostalgia that is so endemic to Tamil labors of loss around Lemuria. Here, for example, is Churchward lamenting the loss of Mu: "Poor Mu, the motherland of man, with all her proud cities, temples, and palaces, with all her arts, sciences and learning, was now a dream of the past. The deathly blanket of water was her burial shroud. . . . Where man once reigned supreme was now the abode of fishes and the haunt of uncanny, creepy things. . . . Everything was gone! All was lost!"[170]

This, too, facilitates the subsequent New Age transformation of Lemuria, from a continent of the Earth's remote paleo-past into a future utopia of wisdom and well-being for all mankind. Convinced as the mid-twentieth-century occultist is that Lemurians had led a life of perfect plenitude on their Pacific paradise, (s)he seeks to recover not only the memory of their very existence, but also to reinstitute their lives as a model for the present and the future. It is this attempt to live life as it had been led on Lemuria that marks the New Age moment in the occult imagination to which I now turn.

New Age Lemuria: The Abode of Lost Wisdom

There is a double genealogy for Lemuria's New Age incarnation:[171] the suggestion by Cervé and other occultists of the 1930s that descendants of the wise Lemurians are still among us, albeit living secret and hidden lives that only the occultist can apprehend; and also the notion voiced in the early 1880s by Master Koot Hoomi in Blavatsky's Mahatma letters that Lemuria would resurface in the future as the home of the Sixth Root-Race, more spiritually and philosophically "evolved" than us moderns.[172] This clears the ground for fast-forwarding Sclater's Lemuria out of paleo-time to the timeless future of the New Age: "Science acknowledges that virtually every portion of the earth's surface now above water was at one time submerged. . . . And conversely, virtually every portion of the earth's surface now beneath the sea was at one time above water. . . . Since these are facts that cannot be successfully contested . . . a New Earth must eventually replace our present continents."[173]

Thus, for these occultists, the scientific contention that Earth is ever-changing provides an opportunity to ground their critiques of existing institutions and practices—the "Old Order"—in the truths of the paleo-sciences, and to note that the New Age that was dawning would be based on a different configuration of continents from that which had produced the corrupt materialistic civilizations of the present. A New Earth for the New Age: this is the key place-making contention of this occult moment, and Lemuria—the continent predicted by Blavatsky to reemerge from its oceanic grave—assumes renewed significance in this context.[174]

For some, Lemuria occasionally figures as a lost Pacific continent that had once flourished in the distant past and will reappear in the future as the site for a universal human utopia; for other New Agers it is not a phantom continent of Earth's prehistory but a flourishing civilization that exists even today, but in another dimension, "on another frequency." In such imaginings, Lemurians are recast as "higher aspects of our souls," or alternatively, the more awakened among us are seen as reincarnations of ancient Lemurians. But more often than not, as had already happened in the course of its appearance in adventure fiction from the 1940s, Lemuria has increasingly become a resonant place-name with no fixed geographical location.[175] Rather than a specific place, it has instead come to signify a state of wholeness, superior wisdom, well-being, peace, and harmony that once was in man's distant past and to which we can all aspire in the present and future as part of the New Age personal and psychic transformation. It has even become a state of mind. Liberated from any specific geographical referent or historical contingencies, today Lemuria lends its name—rendered mysterious, alluring, and wondrous through a century of occult labors of loss—to New Age bookstores and healing programs, to meditation centers and utopian communes, and writers' retreats.[176] Its off-modernity usefully serves the countercultural agendas of the New Age.

The New Age Lemuria bears little resemblance to Sclater's lost continent, or even Blavatsky's, for that matter. Although the occasional New Age place-maker remembers the continent's ancestry in the speculations of Victorian natural history or in the spiritual and epistemological struggles against Darwinism, this does not really matter, thanks to the many occult labors of loss since that have fashioned for it a new identity as an abode of lost wisdom. The promise of Lemuria for the New Agers is the promise of the recovery of its lost wisdom and the reestablishment of a new world order based on "on an entirely new geographical allocation of continents."[177] As one such place-maker declared in 1939: "A history of the world is a history of man's divergence from the primal, concordant condition which obtained in his Edenic Paradise. From an Age of Innocence, man has wandered through the wilderness of his own erroneous, finite thinking. . . . He stands [now] upon the threshold of an Age of Virtue which will surely restore to him his long lost, but divinely endowed, birthright of life, liberty and the pursuit of happiness."[178]

It is this premise that led to the establishment by Robert Stelle, in September 1936, of the Lemurian Fellowship, located first in Chicago and then moved briefly to Milwaukee before finally settling, in 1938, in Ramona, California, where it is still headquartered.[179] The founding assumption of this utopian community is that "humanity began its pilgrimage in this world from a now Lost Continent buried beneath the waters of the Pacific Ocean. . . . From this hypothesis, we can conclude only that all peoples, regardless of nationality, are essentially members of one Root-Race, which was Lemurian then, is Lemurian now, and always will be Lemurian."[180] The Fellowship prepares its members for the New Order which will be established with the "submergence of Imperialistic Europe and Eastern North America."[181] At that time, a new Lemurian Continent will resurface off the California coast. In preparation for this grand event, Fellowship members conform to the ideals and principles—the Lemurian Philosophy, or the Lemurian Cosmo-Conception—that had once prevailed on Lemuria (also called Mukalia), which had produced, seventy-six thousand years ago, "the greatest civilization the world has yet known."[182] The Lemurian Cosmo-Conception would have been lost with the catastrophic submergence of the former continent, but foreseeing the cataclysm, the Lemurian Aristocracy had moved out of their Motherland and "stored their accumulated Wisdom of the Ages in secret archives on the Asiatic mainland where They knew it would be safe during the rapidly approaching cataclysm."[183] Today, on the eve of the dawn of the New Age, this lost Wisdom was being channeled through the Lemurian Fellowship and would help rejuvenate a corrupt and materialistic society. The New World Order which will be established by "a New Lemurian Super Race on a New Lemurian Continent" will build super cities with the help of solar energy. These cities will be free of pollution and industrial waste, of crime and disease, of class warfare, greed, and poverty. Out of the ashes of decrepitude, a new Heaven on Earth will be reconstituted with the help of the principles of a lost Lemuria.[184]

Like the I AM movement, the Lemurian Fellowship deploys the same formula of a perennial wisdom, formerly lost but now recovered, combined with a heady patriotism, based on the notion of America as the promised land, and the fantasy of a utopian future where the materialist evils of modern civilization will be transcended under the leadership of enlightened (white) Americans. Thus, an advertisement for a book called *Lemuria the Incomparable,* published by the Fellowship, announces:

> There has come into existence in America an Institution of Higher Learning dedicated to the preservation of the Ancient Wisdom which was responsible for the unrivalled grandeur of two of the world's mightiest prehistoric civilizations—Atlantis and Lemuria. This School teaches that these two super Empires have been reborn in America as a New Order of the Ages (Novus Ordo Seclorum) and that this New Order is today reviving all of the arts, crafts and industries of the ancient world so as to establish a new standard of living for the New Citizen. Does this philosophical, industrial, and racial programme attract you[?][185]

It is difficult to ascertain the membership strength of The Lemurian Fellowship or its popularity, but it produced its own spin-offs, such as the Stelle Group, which was founded, along with its sister society, the Adelphi Organization, by Richard Kieninger (b. 1927). Kieninger started out in the Lemurian Fellowship in the 1950s, but by 1963 went on to set up his own commune in Stelle, Illinois, which later moved to Dallas.[186] Kieninger, too, as did Ballard and Stelle before him, insisted that he had been contacted by "higher beings"—some of whom had lived on Lemuria—to lead select Americans to create a new community out of the ashes of our present decrepit society. Under the name Eklala Kueshana he published his utopian vision in *The Ultimate Frontier* (1953), which enumerated Ten Laws—akin to the Ten Commandments, but also the Bill of Rights—based on the philosophy that had prevailed on the lost Lemuria and that had produced its "master civilization."[187]

More recently, the imperative to establish a utopian society that would mirror the former utopia that had prevailed on Lemuria has also led the Hawaii-based Gurudeva Sivaya Subramuniyaswami to establish a "Lemurian mountain top monastery" on the island of Kauai, itself deemed a remnant of the lost Pacific Paradise. The inmates of the monastery live lives according to the principles laid out in the Lemurian Scrolls, which had once been written within "the great walled Lemurian monasteries" but were lost until Subramuniyaswami recovered them through clairvoyance, very much in the tradition of Blavatsky and her lost Stanzas of Dzyan, Steiner and his Akashic Records, and Churchward and his Naacal Tablets. The narrators of the Scrolls chronicled mankind's journey to Earth from the Pleides, and now with the help of the Gurudeva we, too, can "read in depth about Lemurian culture, its unique perspective on inner and outer worlds, and the Lemurian's approach to manifesting our divine nature."[188] Subramuniyaswami's aim is to reproduce a life of harmony and well-being in his new commune as close to the original ideal that had prevailed on Lemuria, as a model for the world more generally. So, he invites everyone to "Come to Lemuria!"[189]

Lest we think that it is only the fringe elements of the New Age that have seized upon Lemuria—probably because, unlike Atlantis, it was more appropriatable, given its eccentricity even within the occult world—it is worth noting that Sclater's continent plays a key role in the activities of one of its most celebrated channels, J. Z. Knight (née Judith Hampton, b. 1946). A former housewife living in Tacoma, Washington, Knight was contacted by an entity called Ramtha, who had resided on Lemuria (along with two million others) about thirty-five thousand years ago and who had fled to Atlantis when the continent had submerged. In 1977, when Ramtha first contacted Knight, he told her, "I am Ramtha, the Enlightened One, and I have come to help you over the ditch."[190] Knight subsequently established Sovereignty, Inc. and for a decade or so in the 1980s had quite a large following of New Agers who were convinced that adhering to Ramtha's teachings about a Lemurian order based on the worship of the One Unknown God and communication by thought would put them on the path to Self-realization.[191]

It is thus with the New Age that the channeling of the thoughts of Lemurians becomes something of an occult fashion, Knight/Ramtha being among the more famous of these manifestations.[192] The phenomenon of channeling suggests that the entity being channeled continues to exist, albeit in a different plane of consciousness. In New Age place-making, therefore, in contrast to other occult moments, Lemurians are not dead beings of the past, lost forever with the disappearance of their homeland, leaving behind their records for the occultist to recover. Instead, they flourish in altered states, and are available as guides and mentors for those who are fortunate enough to sense and recognize their existence. In particular, these Lemurian guides and their Lemurian wisdom have been incorporated into New Age healing regimes, and all manner of holistic therapies that have found their way to the metropolitan spiritual marketplace (from "ear candling" and "organic cures," to "meditation," "dream therapy," and "relaxation techniques") invoke their connection to Lemuria.[193] So the New Age labor of loss is also a labor of healing.

Not least, New Age labors of loss have found particularly attractive the claim advanced since the 1930s that Lemuria had been a site of both spiritual and technological perfection. The New Age utopia is not one which subordinates science to spirit, or spirit to science, but one which envisions the balanced coexistence of the two, as prevailed on lost continents like Lemuria and Atlantis. The good life on Lemuria had been possible because Lemurians tapped into the energies stored in Earth's very depths, which helped them fashion a civilization far beyond the capability of moderns. The fact that Lemurians and Atlanteans traveled intergalactically, used ray guns, deployed space communicators and teleporters, and harnessed solar energy, even while they led lives of spiritual perfection, enables the New Age place-makers—used to modernity's technological conveniences and yet anxious over their consequences—to dream of a utopian future when science and spirit might be reconciled after their catastrophic disjuncture in the present.

So, in the final analysis, New Age labors of loss take the form of a countercultural and utopian critique of the present and consequently "are a political intervention in contemporary politics."[194] Humanity had, once upon a time, lived a life of peace, love, harmony, and prosperity, even while pursuing a life of scientific and technological perfection. All this was lost, just as the worlds that had supported such wonderful lives—like Lemuria and Atlantis—had been lost and forgotten. Remembering these worlds and their achievements would remind humanity today of what had been lost, and lead us to repair the degenerate present in order to engineer a future where, once again, peace, love, and harmony would coexist with material prosperity and technological advancement, as they once had on the lost continent of Lemuria.

THE POETICS AND POLITICS OF OCCULT LOSS

Lemuria thus has no singular identity in Euro-American occult place-making, as it moves from the Indian to the Pacific Oceans to the subterranean or astral realms, and to nowhere at all. Lemurians, too, are variously configured: as barely human progenitors of mankind whose direct descendants are the "black races" of today's world, but who are by and large transcended and replaced by the more advanced Atlanteans; as the civilized ancestors of all humanity who first appeared in the Americas; and as ethereal inhabitants of other dimensions, alternate realities, and astral realms whose thoughts may be channeled into the here-and-now for their superior wisdom. Even the labors of loss vary, from the Theosophical imperative to reconstruct a lost geography of Spirit on its travels through the history of Earth to return to its source, to those interested in uncovering the secrets of a lost civilization and lost race, mankind's first at that, to the New Age preoccupation with recuperating lost wisdom and lost harmony in the name of a utopian future. But varied though these projects might be, they are all united by the singular concern with reintroducing god back to an Earth that had become a-theized through the runaway triumph of a positivist materialist science that has reduced "nature's majesty" to matter, and matter alone.[195]

Scholars have convincingly demonstrated that post-Enlightenment esotericism has not been opposed to science per se as much as it has been critical of its mechanistic and materialistic excesses. These, it is contended, transformed the universe, the abode of spirit, into a vast machine of matter and motion impersonally obeying mathematical laws. The esotericist project in such an a-theized world is to insist that spirit exists and functions in the universe as surely as matter, and perhaps more so.[196] Paradoxically, at its most imperial moment, mechanistic and materialistic modern science had narrowed the reach of human inquiry, limiting it to objects that were immediately ascertainable by the senses and that could only be verifiable through its positivist methodology. For the committed esotericist, however, there are worlds and experiences outside and beyond science's ken that are available to those who cannot find contentment in the finite, the evanescent, the incomplete. Thus, Rudolf Steiner insisted, "Everything belonging to the outer world of sense is subject to time, and time destroys what in time arises."[197] In contrast, in spiritual science—or "the science of the invisible"—the investigator "presses on from evanescent history to that which does not pass away":

> He who has won for himself the power to observe in the spiritual world, there recognizes bygone events in their eternal character. They stand before him, not as dead witnesses of history, but in the fullness of life. In a certain sense, the past events are played out before him. Those who have learnt to read such a living script can look back into a far more distant past than that which external history depicts—and they can also, by direct spiritual perception, describe those matters which history relates, in a far more trustworthy manner than is possible by the latter.[198]

So, occult thought clears for itself what Ashis Nandy in another context refers to as "some moments of freedom" from the "shackles" of history.[199] Steiner therefore insisted, "It is but a small part of prehistoric human experience which can be learnt by the methods of ordinary history. Historic evidence throws light on only a few thousand years, and even what archaeology, paleontology and geology can teach us is very limited."[200] Occult history perforce sees its mission as filling in the blanks that could never be filled by disciplinary history, archaeology, or any of the materialist sciences.

And this, indeed, was Blavatsky's innovation. Up until the time of *The Secret Doctrine*, as the English Theosophist K. Browning noted, "the question of the existence of lost continents had been dealt with from the viewpoint of ordinary scientific knowledge."[201] Blavatsky changed this by showing that it was possible to gain extra-ordinary knowledge of these lost worlds through trained clairvoyance that allowed her to read "the Archaic Records" of peoples and places long forgotten by material science and indeed incapable of ever being discovered by it. As Mahatma Koot Hoomi observed to Sinnett:

> No doubt your geologists are very learned; but why not bear in mind that, under the continents explored and fathomed by them, in the bowels of which they have found the "Eocene Age" and forced it to deliver them its secrets, there may be, hidden deep in the fathomless, or rather *unfathomed* ocean beds, other, and far older continents whose stratums have never been geolog-

ically explored; and that they may some day upset entirely their present theories, thus illustrating the simplicity and sublimity of truth as connected with inductive "generalization" in opposition to their visionary conjectures.[202]

And, in fact, Blavatsky's innovation has proved to be immensely fruitful, as occultist after occultist who followed in her wake turns to all manner of "lost" sources which show up the incompleteness of the positivist evidence of modern knowledge practices rooted in materialism. These range, as we have seen, from Steiner's Akashic Records, which are a repository of everything that ever happened in collective spiritual history, to Churchward's Naacal Tablets, written in a long-forgotten language, to the newly discovered Lemurian Scrolls of the Hawaii-based Himalayan Academy, once written in a Lemurian monastery. They also include Edgar Cayce's psychic "readings" and New Age "channelings" of the lost thoughts of Lemurian entities. These "living scripts" enable the occultist to fill up the landscape of Lemuria rendered empty and bleak by the work of the materialistic paleoscientist with his "ordinary" methods and "limited" archive. So, for the Theosophist Eugen Georg, such "racial memories" lay the foundation "for a greatly extended conception of prehistory": "Behind the disguises of fancy the essence of truth lies hidden: the testimony as to the fate and experience of mankind, of lost cultures, of the evolution of vanished races of men in early geologic times, and the magic union of man and the elements during the Secondary and Tertiary periods."[203]

Thus, in clearing a space for itself from the shackles of history, occultism generates its own eccentric methodology based on clairvoyance, telepathy, communion with spirits, racial memories, and so on. It also resorts to an astral archive that no self-respecting disciplinary historian would consider legitimate but which to the occultist is the repository of everything that has ever happened from Creation, not just on Earth but on other planets as well. So, the occultist triumphantly compares her "imperishable" archive with the mundane bits and pieces—"the testimony of things"—with which the materialist scientist struggles.[204] Fittingly, the occultist's astral archives are typically located in secret, hidden, buried, or submerged places—the so-called "power spots" of the New Age—far outside the reach of the materialist scientist, such as Tibet,[205] Easter Island, Polynesia, Hawaii, the depths of Mount Shasta, or the bowels of Earth, where survivors from Lemuria (and other vanished place-worlds) had secreted away what they could salvage from their disappearing continent(s). Also appropriately, those who had been dismissed by metropolitan disciplines as "people without history"—Tibetans, Eskimos, Mayans, Native Americans, not to mention Indians and other "Orientals"—are the "store keepers" of these secret archives, guarding the ancient wisdom and knowledges of the lost Lemuria that modernity has discarded till the occultist learns about them through esotericist practices.

Thus, the occult imaginary introduces a completely new—and enchanted—evidentiary base into metropolitan place-making around the lost Lemuria. As we have seen, the paleo-scientist summoned Lemuria into existence through studying the fossil remains of prehistoric flora and fauna, by scrutinizing similarities in rock formations across continents or on ocean floors, or through considering the geography of mankind's races. These are all positivist practices painstakingly established over the course of the professionalization of the paleo-sciences from the nineteenth century. But the occultist bypasses these, and instead deploys "records" of a totally different sort which need a completely different set of powers that only a select few are fortunate enough to possess. The very loss of Lemuria required novel technologies and novel archives to recover and reconstruct the former existence of the continent, and only occult practices can meet the challenge of responding to this novel need. The record of the rock was to be transcended by the record of the spirit.

Freedom from the terror of history there might well be for these occultists, but there is no freedom from the tyranny of the natural sciences. As moderns, they cannot bypass the knowledge-making claims of these sciences. But, more importantly, they have no interest in doing so, intent as they are in showing them up for what they are—evanescent, incomplete, limited—even while going beyond to clear the ground for establishing a "science of the spirit," a science of the invisible and hidden.[206] "Esoteric Philosophy, let us remember, only fills the gaps left by Science and corrects her false premises."[207] Materialist science itself contributes to this new project of establishing a "science of the spirit" in its revelations about submerged continents, which are appropriated into occultism as the productive sites for its own secret archives and lost records. And Lemuria lends itself to this project precisely because, as Blavatsky noted at the beginning of the long occult century, it is "half the creation of Modern Science."

But Lemuria is only half the creation of Modern Science, precisely because occult labors of loss had demonstrated that its existence was already known to the Archaic Knowledges secreted away in hitherto-lost archives. Lemuria is also only half the creation of Modern Science because occult labors of loss contended that, with modernity's disenchantments, humanity had also lost the all-important powers of clairvoyance and psychic communication which had once prevailed among Lemurians and Atlanteans. In their absence, Modern Science had to fall back on the limited materialist archive of rocks and fossils to reconstruct the history of Earth and of man. In the foreword he wrote in English Theosophist Scott-Elliot's 1896 book, *The Story of Atlantis,* Sinnett insisted that literary memoranda, stone monuments, and fossil remains have given us only "a few unequivocal, though inarticulate assurances concerning the antiquity of the human race; but modern culture has lost sight of . . . [other] possibilities connected with the investigation of past events." He went on to write: "The world at large is thus at present so imperfectly alive to the resources of human faculty, that by most people as yet, the very existence, even as a potentiality, of psychic powers, which some of us all the while are consciously exercising every day, is scornfully denied and derided."[208]

Astral clairvoyance, on the basis of which Scott-Elliot's "pioneering essay" on Atlantis was written, allowed the occultist "to get touch [*sic*] with . . . other records in the vast archives of Nature's memory."[209] For,

> there is no limit really to the resources of astral clairvoyance in investigations concerning the past history of the earth, whether we are concerned with the events that have befallen the human race in pre-historic epochs, or with the growth of the planet itself through geological periods which antedated the advent of man, or with more recent events, current narrations of which have been distorted by careless or perverse historians. The memory of Nature is infallibly accurate and inexhaustibly minute.[210]

So, a science of the spirit, which the occultist insists is necessary to complete the incomplete claims of the materialist sciences, is only possible through a renewed effort to recover the clairvoyance and psychic abilities that man once possessed but had since lost. The occultist was aware that his use of clairvoyance to rewrite the history of the past would be dismissed as "fantastic, or accused of groundless speculation," or, worse still, as "a work of imagination."[211] But as Rudolf Steiner smartly retorted, "All we ask of the ordinary scientist is that he shall accord to the student of the Higher Science the same toleration as the latter shows to the mode of thought of Physical Science."[212] The methods of clairvoyance and psychic readings were not without their problems, but "the trustworthiness of such observations is certainly far greater than in the outer world of sense."[213]

Occult technologies thus enable the redemption of that which had been deemed by the materialist sciences as lost for ever. They may be lost forever to the world of material senses, but not so to the occultist who is able to find them in other realms—the astral and the subterranean—and recover and rehabilitate them. So, in occult place-making, the loss of Lemuria is not irrevocable, as it is for the paleo-scientist or even for the Tamil devotee. In the final analysis, this is what sets apart occult labors of loss from others around Lemuria.

LEMURIA AND OCCULT LABORS OF LOSS

The last chapter dealt with the dispassionate and disinterested quest by men of science for the lost worlds of deep time. I characterize the labors of loss undertaken in the name of science as dispassionate enchantment, partly because of the imperative to ground the paleo flights of imagination about lost worlds and submerged continents in the mundane evidence provided by rocks and fossils, earth movements, and ocean floor configurations. As occultists repeatedly insisted, what material science could recover about Lemuria was limited by this imperative. In contrast, occultism's labors of loss are more transparently enchanted because they are liberated from this constraint of grounding their claims in evidence that is subject to measurement, verification, and the other mundane demands of legitimate scientific methodologies. Occultism counters paleo-science's record of rock and fossil with the intangible and immeasurable record of Spirit, and the record of the memories of former Lemurians and other astral beings. The techniques used to "read" this record through clairvoyance, telepathy, and psychic communication are also enchanted, in the sense that they fall outside the realm of normative scientific procedures and protocols.

Occult labors of loss are also enchanted because they are driven by the imperative to re-theize a world from which god had been dismissed by the material sciences. The restoration of the lost unity between spirit and matter, and of the lost nexus between religion and science, is what occultism seeks. The paleo-scientist's labors of loss around Lemuria, I have suggested, have been conducted for the sake of completing the human stock of knowledge. They are propelled by a will to wholeness. Occultism's labor of loss, on the other hand, is compelled by a will to unity: the wholeness that modern science was seeking had always already existed in humanity's distant past on lost continents, like Lemuria, but had been shattered and fragmented. For the occultist, science's quest for wholeness would never be successful precisely because spirit had been cast out and discarded. It was only a science of the spirit that would restore mankind's lost unity and wholeness, lost wisdom and lost powers of clairvoyance.

Finally, occult labors of loss are enchanted precisely because of the claim that the "imperishable" record of the extra-ordinary world beyond the senses allowed the occultist to recover all that has been deemed lost or missing. Yes, one could never really return to the former Lemuria, at least not in this body, but it was possible to remember everything that had transpired on the now-submerged continent, and to even recreate a life and community in the here-and-now, as the Lemurian Fellowship sought to do in the middle decades of the last century, based on this reconstructed memory.

Over more than a century of occultism's interest in Lemuria, its labors of loss have varied, as has the nature of its preoccupation. The occultist has used Lemuria for a variety of purposes, including reconstructing a lost geography of Spirit as it travels through Earth to be reunited with its source. Lemuria, literally, is one of the grounds on which Being manifests itself. It is also the realm of mankind's first civilization and first race, the "motherland" of humanity, for some. Not least, as the repository of the lost Word and lost wisdom, it is the site for the future utopia in which the lost unity would be recovered, as humanity itself would move to a New Age where modernity's disjuncture between spirit and matter, and between science and religion, would be overcome and undone.

And what of "India" in all of this? India was central to the paleogeographer's place-making precisely because its faunal and floral remains, its fossil rocks, and its unique tectonics intrigued many a man of science from the early years of the nineteenth century. The subcontinent is crucial to the occult imaginary as well, but this is the India of the Orientalist rather than of the scientist, hallowed for its spiritualism, its great repository of Sanskrit philosophy and traditions, and its reputed antiquity. For Blavatsky, initially, India is "the cradle of the race," and even after the cradle shifted further West and to the Pacific, she invokes "the most archaic Sanskrit and Tamil works" as "teem[ing] with references to both Continents [Lemuria and Atlantis]."[214] As others have noted, the Theosophical moment in modern occultism would not have been possible without (Hindu and Buddhist) India, which is recast as the repository of all manner of secret knowledges and archaic wisdom, rivaled only by Tibet. Although it has been much more Western in orientation, this recasting is particularly noteworthy in post-Theosophical occultism, across whose labors of loss "India" continues to flicker as a hallowed land. James Churchward found his Naacal Tablets there, hidden in the vault of an old temple in Burma, then a part of British India. For Robert Stelle and his Lemurian Fellowship, after the disintegration of Lemuria/Mu, India was one of the lands that was colonized by those Lemurians who were more spiritually inclined, the materialist among them heading West out of the distressed continent. Knight's Ramtha is none other than a former incarnation of the Hindu god Rama, for once his Lemuria had submerged, he and his fellow survivors headed toward India. In all such place-making, the "spiritual" Indian is imagined as a descendant of the Lemurian, and, correspondingly, "Indian" wisdom is turned to, again and again, as a source of inspiration on the lost Lemuria.

A telling reminder of this may be found in Paul Brunton's *The Message from Arunachala* (1936), in which the author, a leading conduit for channeling "Eastern wisdom" to the disenchanted West in the middle decades of the last century, revisits Tiruvannamalai ("Arunachala") to remember his erstwhile guru, the mystic-sage Ramana Maharshi.[215] He recalls that an American geologist had told him that Arunachala was as hoary as our very planet, and this is because "it was indeed a remnant of the vanished continent of sunken Lemuria." In addition, like Blavatsky before him, Brunton refers to "Tamil traditions" that mention the lost continent, and observes that he had heard from Ramana Maharshi himself that "the lost continent of Lemuria had once stretched all the way across the Indian Ocean, embracing Egypt, Abyssinia and South India in its confines."[216] Such is the power of the hallowed mountain that it continues to lure Brunton back, again and again. After his visit to the top of the mountain, Brunton returns to "the world beyond its Lemurian height" with tablets that had been written "at the bidding of a strange Messenger who gazed at me commandingly and said: "'Give ear, my son, take up your pen and write. Search deep within your mind for its most vital thoughts. . . .' Not many days passed before it seemed to me that I must unburden myself of those charactered Mosaic tablets, those indictments and commandments which I had carried down from my strange Sinai. . . . So, I present these souvenirs of my last wanderings upon Arunachala, and submit them to the wizardry of the printing-room."[217]

In the tradition of Blavatsky's *Book of Dzyan,* Churchward's Naacal Tablets, or, more recently, Subramuniyaswami's *Lemurian Scrolls,* Brunton's Mosaic tablets, with their "message from Arunachala," derive their resonance in the Euro-American spiritual marketplace precisely because of that last half a century of occult labors of loss which had come to indelibly associate Sclater's Lemuria with the lost Word and lost Wisdom.

Chapter 4

Living Loss at Land's End

Alas! What can we say about the suffering heaped upon the ancient Tamil land, which gave birth to the arts and to human civilization?[1]

INTIMATE ELSEWHERES

By the opening years of the twentieth century, far away from the metropolitan sites where it had hitherto largely circulated, Sclater's Lemuria found a new and enthusiastic following in the Tamil-speaking region of colonial India in the context of the upsurge of language consciousness and mobilization that I have characterized as "Tamil devotion."[2] A complex network of praise, passion, and practice centered on the adoration of the Tamil language, Tamil devotion mobilizes metropolitan labors of loss over drowned continents and submerged landbridges toward a brand-new narrative of origins in which Lemuria is recast as the birthplace of the Tamil people, their ancestral homeland lost catastrophically to the ocean. As a consequence, Lemuria acquires a commemorative density in the Tamil country over the course of the twentieth century that it does not command anywhere else in the world.[3] Indeed, for its Tamil place-makers, everything in the known world—be it (the Tamil) man or music, medicine or the martial arts—had its origins in Lemuria. Convinced as they are about the utter state of humiliation and neglect in which their beloved Tamil languishes in colonial and postcolonial India, its devotees have lived out their own lives in the shadow of decline and loss, as I have documented in my *Passions of the Tongue.*[4] It is their everyday experience of despair and yearning that powerfully anchors their labors of loss as they fantasize about Lemuria as their former homeland, a place of promise, plenitude, and perfection that had once existed elsewhere but no more. Hence also their intense preoccupation with its catastrophic loss, which leads them to teach it in schools and colleges and has secured for Sclater's lost world, for the first time, anywhere in the world, the patronage of the modern state.

Indeed, in January 1981, during the Fifth International Conference of

Tamil Studies held in historic Madurai, a short documentary titled "Kumarikkaṇṭam" was screened in Tamil and English. Produced with the financial support of the Tamilnadu government and the personal backing of Chief Minister M. G. Ramachandran, the documentary recounts an ancient tale of origins in the paleo-scientific language of modernity. It traces the birth of Tamil and its literature to the very beginning of time on Lemuria, referred to also by its Tamil name Kumarikkaṇṭam. In the documentary's recounting, the paleo-history of the earth turns around the Tamil land, language, and literature. In such a planetary vision, the history of Tamil and its modern speakers is both deeply temporalized and ambitiously spatialized: the entire world was Tamil's domain, once upon a time, millions of years ago.[5] With the making of this film, and with the pedagogical circulation of Lemuria in schools and colleges, Sclater's lost continent has been officially installed in Tamil collective memory at the heart of a catastrophic narrative about the loss of the prelapsarian Tamil past and self. This is, of course, in striking contrast to its presence in Euro-America, where it has been largely confined to the occasional scientific conjecture or to place-making on the occult fringe.

In what follows, I document how and why Lemuria accumulates this density of commemorative meaning in the Tamil country by first considering its transformation from the homogeneous paleo and occult place-worlds of Euro-America into an intimate Tamil *home-place* that is catastrophically lost to the ocean. Labors of loss accompany paleo-scientific and occult place-making around Lemuria in Euro-America, as we have seen, but the loss experienced by its Tamil place-makers appears much more profound and personal precisely because it is not some remote paleo land or occult domain that vanishes, but the very birthplace of the Tamil language, literature, culture, indeed, the Tamil man—the Tamil prelapsarium. Thus, it is under the sign of the catastrophic disappearance of everything that belongs to Tamil, and the impossibility of their return, that labors of loss around Lemuria take place in Tamil-speaking India. Hence also the fascination with it among so many.

COLONIAL LABORS OF LOSS

As was so often the case in colonial India, but for a passing interest expressed by sundry British scholars and administrators in Sclater's vanished continent, the transformation of Lemuria into an intimate Tamil place might never have happened. Fleeting though the references to Lemuria in the colonial archive may be, they are accorded a significance in Tamil labors of loss that far exceeds their original objectives. In fact, colonial pronouncements on Lemuria are frequently misquoted as the Tamil place-maker hastens to incorporate their limited speculations into his own certitudes. In this process, what is evanescent in the imperial archive assumes a hard materiality beyond it, as marginal statements and conjectural footnotes come to be (mis)cited over and again in the cause of Tamil devotion. Most consequentially, it was primarily through the colonial place-making of Lemuria that metropolitan labors of loss around it filtered through to the Tamil life-world, where the words (and even the names) of a Philip Sclater or an Ernst Haeckel are fetishized, sometimes beyond recognition.

In the colonial archive, Lemuria appears in two somewhat contrary incarnations which are extensions of the Euro-American labors of loss I earlier identified as the paleogeographical and the ethnological. In the former, as we have seen, Lemuria is summoned into existence as a Mesozoic continent (or "landbridge") that sprawled across the Indian Ocean. It is in this guise that it puts in an appearance in the subcontinent for the first time in 1873—just a few years after speculations about its existence had begun in the metropole—in a physical geography schoolbook by Henry F. Blanford (1834–93).[6] While not naming Lemuria as such, Blanford informed Indian student-readers that "at a very early period," their India was connected with southern Africa, and that faunal and geological evidence favored "a communication between the two regions." However, as a result of "the enormous outburst of volcanic activity" at the end of the Cretaceous, this land-connection submerged and India as we now know it today from modern maps took its shape.[7]

Soon after, in 1879, the Geological Survey of India (GSI) published its first synthetic *Manual* of the geology of the subcontinent, which cautiously but unambiguously discussed the Mesozoic land-connection between southern India and Africa.[8] With this publication, Lemuria's presence in Earth's paleo past was officially authorized by the colonial state, and indeed H.B. Medlicott and William T. Blanford's statements in the *Manual* remained the GSI's official position on the paleogeography of India into the 1950s.[9] So much so that D.N. Wadia (1883–1969), a college professor who had also worked for the Geological Survey, declared in his widely prescribed college textbook in 1919: "The evidence from which the above conclusion regarding an Indo-African land connection is drawn, is so weighty and so many-sided that the differences of opinion that exist among geologists appertain only to the mode of continuity of the land and the details of its geography, the main conclusion being accepted as one of the settled facts in the geology of this part of the world.[10]

This "settled fact" of the subcontinent's paleo-past appears, if only in passing, in many general works on the subcontinent by British as well as Indian authors,[11] as it increasingly does in school and college textbooks.[12] That this paleogeographic imagination about a lost Indian Ocean continent had become part of a colonial common sense is also apparent from the following statement in E.M. Forster's acclaimed novel *A Passage to India* (1924), whose celebrated section on the Marabar caves begins thus:

> The Ganges, though flowing from the foot of Vishnu through Siva's hair, is not an ancient stream. Geology, looking further than religion, knows of a time when neither the river nor the Himalayas that nourished it existed, and an ocean flowed over the holy places of Hindustan. The mountains rose, their debris silted up the ocean, the gods took their seats on them and contrived the river, and the India we call immemorial came into being. But India is really far older. In the days of the prehistoric ocean the southern part of the peninsula already existed, and the high places of Dravidia [*sic*] have been land since land began, and have seen on the one side the sinking of a continent that joined them to Africa, and on the other the upheaval of the Himalayas from a sea. They are older than anything in the world.[13]

So, how did Tamil place-makers benefit from such colonial labors of loss? First, although the paleo land-connection to Africa was not always called "Lemuria," and although a continental status was not always granted to it, nonetheless the very fact that the physical sciences, especially geology, sanctioned its presence meant that Tamil labors of loss around Kumarikkaṇṭam, based on literary tradition, could receive immediate credibility merely by citing their findings.[14] Not surprisingly, even when pursuing contrary agendas, Tamil place-makers repeatedly invoke European "scientists" and their utterances to confirm the "scientific" reality of their ancestral home-place.

Moreover, Tamil place-makers are also enormously gratified that the colonial scientist favorably contrasted the great antiquity of the peninsula with the relative geological youth of the northern plains, the so-called heartland of India. "In the Deccan we are . . . in the first days of the world."[15] Here is Wadia in his widely quoted college textbook on the geology of India: "The great alluvial plains of the Indus and the Ganges, though, humanly speaking, of the greatest interest and importance as the principal theater of Indian history is, geologically speaking, the least interesting part of India. In the geological history of India, they are only the annals of yester-year."[16]

Indeed, the geological proposition that the northern plains might have come into existence at the expense of the former great southern landmass meant that the putative violence perpetrated by "the north" over "the south" in historical time had deep geological precedent as far as fervent devotees of Tamil were concerned. In fact, as many of them frequently insisted, the mighty Himalayas had not even existed when the first Tamil speakers had lived out their lives on their antediluvian homeland before it drowned.

Finally, most members of the colonial geological establishment were loyal Lyellians in their conviction that the earth as we know it today had been dramatically different in the past. Although they did not necessarily entertain catastrophic conceptions of Earth's deep history along the lines of Suess and some of his continental colleagues, the language of their place-making was punctuated with allusions to turbulent oceans and violent earth upheavals that caused the subsidence and loss of land. This, in turn, as I document later, bolstered Tamil labors of loss around the "cruel sea" and "the mischievous ocean" whose destructive work resulted in the disappearance of a beloved homeland.

Helpful though such colonial labors of loss were to Tamil place-making, there is one important regard in which they were a letdown. For colonial paleogeographers, as for their Euro-American counterparts, Lemuria was a Mesozoic landmass whose existence and disappearance far pre-dated the arrival of the first humans on Earth in the late Pliocene or early Pleistocene. It had nothing to do with the human, let alone Tamil, past. Yet, as in Europe, other colonial bureaucrat-scholars were not entirely convinced of this, and it is in their place-making that the important proposition that Lemuria might possibly be the lost Dravidian homeland is first aired in India, proving to be immensely fruitful for the Tamil agenda.

The foundational text for this all-important claim was another official text, *The Manual of the Administration of the Madras Presidency*, published in 1885 by the Madras government. Its author, Charles D. Maclean (1843–1916), worked for the Indian Civil Service from 1865 to 1893, holding numerous administrative positions in the Madras Presidency.[17] In the *Manual*'s "Ethnology" chapter, where he discussed the racial constitution of the Presidency in some detail, Maclean included a long footnote—one of the most repeated in Tamil labors of loss—entitled "Sketch history of race movements as they may be inferred for Southern India." He began by questioning the prevailing tendency "to suppose that all population questions in India are to be explained by migrations, and those in a southerly direction." In particular, he wondered about the wisdom of assuming that the Dravidians came into India from the northwest and rushed "in a mighty body to the south." Having raised such doubts, he introduced the Tamil reading public to Ernst Haeckel, whose labors of loss around Lemuria as "the primeval home of man" he helpfully recycled. He also consolidated the German biologist's theory of the origin of the various races of mankind on the drowned Indian Ocean continent by reiterating that it was the "primeval home" of the ancestors of "the Dravidas of [India] and Ceylon."[18] Further, Maclean insisted that the Dravidians are indigenous to India, and that "there are no living representatives in Southern India of any race of a wholly pre-Dravidian character."[19] Indeed, reversing the terms of migration theories then popular among Victorian and colonial ethnologists, he even suggested that "Southern India was once the passage-ground by which the ancient progenitors of northern and Mediterranean races proceeded to the parts of the globe which they now inhabit" from Lemuria.[20]

Maclean's *Manual* was published in the heyday of the colonial ethnological and bureaucratic preoccupation with classifying the diverse populations of the subcontinent into the master racial categories of "Aryan," "Dravidian," and "Kolarian," with determining their original homelands and subsequent migrations, and with ranking them on an evolutionary scale ranging from "savage" to "civilized." By the 1880s, while there was general consensus that "the Aryans" were later arrivals to the subcontinent, and undoubtedly more civilized, there was much debate over the "aboriginal" population of India, as there was over the original homeland of one of these potential autochthones, the Dravidians. While Maclean's *Manual* did not end these debates, it added a new dimension by proposing a Lemurian homeland as the solution to "the riddle" of Dravidian origins. This was a theory that was embraced—if only fleetingly and tentatively—by other colonial bureaucrats and ethnologists who wrote about Dravidian origins in the next few decades, including influential men like Edgar Thurston (1844–1935) and Herbert H. Risley (1851–1911).[21] The decennial census reports of the government of India of both 1891 and 1901 briefly aired the theory—not without disapproval—in their discussion of the origins of the Dravidians.[22] And from around 1909 well into the present, history schoolbooks incorporate Maclean's theory in their own speculations on the prehistoric movements of various races into the subcontinent as they discuss the possibility that Lemuria—which they frequently do not name, but refer to as "a land that existed once upon a time to the south of India" or as "a vast landmass in the Indian Ocean that submerged"—*could* have been the ancestral homeland of the Dravidians.[23] Indeed, as recently as 1981, a history textbook in Tamil, published under the auspices of the government of Tamilnadu for sixth-grade students, tells them in the course of a discussion of prehistoric Aryans and Dravidians that "some say that in the Indian Ocean several thousand years ago there was a vast landmass called Ilemūriākkaṇṭam [Lemuria continent], and that was the birth place of Dravidians."[24]

Such colonial pretexts for the Tamil project around Lemuria notwithstanding, its place-making is not accompanied by the deep, wrenching sense of grief that is so characteristic of Tamil labors of loss. The "primeval" inhabitants of Lemuria—soon to disappear into the ocean—are barely human, and certainly not the acme of world civilization that they appear to be in Tamil place-making. They might have been Dravidians, but in turn, Lemuria is only one of their hypothetical homelands, and it was just as possible that they might have come into India from the northwest, and just as possible that they might not be the oldest peoples of the subcontinent after all. Most importantly, nowhere in the official colonial archive is it even hinted that Lemuria is the *Tamil* homeland, the birthplace of the Tamil language, the "cradle" of Tamil civilization—the Tamil prelapsarium.

As we will see, Tamil place-makers do not take aboard everything that the colonial ethnologist had to say about Lemuria, and they even question the limited colonial labors of loss around the vanished land in some critical respects. Nevertheless, by recirculating metropolitan speculations about Lemuria as a possible *Dravidian* homeland, colonial ethnology cleared the ground for its appropriation for the *Tamil* cause. The Dravidian category gained salience after 1856 with the publications of the missionary-grammarian Robert Caldwell (1814–91), which established a clear distinction between Tamil and the larger family of "Dravidian" languages of southern India to which it belonged. Yet from the 1880s, with the increasing use of the category in popular and political discourses in Madras, the original distinction between "Dravidian" and "Tamil" comes to be blurred. Because Tamil is deemed the oldest and most "cultivated" of this family of languages, it metonymically represents Dravidian, even among those Tamil intellectuals who did not wholeheartedly embrace Caldwell's theories for a variety of reasons.[25] For a large majority, Dravidian is always already Tamil.

The ground was thus cleared for the entry of Lemuria into the Tamil lifeworld with the colonial labors of loss around a drowned Dravidian homeland, and indeed, from the late 1890s, the very first murmurs about the presence of a lost *Tamil* continent in the Indian Ocean begin to circulate in Tamil-speaking intellectual circles. Thus, in 1898, J. Nallasami Pillai (1864–1920) wrote an editorial in the philosophical-cum-literary journal *The Light of Truth or Siddhanta Deepika* in which, for the first time, place-making around Lemuria in colonial India shifts to a clearly Tamil terrain: "If we can believe in the tradition of there having been a vast continent south of Cape Comorin, whence all humanity and civilization flowed east and west and north, then there can be nothing strange in *our regarding the Tamilians as the remnants of a pre-diluvian race.* Even the existing works in Tamil speak of three separate floods which completely swamped the extreme southern shores and carried off with them all its literary treasures of ages."[26]

Writing very much in the vein of the cautious colonial, he concluded, "However, this theory stands on no serious historical or scientific footing."[27]

Fifty years later, the well-known Tamil scholar Maraimalai Adigal (1876–1950) similarly invoked ancient Tamil literature to confirm the findings of colonial and modern science, but now without the ambivalence of Nallasami:

> From the time I read the ancient Tamil classics I came to have a dim notion that the present Tamil country in the South of India could not have been more than a remnant of a vast continent now sunk in the Indian Ocean. . . . Possessed with this idea of a submerged continent in the south, I was kindled with a strong desire to know whether this could be proved by the sciences of geology and Physical Geography, Biology and Ethnology and began in my nineteenth year to apply myself to a careful study of authoritative works on these sciences.[28]

After his years of careful study of these sciences, Maraimalai Adigal was delighted to observe: "It is manifest that scientists are unanimous in holding that the submerged continent called Kumari Nadu by the ancient Tamils and Lemuria by the modern scientists, constituted the only primeval home of man when he first made his appearance on this globe. . . . The language spoken there by the first man was Tamil."[29]

Between Nallasami's cautious speculation, tinged with skepticism, of 1898 and Maraimalai Adigal's jubilant pronouncement, untouched by equivocation, of 1948 looms the dense work of Tamil devotion. It is this work which transforms the paleo and occult place-worlds of the Euro-American imaginaries, and the possible Dravidian homeland of the colonial labors of loss, into an intimate Tamil home-place.

ACTS OF NAMING

Critical to this transformation is the re-naming of Lemuria in order to claim it for a distinctly Tamil project. Following Keith Basso's invitation to attend to "place-names and the full variety of communicative functions served by acts of naming in different social contexts,"[30] I consider the many different names bestowed upon Lemuria by Tamil labors of loss to suggest that through such nominations, Sclater's lost continent comes to reside within a Tamil horizon of memory and meaning. As Paul Carter effectively demonstrates, naming is paradigmatically an act of possessing and making one's own, but it is also a summoning into existence. Moreover, to name is to know.[31] Invested with "Tamil" names, Lemuria is nominatively transformed into a Tamil home-place *(akam, nāṭu)* and launched into the Tamil life-world for circulation, contemplation, consumption, and action.

In the late 1890s, when Tamil place-makers first started to reconcile the story of the earth and the formation of its continents narrated by modern geology and natural history with the prediluvian history of their own land and language as they learned it from recently published ancient Tamil poems, they used phrases such as "the vast continent south of Cape Comorin" or "the land that had extended further south" of Cape Kumari to refer to the territorial entity in the Indian Ocean that Euro-American labors of loss had designated as Lemuria, the Indo-African continent, even Gondwanaland.[32] And, indeed, through much of the next century as well, this practice of non-naming continued. Such acts ought to be noted as well, for leaving something unnamed suggests it cannot be captured through human utterance, pointing in turn to its wondrous, even awe-inspiring, nature.

All the same, it is a measure of the conflicted intimacy between Tamil labors of loss and European science that the name Lemuria enters the Tamil-speaking life-world in 1903.[33] Within a few years, it gains in popularity, and remains widely in use today, sometimes as such, and at other times Tamilized as "Ilemūria."[34] But from the start, Tamil names were also sought, the most common of which was Kumarināṭu (lit. Kumari territory), used for the first time in 1903 by V.G. Suryanarayana Sastri (1870–1903) in his pioneering *Tamiḻmoḻiyiṉ Varalāṟu* (*History of the Tamil Language*).[35] "Kumarikkaṇṭam" (lit. Kumari continent) becomes a part the lexicon of Tamil devotion only much later, from the 1930s.[36] The imperative to christen the ancestral homeland with such names is not surprising. For one thing, while the name Lemuria lent the authority of European science to the Tamil ancestral place, it had picked up some embarrassing associations as well. As we have seen, the paleo continent was named as such in 1864 to account for the geographical distribution of the lemur, the most primitive of primates. Similarly, in Theosophical labors of loss, with which many Tamil place-makers were undoubtedly familiar because of the visibility of these occultists in the Madras Presidency, the name had come to be associated with a continent inhabited by proto-human beings of doubtful sexual orientation.[37] At a time when the Dravidian category, as it was used in colonial *and* modern Indian ethnological discourses, was so closely associated with primitive savagery, even simian bestiality, such characterizations were potentially damaging to a project that sought to transform Lemuria into an ancestral Tamil homeland.[38]

In contrast, Kumarināṭu and Kumarikkaṇṭam had the (apparent) virtue of being old Tamil names that linked the paleo place-world of the European material sciences to the hallowed knowledges of ancient Tamil tradition. Thus, T.V. Kalyanasundaram (1883–1953), the famous Congress nationalist, influential journalist, and Tamil scholar noted, in an essay published in a Tamil-language textbook for ninth-grade students, that the Lemuria of "western scholars" like Ernst Haeckel and Scott-Elliot was none other than the Kumarināṭu of Tamil literature.[39] Although he (and others) may so insist, I have found no use of either Kumarināṭu or Kumarikkaṇṭam until the opening years of the twentieth century, when they came into vogue precisely because in their very semiosis the names are very suggestive. The primary denotative meaning of the Tamil word *kumari* is "virgin" or "pre-pubescent girl," while the secondary connotative meanings include "pristine chastity," "sexual purity," and "everlasting youth." And, indeed, the reasons that Tamil place-makers themselves offer for why their ancestral homeland was named Kumari draw upon both sets of meanings which allow them to summon into existence a prelapsarian place. Some suggest that the antediluvian homeland was so named because "for centuries it was ruled and presided over by Queens."[40] In turn, this spawns the frequent claim of gender egalitarianism that is said to have prevailed in the prelapsarian Tamil homeland—yet another hallowed virtue that was lost with the submergence of Kumarināṭu. So, D. Savariroyan (1859–1923) notes, "The dame of the ancient *Tamilagam* (Home of the Tamils) had the right to select her husband at [her] own choice. She was the mistress of the house and the heir to and owner of all property. *Tamilagam* was known in ancient literature as *Kumari-nadu,* 'the land of the maiden,' and among other nations as the 'Land of Queens.'"[41]

If the naming of the ancestral homeland as Kumari enables some to make such claims of gender egalitarianism—always useful for a modern project, especially in colonial Madras—others borrow from the power and popularity that had for centuries adhered to the goddess called Kanyakumari, whose well-known shrine stands at land's end.[42] Indeed, the Jaffna-based Tamil scholar N.C. Kandiah Pillai (1893–1967) fashioned a new history for the goddess and her shrine when he noted in a book intended for young children that the inhabitants of "the vast land that extended from Kanyakumari" were worshippers of the mother goddess. "The Mother Goddess is called Ka<u>nn</u>i or Kumari. Hence the land came to be called Kumarināṭu." Those who survived the ocean floods subsequently established a shrine for the goddess at the Cape, which is named after her.[43] Although Tamil labors of loss around Lemuria are strikingly secular—not the least of the reasons that they stand out in colonial and postcolonial India—this is one of the few instances in which a religious element does come into play. In her mythography, Kanyakumari is associated with patience, endurance, and penitence as she performs austerities to win over the divine Siva, who she is destined to never secure since the gods themselves conspire to keep her a virgin so that she can combat the demons. There is little doubt that in the popular religious imagination about the goddess, her power stems from her everlasting virginity.[44] Other female deities in the Tamil religious world are associated with virginal maternality, but none more poignantly so than the goddess who bears the name Kumari and waits patiently to unite with her divine lover. For some, it is telling that such a goddess presides over land's end, for they imagine her shrine as the surviving outpost of their beloved homeland lost forever to the ocean's ravages.[45] In the words of K. Anbazhagan (b. 1922), a Tamil professor-turned-politician (and minister in several Tamil nationalist governments since 1967) whose atheistic Dravidianist predilections did not preclude him from drawing upon the rich reservoir of religious imagination surrounding the goddess: "[At Cape Kumari] stands the temple in which resides our guardian goddess Kanyakumari. This is also the temple of our ever-virginal Tamil who has survived, without destruction, numerous ocean floods. It is as if the ever-virginal Kanyakumari—a woman who never marries—is the very embodiment of our virgin Tamil language."[46] Indeed, for some others, Kanyakumari stands in perpetual penance in order to prevent further loss of the Tamil homeplace over which she had formerly presided, her tears of anguish producing the Indian Ocean itself.[47]

For all these reasons, I suggest that Kumarināṭu and Kumarikkaṇṭam are the spatial equivalents of the symbolic notion of Tamil as *ka<u>nn</u>ittami<u>l</u>*, "virgin Tamil." As I have discussed elsewhere, and as Anbazhagan's statement reminds us, this is one of the most popular ways in which the language has been characterized by Tamil's devotees since at least the latter decades of the nineteenth century, for Tamil had been pure, untouched, and self-sufficient before the arrival of the Aryan hordes from the dreaded north.[48] Similarly, the (re)naming of Lemuria as "virgin territory" (Kumarināṭu) or "virgin continent" (Kumarikkaṇṭam) suggests that it, too, had been a pure and unpenetrated Tamil place, particularly useful for a project concerned with creating an authentic, originary, and sovereign homeland.

Beginning in the 1930s, although all three terms—Lemuria, Kumarināṭu, and Kumarikkaṇṭam—come to be used interchangeably, the name that is most widely used for the antediluvian homeland is Kumarināṭu. It is not accidental that the term *nāṭu* is deployed for this drowned homeland, for it is the most enduring and popular Tamil place-making term used to designate cultivated territory said to be characterized by settled populations, government, and sovereignty.[49] In using the term *nāṭu* in conjunction with *kumari,* Tamil place-makers dramatically shift the meaning of Lemuria away from its dominant Euro-American scientific designation for a remote uninhabited paleo continent to a familiar and real *Tamil* place, inhabited by Tamil speakers governed by Tamil kings and living in a Tamil state.

Crucially, the use of the suffix *nāṭu* also opens up the possibility for designating this antediluvian land as Tami̱lnāṭu, "land/nation of Tamil," from as early as 1903.[50] The term "Tami̱lnāṭu" has been used extensively since the early years of the twentieth century to refer to the Tamil-speaking region (even "nation") of modern India, and it became the official name of Madras State in 1968.[51] The renaming of Lemuria as "ancient Tami̱lnāṭu" or "our Tami̱lnāṭu" implies that the paleo continent has always already been a Tamil place, besides establishing a proprietary claim to it in the name of modern Tamil speakers. Indeed, a 1949 Tamil schoolbook put it unambiguously when it informed its young readers that "those who lived in antiquity in that vast Tamil home-place *(tami̱lakam)* called Kumarināṭu were called Tamil people *(tami̱l makkaḷ).* Since they were our ancestors, we also call them Tamilians (*tami̱lar).* Like Tami̱lnāṭu itself, the people who lived on that land were very ancient."[52] Statements like these, of which there are plenty, imply that the geo-body of modern Tamilnadu had existed for hundreds and thousands of years, a point of great significance for Tamil labors of loss around Lemuria, with their political agenda of claiming first rights in the subcontinent, indeed, in the entire world.[53]

Yet another important nominating act that follows from renaming Lemuria as Tami̱lnāṭu is calling it Tami̱lakam, Tamil home-place. In contrast to Tami̱lnāṭu, which has largely acquired political saliency in the twentieth century, "Tami̱lakam" has an ancient presence which can legitimately be traced to the earliest extant Tamil poems, dated by disenchanted academics to the opening years of the first millennium C.E. Literally meaning the "Tamil home" or "Tamil abode," Tami̱lakam was also the traditional domain of the three royal dynasties of the Chola, Chera, and Pandya celebrated in these poems, and was imagined to have extended south from Mount Venkatam (modern Tirupati) to Cape Kumari. Although Tamiḻnāṭu is the term that has assumed geopolitical significance in the past century, Tamiḻakam continues to have historical and symbolic resonance because of its association with the royal exploits of these kings and with the literary achievements of the poets whom they patronized. Not surprisingly, Tamil place-makers also deploy this term in their attempts to turn Lemuria into a Tamil homeland, beginning in 1912, when Somasundara Bharati (1879–1959) used it in a much-quoted text called *Tamil Classics and Tamilakam.*[54] Since his time, numerous others have followed his lead, enabling the imagination of Lemuria as an intimate Tamil abode and home.

Renaming Lemuria as Tamiḻakam also allows it to be occasionally hailed as "Pāṇṭiya nāṭu," or the "land of the Pandyas," after the dynasts who are most associated with the cultivation and patronage of Tamil literature and who are deemed sovereigns of Kumarikkaṇṭam.[55] In contemporary memory, the Pandyas are also remembered as the most ancient lords of the land, their very name synonymous with antiquity. Most importantly, the renaming of Lemuria as Pāṇṭiya nāṭu allows Sclater's lost land to be hooked to an ancient narrative of the Tamil literary academies *(caṅkam)* that had flourished there before their submergence by ocean floods, as I discuss later.

Through such variegated acts of nomination, of which I have only flagged the most persistent, Tamil's devotees have discursively transformed a terra incognita into a familiar Tamil home-place whose very names bring it within the horizon of popular awareness and the historical consciousness of the (educated) modern Tamil speaker. "Nomination," it has been suggested, "is a mode of symbolic appropriation that furnishes virgin territories with a memory, with a gridding that dispossesses space of its alterity."[56] With the production of a "named network," as Paul Carter observes, "certain historical events might begin to occur."[57] Language is the essential medium through which such nominal acts of appropriation, possession, and occupation proceed. The rechristening of Sclater's Lemuria with ostensibly Tamil names extracts the lost continent from the rarified remote abstractions of paleogeography or Darwinian ethnology and reinstates it in a domain where it can evoke associations with the Tamil language, and with places and persons familiar to the average Tamil speaker. In this process, Lemuria is rendered into a place suitable for Tamil commemoration. Tamil speakers can now be at home in Kumarināṭu or Kumarikkaṇṭam in a way that they arguably never can in Sclater's Lemuria.

ACTS OF CLAIMING

As important as acts of naming for the accumulation of commemorative density are acts of claiming. For Tamil place-makers, Lemuria is the ancestral Tamil homeland and motherland *(tāyakam)*, the birthplace of the Tamil person. From the start, Tamil place-making raised the possibility that modern Tamil speakers could be "the remnants of a pre-diluvian race" which lived on Lemuria, commemorated as the home of Tamil ancestors.[58] As Somasundara Bharati insisted in 1912 after a discussion of the prevailing theories of the origins of the Tamils/Dravidians: "Progressive geological research is ready and willing to shake hands with the primeval poems of the Tamil country and establish that the ancestral home of the Tamils was in the far south of the Indian continent now under the sea and not above the snow-clad Himalayan heights, or in the land of the celestials, or in the country of the Hebrews before their dispersion."[59]

By introducing the crucial term "ancestral home," Somasundara Bharati led the way to the increasing characterization of Kumarināṭu as the "Tamil homeland," and even "the motherland" *(tāyakam)*, with all the symbolic and emotional connotations that those words carry, especially in an increasingly nationalist-minded century.[60] That this was a homeland/motherland that was forever lost "to the swollen tides of the southern sea"[61] only added to the poignancy of the Tamil place-makers' dilemma that they could never return to it and be reunited with it again. As such, Kumarināṭu is also aligned with the ancient eschatological spatiality of *te<u>n</u>pulam*, "the southern country," imagined as the realm of the ancestors, "*te<u>n</u>pullatār*." Indeed, for some, this ancient eschatological notion was itself proof of both the truth of Kumarināṭu and of its status as the Tamil homeland, for had not hallowed old texts like the *Tirukkuṟaḷ* mentioned "the south" as the land of the ancestors?[62] This was so much the case that alternate sites for Tamil origins begin to fade away as a consensus emerges by the 1920s around Lemuria as the lost ancestral homeland—a consensus that the government documentary film only officially reiterated in no uncertain terms in 1981.[63]

For all such claims made on Lemuria as the ancestral Tamil homeland lost to the ocean's tyranny, it was enormously important to establish that the postdiluvian Tamil country, and even the greater Deccan plateau, was an intrinsic part of the paleo continent, for otherwise the possibility of claiming that Tamilians were the original peoples of India, its autochthons, would founder. Hence, from the start Tamil place-makers also insist that "a large continent once existed in the Indian ocean which was *connected with or contiguous to* South India."[64] As historian T.R. Sesha Iyengar (1887?–1939) observed in 1925 in discussing Lemuria:

> According to Sclater, the Dravidians entered India from the South long before the submergence of this continent [*sic*]. There are unmistakable indications in the Tamil traditions that the land affected by the deluge was contiguous with *Tamilakam*, and that, after the subsidence, the Tamils naturally betook themselves to their northern provinces. The assertion of the geologists that Lemuria touched China, Africa, Australia, and Comorin will only show the

vast extent of the Tamil country, and can never help to dogmatise [*sic*] that the Tamils came from any of these now far-off regions, and settled in South India.[65]

As I discuss later, Tamil place-makers go to great lengths to delineate the boundaries and limits of the lost homeland and its relationship to the lived Tamil homeland, the Tamil-speaking region of India today. The intent is to ensure that known features of the lived homeland, especially Cape Kumari, were part of the landscape of Kumarināṭu before it disappeared into the ocean. In Somasundara Bharati's baroque formulation from 1912:

> The only conclusion borne in upon us by a reading of the oldest of the old Tamil works is that the Tamils could not have come into southern India from elsewhere. They were here in all the time past. . . . They grew up in the sunny bosom of Tamilakam between the Mahanadi and the submerged Pahruli rivers, and, like the Swiss patriots, clung "close and close to their mother's breast," as the "loud torrent and the whirlwinds roar but bound them to their native mountains more."[66]

Tamil labors of loss around a drowned Kumarināṭu that had been an extension of India and that had formerly been home to Tamil speakers make sense when we place them in the context of early-twentieth-century debates about the original inhabitants of the subcontinent—about who had arrived there first of all from among the many complex peoples and communities who currently live on the land. For most Tamil devotees and nationalists, Tamil speakers are the true indigenes of southern India, its antediluvial spread notwithstanding.[67] Since southern India had in turn been proclaimed by colonial science as the most ancient part of the subcontinent from "the first days of the world," this makes Tamil speakers the original inhabitants of India, and in contrast to others, especially the Sanskrit-speaking Aryans, they are not foreigners in this hallowed land.[68] In terms that echo the logic of linguistic paleontology used by metropolitan philologists, prehistorians, and ethnologists to delineate the Indo-European homeland,[69] Sesha Iyengar wrote thus in 1925 about the original Tamil homeland: "In the oldest extant Tamil classics there are no traditions pointing to a home outside Tamilakam. . . . There is nothing in Tamil to answer to the cold regions of the Asiatic table-lands, to the ice-bound polar plains, or to the vine growing, fig-shadowed Chaldean regions. . . . The Tamils always believed that from the outset they were the aboriginal inhabitants of the great territories bounded by the two seas on the east and west, and by the Venkata hills on the north, and the submerged rivers Pahruli and Kumari on the South."[70]

Through such claims Tamil devotion's place-making recasts many aspects of the fabulous geography of Lemuria in the Euro-American imaginaries. Rather than a remote Mesozoic continent in the Indian or Pacific Oceans, whose links with India and Tamil country are tenuous at best, Kumarināṭu incorporated within its borders a good part of the Indian peninsula, and especially the Tamil-speaking regions of the subcontinent; hence its status as a "Tamil" continent. Consequently, the claim of Western science that Lemuria had been "the cradle of mankind" could only mean that the *Tamil* country had been the birthplace of humanity, and correspondingly, that Tamil speakers were the first humans.[71] So, P.V. Manickam Nayakkar (1871–1931) insisted in 1917 that "Zoologists and Paleontologists tell us that, in the development of species, South India ranks clearly the first. They tell us further that the most probable home of the *homo* or generic man in the world was the submerged Tamilagam or Limuria [*sic*]."[72] Similarly, Abraham Pandither (1859–1919) concluded in 1917, after a long discussion of Haeckel's ethnology, that Lemuria was "the cradle of the human race" and "the habitation of the first Man and the place where the first language was spoken." That language, he insisted, was Tamil.[73] In Kandiah Pillai's unequivocal conviction, "The creation of mankind took place in Tamiḻakam."[74] School and college textbooks in Tamil reiterated such claims throughout the course of the twentieth century.

As I earlier noted, an intense concern with human origins, racial homelands, and with aboriginal populations was a hallmark of Victorian anthropology and colonial ethnology. In turn, this concern fed into a nationalist preoccupation in colonial India, as indeed elsewhere in the world, with ferreting out "the sons of the soil" in whose authentic bodies was imagined to beat the most patriotic heart. Tamil labors of loss around Lemuria emerged and flourished in the shadows of these more hegemonic preoccupations, from whose findings they benefited to some extent, but in whose more sweeping reach there was the danger that Tamil claims to primordiality, authenticity, and originality would be lost. The labors of loss around Lemuria that European science initiated has hence allowed Tamil devotion to insinuate itself into the continuing global quest to crack the mystery of human origins, in order to ensure that the candidacy of the Tamil speaker as the first human, and the Tamil country as the birthplace of mankind, would not be lost in the clamor of similar claims made on behalf of other peoples and other places.

ACTS OF COMMEMORATION

The ethnological preoccupation with proving the origins of the Tamil speaker on Lemuria, the birthplace of all mankind, did not lead Tamil labors of loss to become entirely victim to the evolutionary thinking that was also such a hallmark of Victorian and colonial anthropology. In Tamil devotion's place-making, Kumarināṭu was primordial but not primitive. Human history on the antediluvian continent was not plagued by the long evolutionary march toward "civilization" from a state of primitive savagery that metropolitan ethnologists documented for the (postlapsarian) world.[75] Instead, the lost Tamil homeland is remembered and commemorated as a prelapsarian place of promise, plenitude, and perfection where civilization appears full-blown from the start, as the 1981 government film on Kumarikkaṇṭam only visually recapitulated in its striking images of cities and countryside resplendent with mansions and gardens, the arts and crafts, music and dance.

The transformation of Kumarināṭu into a Tamil prelapsarium began as early as 1903, when Suryanarayana Sastri wrote briefly but eloquently about "the state of civilization" of the antediluvian Tamilians, expert cultivators who lived in an egalitarian and democratic society. They wrote fine poetry, and their merchants traveled far and wide, spreading the Tamil language and the message of Tamil culture.[76] Similarly, for Savariroyan Pillai writing a few years later, the southern continent "was thickly inhabited, was the seat of learning and culture and the centre of the ancient civilisation of the antediluvian Tamil race."[77] Sivagnana Yogi (1840–1924?), a pioneer of the "pure Tamil" movement, insisted that the evil of caste differences that so plagued later Tamil society did not exist then.[78] And Kandiah Pillai informed young children in 1945 that Kumarināṭu was ruled by a mighty and just emperor called Sengon. He was utterly devoted to Tamil and invited Tamil scholars to his capital, where they were members of a large learned assembly. They wrote great Tamil books, "and the fragrance of Tamil wafted across the land."[79] That such claims were pedagogically reproduced is clear from a 1956 Tamil-language instruction book meant for fourth-grade children which presented the following glowing image of the Tamil prelapsarium:

> From ancient times, Tamilians have lead great lives. They are possessors of a great civilization. In the ancient period, our Tamiḻ nāṭu had spread over a vast extent. Because of three oceanic floods, several parts in Tamiḻnāṭu's southern portion disappeared into the ocean. . . . In the ancient Tamil land, everyone was educated. The learned commanded respect everywhere. Kings honored poets. Men and women were educated and virtuous. They were excellent scholars as well. Learning and commerce and handwork excelled in these ancient times. Tamilians were heroic. They went to the west and to the east for commerce. They spread Tamil culture everywhere. Tamil kings ruled without swerving from the path of justice. . . . The people were devoted to language and land. We have descended from such great Tamil people. We have to protect our ancient greatness, and excel once again in learning, commerce and work.[80]

As this textbook statement suggests, place-making of Kumarināṭu as a lost perfect world is frequently didactic and mnemonic: it is meant to remind modern Tamil speakers of the state of plenty, peace, and perfection attained by their antediluvian ancestors, and to cajole them into waking up from centuries of decline and "sleep" to live up to their former potential and promise in the present.[81]

Unlike many other utopian and Golden Age discourses which similarly imagine an elsewhere of plenitude and perfection, Tamil labors of loss are not content with insisting that the prelapsarian Tamil ancestors lived in freedom or as equals in bucolic peace and pastoral harmony. Instead, even more pressing than such claims was the imperative to declare that the antediluvian Kumarināṭu was "the cradle of human civilization," and that Tamilians had become "civilized" thousands of years ago, far before anyone else in the world. Today, Tamil speakers might be living as colonized subjects whose civilizational status was highly suspect, but in the distant past their ancestors had been the benefactors from whom "civilization" had diffused to the rest of the world.[82]

This preoccupation with civilization has to be located in the colonial context of Victorian and Edwardian India, in which it was perceived "as the unique achievement of ethnologically 'advanced' races."[83] A systematic scholarly analysis of the discourse on civilization in colonial India has yet to be produced, but it is clear that the yearning for it among its intellectuals was generated at least partly in response to Britain's own "civilizing mission" that served as a rationale for its empire. As we well know, this latter project only ambivalently conferred the mantle of civilization on the colony, reserving it for certain select peoples and cultures. At least until the 1930s, Dravidian culture and Tamil speakers were not generally counted among the "civilized" few, although every now and then an occasional colonial voice spoke to the contrary in their favor.[84] Predictably, Tamil devotion is haunted not by the fall from a pristine state of Edenic nature, but by the exilic separation from the pure and true Tamil civilization. Hence, chronological primordialism in these labors is not accompanied by cultural primitivism where nature is the norm, as it is in many a narrative of prelapsarian origins. The ancestral Tamilian of Kumarināṭu was most certainly not a Rousseauian man of nature, living a life of rustic simplicity and bucolic innocence. Instead, he led a "great life" devoted to learning and education, travel and commerce, culture and civilization. He had reached the pinnacle of Tamilian, even human, achievement.

Aside from formulaically reminding their fellow speakers of its former existence, however, Tamil place-makers offer them few guidelines on how one may reach—or more to the point, return to—the promise, plenitude, and perfection of the prelapsarian Kumarināṭu. To this extent, Tamil labors of loss are very much like those expressions of utopianism whose principal purpose is the imagination of a perfect elsewhere and not necessarily its attainment; hence, the free reign many utopian projects give to the play of the fabulous and the marvelous, beyond the realm of possibility and outside the shackles of history.[85] On the other hand, in contrast to the classic utopian discourses of the modern West which focus on how to achieve the good life in the future, in Tamil labors of loss the good life had *already* happened in the vanished past of Kumarikkaṇṭam. The imperative to aspire to a new state of perfection and plenitude in the here-and-now or in the future is constantly undercut by the willful hankering after a past time when it had already been achieved, and in the shadows of whose catastrophic loss modern Tamil speakers are destined to eke out their lessened lives, forever. An overwhelming, almost willful, nostalgia for a lost place and time of plenitude and perfection ensnares much of Tamil devotion in melancholic fables of the archaic instead of driving it to aspire to and create a just and equitable society in the present, a point to which I return later.

LABORING OVER LOSS AT LAND'S END

Keith Basso suggests that in addition to naming, claiming, and commemorating, place-making is a way of "constructing history itself, of inventing it, of fashioning novel versions of 'what happened here.'"[86] The principal history of Kumarināṭu is of irredeemable loss—of literary texts and of the original Tamil speech; of Tamil antiquity and purity, sovereignty and unity; and not least, of the ancestral Tamil homeland. It is here that Tamil labors of loss are at their fabulous best, mixing fact with fiction, "outrageously inventing" the impossible and the unbelievable, but mixing it up with the known and the ascertainable, in the course of creating a place-world that can be embedded within a Tamil horizon of meaning and memory.[87] In other words, the Tamil place-maker, like his occult counterpart, flirts with fantasy as he struggles to leaven the ponderous weight of the real and the admissible with the enchanted magic of the fanciful and the marvelous. The result is the production of a "form of 'reality' in which an unstable interplay of truth and illusion becomes a phantasmic social force."[88]

The creation of place-worlds, Basso also suggests, "is not only a means of reviving former times but also of *revising* them, a means of exploring not merely how things might have been but also how, just possibly, they might have been different from what others have supposed. Augmenting and changing conceptions of the past, innovative place-worlds change these conceptions as well."[89] So, how are received notions of the Tamil past revised by Tamil devotion in the process of making the innovative place-world of Kumarināṭu? How does the place-maker persuade his audience that things might *just possibly* have been different from what they had hitherto been led to believe? I begin by looking at the lamentations over the loss of the "gems" of Tamil literature that preoccupies every Tamil devotee.

Drowned Books

In 1955 a Malaysian-Tamil amateur historian named K. P. Sami published a historical work which began with the claim that any history of the Tamil land had to necessarily begin in Lemuria, for "why should we hide the first eleven thousand years of [Tamiḻakam's] history and write only about the last two thousand?"[90] Sami was not alone in making this claim; others well before him had done so. Yet Sami's dilemma, and that of others who want to write the "history" of prelapsarian Tamiḻakam, is that texts and records that the modern historical profession privileges as material "evidence," and as necessary for the legitimate narration of the human past, were lost forever to the ocean in the floods that ravaged Kumarināṭu. When he would ask his Tamil teacher M. Ramalingam about these lost Tamil books, the latter's eyes "would well up with tears," Sami tells us.[91] Undeterred, however, by this catastrophic loss, Sami went on to publish his sixty-four-page "history" of the Tamil country, including its antediluvial past, which ends with an impassioned plea to stop the continued destruction of Tamil books and with a call to restore Tamil to its prelapsarian glory.

From the closing years of the nineteenth century, similar lamentations about the "countless" and "numerous" precious Tamil books and "literary treasures" that met "a watery grave" are a constant refrain of Tamil labors of loss. These works were the creations, devotees insist, of the 4,449 poets who had graced the first literary academy *(caṅkam)* in the antediluvial city of Teṉmaturai, and of the 3,700 bards of the second academy in the drowned city of Kapāṭapuram. When these cities and their learned academies were washed away by the ocean, they took with them—forever—these poets and their works. While most do not venture to enumerate the books that are believed to have been destroyed, settling instead for generic invocations of "innumerable" and "many," some more daringly name names and even discuss their contents. So, in 1903, drawing upon a Tamil commentarial tradition that over the centuries had kept alive some memory of lost treatises, Suryanarayana Sastri recounted that works such as *Mutunārai, Mutukuruku, Māpurāṇam,* and *Putupurāṇam* had been seized by the ocean.[92] Soon after, in 1917, Abraham Pandither, in his massive history of Tamil music, insisted that several of these lost works, such as *Nāratīyam, Perunārai,* and *Perunkuruku,* had been the world's first treaties on music, all of which had disappeared into the ocean along with several rare musical instruments like the thousand-stringed lute.[93] Thudisaikizhar Chidambaranar (1883–1954), a retired petty bureaucrat, pointed to the loss of rare grammatical works, the first in the world, and Deveneyan Pavanar (whom I write about a little later) even printed a list of submerged books for the contemplation of the modern Tamil speaker.[94] Indeed, this general conviction that extant works of Tamil literature are but a small fragment of a much larger whole that is irrevocably lost has licensed many to locate antediluvial origins for everything ranging from medicine to the martial arts whose foundational books are, however, unfortunately drowned in the ocean and hence cannot provide empirical proof for the authors' claims: "Treatises on logic, painting, and sculpture, yoga (philosophy), music, mathematics, alchemy, magic, architecture, virtue, poetry, overcoming the nature of elements, water, soils, metals, (causing of) death, (acquiring of) wealth and many other subjects have been, alas! swept away and swallowed up by the sea, so completely that even their very traditional names have disappeared."[95]

Because of this mystique of loss that cloaks the entire body of Tamil literary works,[96] a special significance is attached to those that are declared survivors of this catastrophic destruction. These include the ancient grammars *Akattiyam* and *Tolkāppiyam,* and the extant poems of the anthologies *Eṭṭutokai* and *Pattupaṭṭu* (the products of the third academy, which succeeded the earlier two that were lost to the ocean). The latter constitute the core of what Tamil's devotees, as well as literary historians, refer to as "Sangam poetry," knowledge of whose very existence became part of the educated modern Tamil consciousness only from the 1890s. Indeed, the very "discovery" and publication of these "lost" anthologies through the hard labor of scholars like C.W. Damodaram Pillai, U.V. Swaminatha Aiyar, and others from the 1870s through the early years of the twentieth century precipitated the intense labors of loss discussed in these pages.[97] This is not surprising because several poems in these recovered anthologies do indeed allude to oceanic floods and consequent loss of land and life, which, as we will see, are quoted over and again in Tamil labors of loss as empirical proof of the catastrophic seizure of Kumarināṭu in the antediluvial Tamil past. So Somasundara Bharati observed in 1935, "Thus the fact of a prehistoric deluge wiping out a wide expanse of the southern Pandyan country is indisputably established by . . . [the] authoritative and well authenticated internal evidences in the earliest Sangam poetry. . . . As such, the statements they record must be accepted as fairly established facts of history."[98]

Further, the recovery of the *Eṭṭutokai* and the *Pattupāṭṭu* also fanned the hope that if these lost works could be found, why not others? This accounts for the great excitement over the publication in 1902 of a slim book, not more than ten printed pages long, entitled *Ceṅkōṉṟaraiccelavu,* which claimed to be some verses of a longer poem of the first academy of antediluvian Teṉmaturai.[99] As one enthusiast wondered, "It is not clear why such an ancient text with such rich information has been not been mentioned by any of the old commentaries. It reeks of antiquity."[100]

The author of the poem styled himself "Mutalūḻi Cēntaṉ Taṉiyūr" (Chentan who lived in Taṉiyūr before the first deluge), and his verses commemorated some of the exploits of his patron, the antediluvian Tamil king Sengon, who ruled the region called Peruvaḷanāṭu that stretched between the rivers Kumari and Pahṟuḷi, now lost to the ocean. Here is Chidambaranar who, years after the publication of this work, is barely able to contain his excitement over—and his own involvement in—its recovery:

> The above said book was discovered by me from some old cudgan [*sic*] leaves and it was copied out and kept ready for publication with some notes. If it comes out, the oldest civilisation of the Tamilians will be known to everybody. Emperor Sengon . . . maintain[ed] battle ships and fought many battles overseas with the help of his dreadnought. Emperor Sengone was reigning in the submerged land called continent of *Lamoria* [*sic*]. His original home was Olinadu which was situated south of Equator, i.e., thousand miles south of present Ceylon. . . . Emperor Sengone conquered lands as far as Tibet and planted his Bull flag on the Himalayas. This was sung in *Sengone Taraichelavu*. Continent of Lamoria was called Tamilagam by the Tamilians. Many pandits and scientists hold the opinion that the human species first evolved in the Great Indo-African Continent. . . . This large continent is of great importance for being the probable cradle of the human race.[101]

Although the *Ceṅkōṉṟaraiccelavu* was declared a forgery by the eminent S. Vaiyapuri Pillai in the 1950s, who noted that "no responsible scholar now takes any serious notice of it,"[102] this has not deterred Tamil labors of loss from invoking its veracity till this day. The 1981 government film even declared it to be the "world's first travelogue," which it details at some length.

While the publication of *Ceṅkōṉṟaraiccelavu* resembles other nationalist attempts to recuperate "forgotten" or "stolen" literatures, such as the Scottish Ossian,[103] I want to focus here upon the imperative to generate *material* proof of loss that underlay this exercise, setting aside the charge of forgery. In an age deeply embedded in empiricism and positivism, the Tamil place-maker is obviously troubled by the fact that the loss of his precious ancestral books has meant that his preoccupation with lost lands, lost culture, and lost civilization could hardly aspire to scientific credibility. So, having noted that "existing works in Tamil faintly speak of three separate deluges which completely swamped the extreme southern shores and carried off with it all its literary treasures of ages," Nallasami Pillai concluded in 1898 that "it stands to reason why, in South India, unlike in ancient Chaldea and Babylon, none of the old records of the pre-historic civilization are absolutely not forthcoming." Hence, for Nallasami, such claims could not stand up to genuine historical or scientific scrutiny.[104] But as the place-making of Kumarināṭu gathers steam through the next century, skepticism gives way to certitude, and different explanations are sought for the lack of material evidence of Tamil antiquity: the frailty of palm leaves on which ancient Tamils recorded their thoughts, the destructive actions of fire and insects, the "ravages of time," and so on. But the favorite explanation of all, as I discuss later, is *kaṭalkōḷ* (lit. "seizure by the ocean"), the catastrophic floods caused by the cruel ocean. Because of *kaṭalkōḷ*, "the various signs of the ancient civilization of the Tamilians are lost. Their books have been lost."[105]

This is not the least of the reasons that nineteenth-century metropolitan labors of loss over submerged continents and disappeared landbridges were seized upon with such alacrity and enthusiasm by Lemuria's place-makers in the Tamil country because, for the first time, "scientific" proof for "traditional" claims of loss was forthcoming. Thus, Savoriryoyan observed in 1907 with some glee: "A greater portion of the land, it is said, has since been claimed by the ocean. This tradition recorded in the ancient Tamil classics has been confirmed by the researches of geologists and naturalists. . . . [And here he quotes Haeckel:] *It is noteworthy that centuries before the birth of the sciences of Geology and Natural History these facts have been recorded and preserved in a more or less accurate form in ancient Tamil classical works.*"[106]

So, for Tamil labors of loss over Kumarināṭu, instead of modern science destroying tradition as it had in so many different parts of the world, it only proved the veracity of ancient Tamil literature, pointing as it were to the latter's inherent "scientific" status, and also in the process establishing the truth of Tamil antediluvial patrimony.

Forgotten Antiquity

If the catastrophic drowning of Tamil's most ancient books had denied their modern inheritors from empirically proving their very existence and that of a flourishing antediluvian Tamil civilization, just as importantly, it had meant that others far less deserving had been able to lay claim to what was really Tamil—and gain materially, politically, and symbolically. Typically, the Tamil place-maker's bête noire in this regard is Sanskrit, the language singled out by colonial Orientalism *and* Orientalist nationalism to be India's most ancient and loftiest of "treasures."[107] So R. Mathivanan noted after decades of discourse on this that "there are persistent traditions of the systematic destruction of ancient Tamil books on various topics. Such destruction and the availability of only secondary Sanskrit versions of many such books has enabled Sanskritists to fob off as their own, the discoveries, findings and thoughts of the Tamil genius."[108] In a similar vein, a few decades earlier K.P. Sami had bitterly observed that Sanskrit had completed the destruction of the antediluvial Tamilian patrimony that the sea had catastrophically initiated.[109] In fact, it is not surprising that Tamil devotion gathered momentum in the early decades of the twentieth century when a heated battle on behalf of "the vernaculars" (like Tamil) was fought in the University of Madras to save them from the tyranny of Sanskrit.[110] As M.S. Purnalingam Pillai (1866–1947) observed in 1925, the failure to recognize Tamil's primordiality, anteriority, and antiquity had allowed Sanskrit to rule the pedagogical and political roost as India's most ancient and classical language. "It is high time for the reformed Madras University to disabuse itself of this unfortunate prepossession [*sic*] and to recognize the ancientness of the Tamil language and literature in the light of what Strabo, Ptolemy and Pliny, and Periplus have said about the Lost Lemuria or Tamilakam as the cradle of the human race."[111] Here, once again, the Tamil place-maker ran up against the fact that whereas Sanskrit could offer empirical proof of its ancient works as testimony to its classicality, Tamil's even more ancient and glorious wealth had been plundered by "the cruel sea," and lost forever.

Inevitably, therefore, the labors of loss around vanished Tamil books prompts a discourse on the failure by the world to recognize the hoary antiquity of Tamil, its literature, and its civilization—by all accounts the most ancient on earth. There is little agreement, however, on how "ancient" Tamil antiquity is, and figures range wildly from precise determinations such as "eight thousand B.C." or "twenty thousand" years, to the fuzzier "thousands of years ago," "at a very remote time," or "from the beginning of Creation." In all such attempts, as I write later, Tamil labors of loss have to take head on the rigorous demands of positivist disenchanted history, a discipline with which many a place-maker is clearly in awe but one which is not on his side when it comes to matters of empirical reckoning, causal reasoning, or material evidence. Some, emboldened by history's complicity with the West and with colonialism, are quite willing to discard that discipline's stubborn requirements, declaring that the life and times of the prelapsarian Tamiḻakam are "beyond historical research."[112] For others, Kumarināṭu's very existence points to the fact that history itself has to be rethought. Not only does the history of the world have to be rewritten, but most especially that of India. As R. Nedunceliyan (b. 1920), a Tamil nationalist who went on to hold key cabinet positions in later Tamil nationalist governments in the state, insisted in the 1950s, echoing a claim that had been made half a century before: "Sir John Marshall declared vehemently, 'The beginning of Indian history should be in the Vaigai river [*sic*].' Therefore, the history of the Indian subcontinent has to begin not in the north but in the southern land, not on the banks of the River Ganges but that of River Cauvery."[113]

The submergence of so much of the Tamil patrimony also means that in his attempts to prove Tamil antiquity, many a place-maker turns to archaeology, the science of antiquities, as holding the key to the lost and forgotten past. This was especially true after the discovery in the distant Indus Valley of the ruins of Mohenjodaro and Harappa in the early 1920s by John Marshall and other archaeologists associated with the Archaeological Survey of India, who soon declared them to be remnants of India's most "ancient" urban "civilization." Within years, the ruins were claimed by Tamil intellectuals as unequivocally "Dravidian," the work of Tamil speakers fleeing from the ravages of the flood-hit Kumarināṭu.[114] As Purnalingam Pillai insisted in 1927, "When one deluge after another overcame Tamilaham, when the

Tamils dispersed in different directions to save their lives, and when the sea beyond the Vindhyas became dry and there was land to traverse as far as the Himalayas which had recently emerged, the Tamil emigrants passed over the jungles and sandy deserts and found their abodes in North India." The "recent discoveries at Harappa and Mohenjodaro" offered testimony to this ancient migration.[115] Years later, Nedunceliyan, by now the state's education minister, reiterated this claim in as august a place as the Tamilnadu Legislative Assembly in 1970: "Father Heras has shown how [Dravidian] civilization moved from Lemuria to southern India, and then reached Harappa and Mohenjodaro and from there later went to the Tigris and Euphrates and Rome and so on. . . . This also I learnt when I was studying in Annamalai University."[116]

About two decades before he made this proclamation, Nedunceliyan had published a pamphlet called *Maṟainta Tirāviṭam* ("*Lost Dravidian Land*"), in which he insisted that scholars, especially Brahman historians with their inherent partiality for Sanskrit, had deliberately kept the greatness of the Dravidian nation "hidden" from the public, and consequently, the knowledge of its antiquity and greatness was "lost," the memories of its glory "forgotten."[117] However,

> Sir John Marshall who excavated and found the lost cities of Mohenjodaro and Harappa that were buried in the Indus valley, and Father Heras who studied the ruined cities buried in the sand, declared that the lost civilization was Dravidian and in those days, this entire continent was inhabited by the people of that civilization. The history of the Dravidian land concealed both by nature and by [its] enemies, and made to be forgotten, is now being dug up. The task of digging up is not over. It is at the very beginning. The horizon of Dravidam's history, hitherto shrouded in darkness, is at long last being lit up.[118]

Nedunceliyan was not alone in placing faith in the possibility that archaeological spade work would unearth the "buried" and "lost" remains of Tamil antiquity. M. Rajamanikkam (1907–67) similarly introduced Mohenjodaro and its excavated remains, this time to Tamil children, by noting that he wrote the book to arouse their consciousness about "their lost and buried past," and to make them realize that "the predominant part of Tamilnadu's ancient history lies hidden in the earth."[119] Ongoing archaeological work was only confirming what Tamil literature had maintained all along, that there had been a Tamil civilization which was lost to the ocean and that Dravidians had lived all over India thousands of years ago: "Some are attempting to conceal this fact. As devotees of Tamil, you should come forth and help establish the antiquity of your mother tongue, the greatness of your Dravidian civilization. May Mother Tamil offer you her grace."[120]

Although such pronouncements have not necessarily translated themselves practically into making archaeology a viable discipline that is rigorously taught in Tamilnadu schools and colleges (as it has in places like Israel where, too, the nationalist stakes are high in the search for buried antiquity), its promise continues to be invoked by Tamil place-makers who in recent years have called upon the state government to commission underwater explorations off Kanyakumari that would finally put to rest the skepticism of those doubting Thomases who scoff at the very idea of the lost Kumarināṭu.[121] Just as geology and biology had demonstrated the veracity of Tamil traditions about loss in the twentieth century, archaeology, too, would come to the Tamil devotee's aid in this regard in the new millennium, it is hoped.

Castaway Tongue, Lost Purity

As already noted, Tamil devotion's interest in Kumarināṭu is the place-making expression of the devoted Tamil speaker's overwhelming preoccupation with his language. When the place-maker mourns the loss of Kumarināṭu and the submergence of its literary treasures, and rants against the global refusal to recognize the Tamil country's ante-diluvial antiquity, he is really grieving over Tamil. So Tamil devotion's labors of loss are very much a response to the perceived decline of the Tamil language, from its glorious beginnings on a prelapsarian Kumarināṭu to its current state of a castaway, ignored by even its own speakers, who are more enamored of rival tongues such as Sanskrit, English, or Hindi.[122]

From as early as 1903 Tamil place-makers insisted that the language spoken in the antediluvian Kumarināṭu was Tamil and only Tamil, and this position is steadfastly adhered to until today. From this foundational claim flows numerous important corollaries: Tamil is the world's most ancient language; indeed, it is the primeval tongue of all of mankind. As such, it is the most "natural" of human tongues, the most original and perfect. It is also, of course, India's most ancient language, with a presence on the subcontinent far prior to anything that Sanskrit could claim. And it is the "mother" of all other Dravidian languages, which emerged from it after the submergence of Kumarināṭu and with the migration northward of Tamil speakers. Separated from their "mother" land and their "mother" tongue, these postdiluvian Tamilians progressively turned into speakers of Kannada, Telugu, Malayalam, and other languages, frequently forgetting their umbilical connection to Tamil.[123]

In and of themselves there is nothing particularly remarkable about such declarations—numerous other language movements have spawned similar fantasies all over the world.[124] What does distinguish Tamil devotion is the spatial and temporal dimensions of its preoccupations, for few language revivalists in the modern age have so systematically and in such a sustained manner developed a theory about the antediluvian origins of their language as do Tamil's admirers, nor have they been involved in such intensive labors of loss. It is also in the Tamil country that notions of decline and disappearance—which numerous other language movements have also experienced—take on such a catastrophic imperative, far in excess of anything claimed on behalf of any other tongue. The writings of Devaneyan Pavanar (1908–81)—a product of the "Pure Tamil" movement—are the most dramatic examples of this catastrophic view of language loss.[125] In a spate of essays and books, Devaneyan wrote agonizingly about the linguistic consequences of the submergence of the prelapsarian Kumarināṭu, which he held responsible for everything from the degeneration of Tamil's originally pure and true grammar to semantic impurities that crept into the perfect language, and most intolerable of all, to its increasing Sanskritization.[126] So, for devotees like Devaneyan, the tragedy of loss of the original homeland is further heightened by the fact that the Tamilian ceased to live in a state of pristine purity, as words from other languages flooded into their perfect Tamil, contaminating it—and its speakers—forever. Further, Tamil ceased to be the sovereign language it had been on Kumarināṭu, and correspondingly, Tamil speakers ceased to be independent as well. Abraham Pandither thus insisted in 1917: "Those who realise the importance of the Tamil language know how it was pure and easy and unmixed and had attained a high state of excellence [on Lemuria]. . . . But after the deluge, as the Tamilians spread over different regions and as people with different languages . . . mixed freely with them, the original grandeur of the language was lost."[127]

As this statement clearly illustrates, Tamil place-makers are not consistent in their labors of grief over the loss of antediluvian Tamil. For one, if Tamil is the primeval language of *all* mankind, on what grounds can Tamil speakers lay a proprietary or exclusive claim to it? How, indeed, can Tamil speakers even define themselves as a distinctive race of people if everybody spoke Tamil, once upon a time, everywhere in the inhabited world of the antediluvium? Further, there is not much clarity on whether, in the thousands of years that Kumarināṭu flourished while Tamilians lived on that hapless continent, the rest of the habitable world was also populated, and if so, by whom.[128] If speakers of other languages did not inhabit regions other than Kumarināṭu, how could one account for the progressive corruption of the Tamil language through admixture that is also such a continual source of grief for the Tamil devotee? Some account for the birth of new languages as a result of the submergence of Kumarināṭu and the dispersal of Tamil speakers away from their "motherland." Its place-makers are divided, however, on whether this is to be celebrated or bemoaned. Some take great pride in insisting that Tamil roots and words can be found in almost every language of the world, from Maori to Maya, including Greek, Sanskrit, and Hebrew, and they undertake elaborate and fantastic etymological exercises to demonstrate this "fact."[129] Others, however, lament that numerous "foreign" words have flooded postdiluvian Tamil, destroying its virginal purity, and they mourn the dispersal and destruction of the pristine tongue of prelapsarian Kumarināṭu.[130]

Regardless of these inconsistencies, which a fabulous imagination can arguably afford to leave irreconciled, the Tamil place-maker is convinced that postlapsarian Tamil is a pale remnant of the glorious and pure language formerly spoken in Kumarināṭu (and in the rest of India as well, once upon a time).[131] Of course, he has to be careful about not pushing this point too hard since he has to get his fellow Tamil speaker to organize and mobilize around the language as it exists and flourishes today, catastrophically flawed though it may be. The devotees' solution to this dilemma is to remind modern Tamilians about the former glories of their language, and to urge them to work hard to save what is left of it. Otherwise, that, too, will be lost forever, destroyed now not by the actions of the sea or by the evils of Sanskrit, but by the indifference of its own speakers.[132] In this logic, the Tamil that is extant today—even if corrupt and contaminated—is one of the few surviving mementos of the antediluvian place-world in which Tamil speakers had once flourished, hence the need to cherish it and guard it with one's life if need be.

Dispersed Unity

The claims of lost purity and integrity are clearly at odds with another preoccupation that grips Tamil devotion's place-making, and that is the peopling of the world by the former inhabitants of Kumarināṭu: "The Dravidian civilization, which emerged on the ocean-seized Lemuria or Kumarikkaṇṭam, was the civilization of southern India. It then spread to Mohenjodaro and Harappa, and to the Euphrates-Tigris river valleys as the Sumerian civilization, and then, moved to Arabia, Egypt, Greece, Italy, Spain and other places. Whenever any Dravidian thinks of all this ancient greatness, there is little doubt that he will swell up with pride."[133]

As early as 1903 Suryanarayana Sastri gave voice to such a claim, and since his time, it has only burgeoned into a full-throated cry about the existence of a Tamil diaspora that is literally constituted by all peoples of the world.[134] In Purnalingam Pillai's evocative words, which cleverly transform a colonial stereotype of Dravidian servility into Tamil global mastery, "in its widest sense Tamilaham at present lies all the world over, wherever the enterprising Tamils have found their home."[135] Hence also the oft-repeated insistence—in a significant recasting of the key metropolitan ethnological contention about Lemuria—that Tamiḻakam is "the cradle of the whole human race," the birthplace of all of mankind.[136] In the gendered vocabulary that is so endemic to Tamil devotion, "all humans are ultimately children of the same mother's womb, and the birth place of humanity is Kumarināṭu."[137]

This insistence produces its own irresolvable contradiction, as is clear from the following statement of Maraimalai Adigal:

> Our observations based on evidence afforded by the ancient Tamil classics and modern Western Science lead us to conclude that the first appearance of man on this globe took place in the vast southern continent that lay on the two sides of the equator . . . that the language spoken by the first man was Tamil; and that the religion he professed was either the worship of Light itself as God or the vehicle of God; *that when the submergence of the southern part of this great land occurred the one great human family that lived there dispersed in different directions going to the north, north-east and north-west* . . . that it is this dispersion of the original human family into different groups and into different situations that accounts for the immense variety which at present prevails in the human features, speeches, customs and manners, and that notwithstanding this variety certain original elements of the Tamilian life and Tamilian speech have left their indelible prints in it to show their primitive Unity to a discerning intelligence.[138]

Again and again, the obvious pride that the place-maker experiences in the peopling of the world by the Tamilian is undercut by the realization of the loss of the primal unity of the Tamil culture of Kumarināṭu, and of "the one great human family" as it fragments and disperses across the globe. This contradiction, which Tamil place-making never resolves, is very much shared, as we know, by nationalist movements which have turned imperialistic, so that claims of national purity have gone hand in hand with projects for global hegemony. In the case of Tamil labors of loss, this fantasy of global hegemony has remained fabulously discursive, for even the most ardent and ambitious of place-makers has resisted chasing the impossible dream of a worldwide Tamil empire that would recover the lost unity of the prelapsarian Tamilians. Instead, the fantasy has been pursued differently, as labors of grief over the various postdiluvian communities and nations of the world who are *really* Tamilians but who have forgotten their Tamil-ness, and yet whose language, customs, and achievements continue to bear its traces.[139]

Such grieving over a lost unity reaches its ironic extreme in the insistence by some place-makers that even the dreaded Aryans, the nemesis of the postdiluvian Tamil-Dravidian world, are after all descendants of those inhabitants of Kumarināṭu who, after its submergence, moved north to Central Asia, settled there for a time while their language got progressively transformed into Sanskrit, and then returned to India as "invaders" and "conquerors." This explains the presence of numerous Dravidian words in even the language of the *Veda*s, primeval ritual texts of the Aryans though they might be. For the Indianists among the place-makers, this also explains the original unity of all "Indians," for whether "Aryan" or "Dravidian," they are all children of the same mother's womb. For the radical, however, it points to the fact that "India" is ultimately and at its core "Dravidian," and that modern Tamilians have been robbed of their patrimony by the upstart Sanskrit-speaking interloper.[140]

Most critically, this claim of the ur-dispersal of the ancestral Tamil peoples out of Kumarināṭu accounts for the dominant metropolitan theory of the nineteenth century that the Dravidians were a Mediterranean race which moved into India, even while it allows Tamil's devotees to hold on to what was after all a fringe speculation about their origins on a lost continent in the Indian Ocean.[141] As a college textbook on Tamilnadu prehistory published by the government in 1975 informed its adult students, the Dravidians of Kumarināṭu moved to the Mediterranean *after* the submergence of the ancient continent. Several hundred years later, they were pushed out of the region by the Semitics, and they moved east and entered India through the Himalayan passes in the northwest. Subsequently, they spread out all over India, and eventually one branch moved along the coast to the south to settle there.[142]

In this regard, Kumarināṭu's place-maker grounds his fantasy in empirical evidence, which he invokes in support of a worldwide Tamil diaspora. The presence of Tamil speakers in Burma and Southeast Asia, Mauritius and southeastern Africa, or in the Americas, is one such proof. Here, rather than recognizing these migrant Tamil populations (as disenchanted historians would insist) as products of commercial flows in the first and early second millennium C.E., or more often, of indentured servitude during the colonial period, the Tamil place-maker sees them as evidence of the ancient dispersal of the population of a drowned Kumarināṭu.[143] Further, some place-makers also draw upon the Victorian and colonial speculations regarding the racial affinity between the Negrito population of Africa, the Dravidians of India, and the Australian aborigines that I noted earlier to insist that descendants of the antediluvial Tamils peopled the entire southern half of the globe. Hence, echoing Searles V. Wood and Andrew Murray a hundred years after these paleo-scientists originally propounded their theory of a prehistoric black diaspora, K.P. Aravaanan (b. 1941), Tamil scholar, teacher, the vice-chancellor of a major university in the state in the late 1990s, and an enthusiastic proponent of this notion, writes:

> Tamil literary evidence allies Tamils with the black races. The three black races—Dravidians, Australians and Africans—ought to have separated from the same ancestor. This separation took place several hundreds of years ago. This primordial ancestor must have lived on the continent of Kumarikkaṇṭam. The Dravidians who lived in the ancestral homeland of Kumari moved north, Australians moved east and the Africans moved south west. They separated from each other several thousands of years ago. Distance as well as separation by the ocean brought several differences between them and undermined their primordial unity. Based on those differences, it is not possible to say that they are separate races.[144]

However misplaced such place-making might seem, the embracing of a black diaspora in this fashion does offer a surprising counterpoint to the much more hegemonic racism of the typical upper-caste modern Indian and his continuing fantasies of an ancestral connection to a "white" and "Aryan" Europe.[145]

Finally, borrowing here from a popular strategy of nineteenth- and twentieth-century diffusionism, some devotees even used maps to show the dispersal of the original Tamil population out of Kumarināṭu and their resettlement in different parts of the world. The earliest of such maps that I have seen is entitled "Descendants of Tamilians" (1943) and shows those parts of the globe that we today identify as the Middle East, northeast Africa, the Mediterranean, and southern Europe.[146] Its author, Kandiah Pillai, illustrates the map with an extensive discussion of such "descendants" as Sumerians, Egyptians, Elamites, Babylonians, Cretans, ancient Britons, Phoenicians, Assyrians, Hebrews, Arabs, and Chinese, among others.[147] A few years later, in a book intended for young children, Kandiah Pillai included a similar map that used arrows to show the migration of the inhabitants of "Kumarināṭu" to the Indus Valley, Sumeria, Phoenicia, Egypt, even England and elsewhere.[148] Such maps graphically illustrate the verbal labors of loss over the dispersal or the ur-Tamilian people, even while confirming the facticity of these labors by relying on scientific cartography's monopoly on place-making truths in modernity, as I discuss later.

I have insisted from the start that Tamil devotion's place-making is very much a product of colonial and Victorian modernity, and as such it becomes prey to many of the latter's preoccupations. Of these, diffusionism ranks high, a theory that informed everything from the nationalistic search for the Indo-European homeland, to philological notions of language change, to ethnological speculations about the distribution of archaic artifacts like the boomerang and the blowpipe.[149] Tamil devotion's preoccupation with the primal unity of antediluvian Tamils and their dispersal throughout the world accompanied by the diffusion of their "culture" and "civilization" has to be necessarily placed in the context of such prior and ongoing fantasies. As in other regards, however, the difference in the Tamil country lies in the extent to which such fantasies linger on into the new millennium, reinforced by the alienating and disempowering experiences of modernity and globalization, by official memorialization, and by the sheer weight of what is by now a deeply embedded cultural habit of thinking in terms of an originary loss.

Forfeited Sovereignty

The devotee's sense of self as a colonized subject living in the debris of former glory and splendor is most apparent in the fantasies about prelapsarian dominion and overlordship that run through the Tamil labors of loss around Kumarināṭu. In the beginning, as the story goes, when Kumarināṭu was the center of the known world, the birthplace of mankind, and the cradle of civilization, it was ruled by the Pandyan kings, mighty and powerful. They were magnanimous and courageous, dedicated to the welfare of their subjects, enterprising patrons of the arts and literature, and most importantly, ardent devotees of Tamil.[150] That the Pandyas came to be so intimately associated with this antediluvian world in Tamil labors of loss is perhaps not surprising, although it is a major departure from the colonial historicist position on this ancient dynasty.[151] Centuries of remembered traditions enabled this, including the belief that their very name, as one devotee was quick to note, was bestowed upon them by virtue of the fact that they had reigned over an "ancient" land.[152] The Pandyas are one among the three dynasties celebrated in the earliest extant Tamil poetry, and hence can indeed lay claim to being among the oldest royals of the region whose antique presence can be vouched for even by the disenchanted technologies of disciplinary history. These poems, of course, celebrate Pandyan heroism and courage on the battlefield, but significantly for the Tamil cause, they also recount encounters between these monarchs and the ocean. As the literary scholar Kailasapathy observes, "A mighty Pandyan potentate had once vanquished the southern sea by hurling a lance at it. Smarting under this humiliation, the sea wrought vengeance by swallowing the Pahruli river and the Kumari hill in the southern region of that primeval kingdom. To compensate for this loss of territory, another Pandyan monarch conquered the Ganges river and the Himalaya mountain."[153] Equally important, a rich commentarial tradition, as well as religious narratives passed down over the centuries from at least the later decades of the first millennium C.E., associated the Pandya with the city of Madurai, with the tradition of a Tamil academy there, and with patronage of Tamil, as I discuss later.[154]

All such traditions feed the mystique that surrounds the Pandyas in Tamil place-making, which raids the vast precolonial archive of historical accounts and royal genealogies for its labors of loss over Kumarināṭu. This archive furnishes Tamil place-makers with the names of a few antediluvial kings—Kalcina Valuti, Kadunkon, Venterceliyan, Nediyon, Mutatirumaran—who are credited with inaugurating Tamil academies, digging irrigation canals, and conquering distant lands, but even their fabulous imagination is hard pressed to come up with specific details about their lives and times.[155] In place of singular histories, Tamil labors of loss typically resort to a formulaic model of ideal kingship as it had developed in the region over the centuries in order to present the prelapsarian Pandyan as the paradigmatic Tamil sovereign of them all. The 1981 documentary film is the most elaborate in this regard. The camera draws us into the antediluvial Pandyan capital of Teṉmaturai, filled with luxurious mansions and pleasant playgrounds, and zooms in on the court of Kalcina Valuti, lord of Kumarikkaṇṭam. The commentator introduces him as the king who "established the first Tamil *caṅkam,* an academy to promote Tamil, the first of its kind, a concept so far ahead of the times, a rare new thought indeed. Great poets and scholars served the academy, to promote Tamil in its many art forms like literature, music, and drama." After a musical interlude, the monarch is shown bestowing generous gifts on scholars and poets. The documentary goes on to present Kalcina Valuti's successors as helping their subjects make their land fertile for cultivation, and as sailing with their merchants "on voyages to distant lands" from where they bring back all manner of riches.

After the deluge that destroyed Te<u>n</u>maturai and its learned academy, the commentator notes that Pandyan kings founded Kapāṭapuram, the new capital of Kumarināṭu:

> In Kapāṭapuram, the second Tamil *caṅkam* was established by Pandyan king Venterceliyan. Succeeding kings nourished the *caṅkam* that flourished for 3700 years.
>
> One of the great Pandyan kings, Tirumaran [was] a great lover and patron of Tamil.
>
> He respected his language more than his regal status.
>
> Great works were written during this period on a variety of subjects like logic, medicine, mathematics, astronomy, and navigation.
>
> Foreign merchants visited Kumarināṭu, not only to buy pearls, ivory and the like, but also to acquire works of Tamil literature whose fame had crossed the ocean.

This idyllic life was brought to a cruel end by yet another burst of oceanic fury, and the city of Kapāṭapuram was flooded. The camera zooms in on the agonized king Tirumaran, and the commentator intones:

> Human beings, however high, were helpless.
>
> The king was losing his riches, yet he did not care.
>
> Come what might, he was determined to save the precious gems of Tamil literature, for the benefit of prosperity.

The commentator pauses here, as the film shows the Pandyan king, oblivious to the crashing waves and flooding waters, gathering up some palm-leaf manuscripts. The commentary continues:

> The king has lost his kith, kin and all, yet he fought to live, not to save himself but something more precious than his life, the gems of Tamil literature.
>
> Alone, anguished, but not awed, bereft of all pomp and circumstance, he struggled, unmindful of injuries and hurdles, to save his language, Tamil.
>
> His body was injured, but not his spirit.

Here, the film resorts to some dramatic footage of the Pandyan king running across the countryside barefoot and injured, his clothes and jewels in disarray. He finally reaches the city of (postdiluvial) Madurai, tired but happy. The commentator continues, as the film draws to a resounding end:

> The history of the Tamil *caṅkam* began on the banks of River Pahṟuḷi, flowed beside River Kumari, and reached Madurai of today on the banks of River Vaigai. . . .
>
> King Tirumaran offered at the feet of Mother Tamil the rare works of Tamil literature he had salvaged.
>
> Tears of joy brightened his happy eyes.
>
> He continued to serve Tamil, and founded the third Tamil *caṅkam*. . . .
>
> Long live Tamil!

Although the documentary appears astonishing in its unabashed nostalgia for kingship and monarchical rule, this filmic fantasy only builds upon similar labors of loss around prelapsarian Pandyan rule from early in the twentieth century. In and of itself, this fantasy seems utterly at odds with the demand for democratic forms of rule and for the abolition of hierarchies of all manner that have been such staples of the political culture of the region through much of that same century.[156] All the same, this yearning for a time when those in power were imagined as selfless, and as devoted to their subjects and to Tamil alike, makes sense when we locate it within the symbolic economy of loss in which Tamil devotion flourished, especially in the colonial period. Colonial rule had not only meant the end of Tamil political sovereignty, but it had also ushered in a modern state that was indifferent, even hostile, to the continued progress of Tamil as the hegemony of English gained ground. But the typical Tamil devotee is not just concerned with British colonial rule, but with a more enduring colonization—that by the Brahmanical, Sanskritic and Aryan culture of the North. Freedom from British rule only meant that the Tamil region was incorporated into the Indian nation-state dominated by what were clearly seen as anti-Tamil interests. Although Dravidian and Tamil nationalisms which underpinned so much of Tamil devotion relinquished their separatist intentions by the 1960s, and although the average Tamil speaker is quite willing to demonstrate his patriotic adherence to the Indian whole today, nevertheless, a powerful sense of Tamil pride and autonomy continues to be the dominant feature of the political culture of the region to this day, as many have noted. This sensibility is kept alive by the circulation of memories about a place and a time when Tamils had been in power, not just in Tamiḻakam but all over India as well—when Tamil speakers had been ruled by their own. Further, today they might be confined to the small region known as Tamilnadu because of the periodic oceanic floods which had robbed them of their homeland to the south, and because of the Aryan invasion which had put an end to their dominion over all of India. But, once upon a time, long long ago, their ancestors had ranged over the vast continent of

Kumarinātu, free and sovereign. The memory alone of this former glory sustains many a Tamil devotee as he deals with the indignities of colonial rule, and with (the imagined) postcolonial marginalization within an independent nation-state.

Loss of Territory

> Today, the Tamilnadu that we inhabit consists of 12 districts within its limits. A few centuries ago, Cēranāṭu and a part of the Telugu land were part of Tamiḻnāṭu. Some thousands of years ago, the northern limit of Tamiḻnāṭu extended to the Vindhya mountain, and the southern limit extended 700 *kāvatam* to the south of Cape Kumari which included regions such as Paṉai nāṭu, mountains such as Kumarik Kōṭu and Maṇi Malai, cities such as Muttūr and Kapāṭapuram, and rivers such as Kumari and Pahṟuḷi. All these were seized by the ocean, so say scholars. . . . That today's Indian Ocean was once upon a time a vast landmass and that that is where man first appeared has been stated by several scholars such as Ernst Haeckel and Scott Elliot in their books *History of Creation* and *Lost Lemuria*. The landmass called Lemuria is what Tamilians call Kumarināṭu. That which is remaining after this ancient landmass was seized by the ocean is the Tamil motherland in which we reside today with pride.[157]

So wrote K. Annapoorni, a rare female place-maker and a Tamil teacher in a Salem high school in 1935. In Tamil place-making, all the other losses that I have detailed so far are ultimately grounded in the loss of something deeply material and tangible, namely, the loss of land—and not just any land, but, as I have repeatedly emphasized, the ancestral Tamil homeland, the birthplace of Tamil and its literature, the nursery of foundational Tamil values, the Tamil prelapsarium. As I detail in the next chapter, Tamil placemakers summon into existence an elaborate landscape of loss, filling up Kumarināṭu with named topographic locations—regions, rivers, mountains, cities—that are derived from fabulous readings of precolonial allusions to disappearance of land and of dispossession. Through such readings, Sclater's lost place-world is converted into a known and familiar Tamil home-place. For now, I focus on the several ways in which Tamil place-makers recall the magnitude and enormity of the territory that is lost, and convey this to their fellow speakers.

For one, they discuss the size of Kumarināṭu in terms of the terrestrial formations with which Tamil speakers, or at least the educated among them, would be familiar. In 1903, Suryanarayana Sastri suggested that the antediluvian Tamil land "extended in length from today's Cape Comorin southward to Kerguelen Island, and in breadth from the island of Madagascar to the Sunda Islands which include Sumatra, Java and other islands."[158] Somasundara Bharati proposed in 1912 that the continent "touched China, Africa, Australia, and Comorin on its four sides."[159] In 1948, Maraimalai Adigal observed that "the vast southern continent . . . lay on the two sides of the equator, the greater portion of it stretching southwards as far as the South Pole and towards east and west as far as Australia and even farther and as far as South Africa."[160] This suggestion has been reiterated recently in the insistence that "before the deluges, Lemuria extended from the South Pole regions to the foot hills of the Himalayas, including Afghanistan."[161]

For those unfamiliar with modern geography, maps, and atlases, the place-maker resorts to numbers to quantify the enormity of the "Tamil" land that was lost. In particular, he turns to the figure of 700 *kāvatam* provided by the medieval commentator Adiyarkunallar, as the size of the land south of Kanyakumari seized by the ocean. While it is acknowledged that the modern equivalent of the *kāvatam* is not known, this has not stopped many from computing the area lost to the ocean using this conventional figure sanctified by Tamil's literary tradition. The earliest and most popular estimate conjures up 7,000 miles of lost "Tamil" land.[162] Others concede that this might be too extravagant, as this would mean that Kumarināṭu extended all the way to the South Pole, and they instead offer more modest figures ranging from 1,400 to 2,100 to 3,000 miles.[163] Still others nuance such figures by suggesting that perhaps an area of 6,000–7,000 *square* miles was lost.[164] Some, including U.V. Swaminatha Aiyar, are satisfied that land amounting in area to only a few villages (equivalent to the Tamil measure of two *kūṟṟam*) was lost.[165] Regardless of these differences, it is worth noting the imperative to quantify the extent of the drowned land in order to convince the modern Tamilian of the enormity of his lost patrimony, an imperative that also speaks to the hold that statistics and numbers had on popular and official imagination in colonial India.[166]

But most importantly, the lost homeland of Kumarināṭu is again and again compared, contrasted, and juxtaposed with the lived homeland of Tamilnadu/India to which modern Tamilians are today confined. Both discursively and cartographically (as I discuss at greater length in later chapters), Tamil speakers are reminded, again and again, that all they have left today is a mere remnant of a much vaster former domain. The poignancy of catastrophic dispossession is heightened by the realization that there is no possibility of return to the lost homeland, now lying at the bottom of the Indian Ocean. Reunion with the lost motherland is therefore an impossibility that every Tamilian has to live with, forever.

In this place-making, therefore, modern Tamil speakers are doomed to live in a state of perpetual exile. It is this state of being-in-exile that runs as an undercurrent through much of the twentieth century, as Tamil devotees and nationalists struggle, first as colonial subjects and then as sovereign citizens, to accommodate their own regional sentiments and aspirations with a national whole that is "India." As many have noted, this process of accommodation has been tendentious at numerous moments, especially during the middle decades of the last century when a powerful separatist surge distinguished Dravidian and Tamil nationalism from other movements of regional assertion in the subcontinent. Even after the 1960s, when the separatist impulse had run its course and the Tamil country ambivalently settled down to being a part of the Indian union, the yearning for a prelapsarian elsewhere in the long lost past, where Tamil and its speakers had been born and had led a life of perfect plenitude, has never really abated. Lemuria lingers tantalizingly as a rival home/land that is the object of unrequitable longing, even while Tamil's devotees call upon their fellow speakers to cultivate love and loyalty for their lived homeland of Tamilnadu/India.

LIVING WITH LONGING AND LOSS AT LAND'S END

There are many reasons that make Lemuria's candidacy as the ancestral Tamil homeland so attractive to Tamil devotion's place-making. It had been identified by metropolitan scientists as a vast ancient continent and as the birthplace of mankind: if Tamil speakers were the original inhabitants of Lemuria, this made them, ipso facto, the most ancient peoples of the world and the ancestors of all of humanity. As many insisted, with the sinking of the ancestral homeland, Tamilians fanned out to the rest of the world, settling and planting the seeds of civilization in the Indus Valley, Mesopotamia, Egypt, China, the Americas, even Europe. Consequently, Lemuria allows Tamil labors of loss to fashion a diaspora for Tamil and its speakers that was as widespread as it was ancient, stretching back to the very beginning of time, in fact. It allows modern Tamilians to assume the status of global peoples, if only vicariously, in an age of global empires and the exercise of global power.

Furthermore, the appropriation of Lemuria for the Tamil cause also enables the recasting of modern Tamil speakers as descendants of an antediluvian people. This allows Tamil labors of loss to tap into all the symbolic potency—innocence, purity, and singularity—associated with prelapsarian virtue and bliss. Lemuria thus provides a context for summoning into existence a Tamil prelapsarium, further deepening the antiquity of the language and its speakers, the first in the subcontinent as well as in the entire inhabited world.

But most significantly, it is the fact that Lemuria had been declared by science itself as a lost continent, forever vanished in the depths of the ocean, that makes it so useful for the Tamil project. The collective yearning for an unreclaimable past plenitude holds together a people-in-exile otherwise riven apart by caste, class, and religious differences. Even within the community of Tamil devotees, which I have shown elsewhere has been radically divided over the meaning of Tamil and its relationship to the other languages, cultures, and speakers of the subcontinent, most unite around the contention that Tamil is an antediluvian language whose origins can be traced to the lost homeland of Lemuria. Loss, therefore, is powerfully enabling in southern India, and is a sentiment, therefore, that has to be continually fed and stoked. As Marilyn Ivy has noted in another context, "the loss of nostalgia—that is, the loss of the desire to long for what is lost because one has found the lost object—can be more unwelcome than the original loss itself. . . . Modernist nostalgia must preserve . . . the sense of absence and lack that motivates its desire."[167] Thus, Tamil devotion's complex and variegated labors of loss cannot actually afford to find the lost homeland, or to close the gap, to present the absent. To do so would mean the end of its project, its own logic for existence. As Susan Stewart writes perceptively: "The nostalgic is enamored of distance, not of the referent itself. Nostalgia cannot be sustained without loss. For the nostalgic to reach his or her goal of closing the gap between resemblance and identity, lived experience would have to take place, an erasure of the gap between sign and signified, an experience which would cancel out the desire that is nostalgia's reason for existence."[168] Yearning for—and mourning—the lost homeland becomes an end in itself, therefore, for a people who imagine themselves in perpetual exile.

Loss in this sense may be empowering, but there are also dangers in pursuing a project whose object of desire is an elsewhere that can never be attained. This is especially so when the elsewhere falls outside the territorial boundaries of the nation-state which is the lived homeland, not to mention the ground of practical politics.[169] From the start, the relationship of Lemuria to the lived homeland of Tamilnadu, itself a small part of "India," is plagued by a strategic ambiguity in Tamil place-making. Some suggest that all of what we know of as India today had been part of Lemuria, spatially reiterating the claim that Dravidians (and hence Tamilians) had been the original peoples of the subcontinent before they had been driven south by "the Aryan invasion."[170] Others are content to include only peninsular India—that region of the subcontinent that colonial ethnology proposed was populated by Dravidian speakers—in Lemuria. Still others insist that only that part of India extending from Mount Venkatam to Kanyakumari had ever been part of Kumarināṭu, the historic Tamiḻakam.

And yet the fact remains, as I have already noted, that loyalty and attachment to the lived homeland of Tamilnadu, itself a part of India, has also to be generated, especially if Tamil's devotees want their fellow speakers to rally together to protect and nurture their language and land—or what was left of these after the ravages of time and after the onslaughts made on them by rival languages and their patrons. Here is where the naming practices I wrote of earlier come in useful. The rechristening of Sclater's Lemuria as

Kumarināṭu and Kumarikkaṇṭam enables the establishment of a direct connect with the remnant of Kanyakumari, which also serves as the land's end for India today. Even more directly, the naming of Lemuria as Tamiḻnāṭu suggested not only that the lost continent was an intimate Tamil territory, but also that the modern state of Tamilnadu, itself a member of the Indian union, had formerly been a part of the lost homeland. Those Tamil devotees who are Indianist in inclination even name the lost homeland as "India," establishing a nominal connection between the nation-state and Lemuria. The lived homeland is thus nominally subsumed by the lost homeland in a variety of ways.

As importantly, the very structure of loss in which Lemuria is embedded proves to be critical. For, as Tamil speakers are reminded again and again, because what they are left with is a remnant of their lost homeland, this is all the more reason that what is remaining—fragmentary though it might be—has to be guarded and cherished. This argument was especially made over and again—to everyone from Tamil schoolchildren to the adult citizen—during the 1940s and 1950s, the very decades that saw the birth of independent India and the reconfiguration of its internal political geography. As Tamil speakers were repeatedly reminded during these years, even though they could never have Lemuria, they should work hard to ensure that what was left of it—namely, the remnant of (postlapsarian) Tamilnadu—should not be lost to the neighboring states of Kerala and Andhra Pradesh. The lived homeland, in other words, functions as "a memorative sign" of the lost homeland. In Jean Starobinski's suggestive reading, the memorative sign "is related to a partial presence which causes one to experience, with pleasure and pain, the imminence and impossibility of complete restoration of [all our former life] which emerges fleetingly from oblivion. Roused by the 'memorative sign,' the conscience comes to be haunted by an image of the past which is at once definite and unattainable."[171]

The politics of loss thus works to establish a productive, even pleasurable, tension between lived and lost homelands in the Tamil country. The two are intimately linked precisely because the lived is imagined as the fragment of the former whole that is lost. Because that which is lost is lost forever, this means that the fragment becomes even more precious and has to be protected and cherished, to the point of giving up one's life for it, if need be. A critical division of labor over sentiment animates this productive nexus between lived and lost homelands. The lived homeland becomes the object of loyalty and love—of patriotism. Meanwhile, the lost homeland is the focus of unsatiable yearning or unrequited longing, in other words, of nostalgia, that most elusive of sentiments that hovers "at the crossroads of imagination and memory."[172]

In recent years, much has been written on this intangible structure of feeling which some have argued "is at the very core of the modern condition."[173] In her recent book *The Future of Nostalgia,* Svetlana Boym reminds us that nostalgia "is a longing for a home that no longer exists or has never existed. Nostalgia is a sentiment of loss and displacement, but it is also a romance with one's own fantasy. Nostalgic love can only survive in a long-distance relationship."[174] Not surprisingly, nostalgia has found a flourishing home in the modern Tamil country, whose relationship to the lost Lemuria is, irrevocably, long-distance. Similarly, the feminist scholar Iris Young insists that "nostalgia is a flight from having to come to terms with [an originary] loss, by means of constant search for a symbolic substitution for the lost home."[175] The symbolic substitution for the lost motherland of Kumarikkaṇṭam, I have noted, is the lived homeland of Tamilnadu/India. The latter comes to stand in for the lost home/land. Young also distinguishes usefully between remembrance and nostalgia, suggesting that while nostalgia accompanies fantasies of a lost home, remembrance preserves and nurtures what one owns, possesses, holds. "Nostalgic longing is always for an elsewhere. Remembrance is the affirmation of what brought us here."[176] In other words, along with a division of labor over the exercise of sentiment, there is also a division of labor in memory-work, so that the lost homeland returns to present consciousness through nostalgia (and a narrative modality that flirts with fantasy, albeit not very successfully, as we will see), while the lived homeland is remembered through linear recollection and the realist genre of history.

Not least, following Edward Casey, I want to suggest that the lost homeland inhabits "an absolute past, a past that was never a present. Such a world with such a past . . . cannot be recollected as such."[177] It can only be recollected, instead, as longing, yearning, and mourning for the home-place where it all started. The lost homeland is a "world-under-nostalgement," to use Casey's terms, "a world vanished in the shadows of an absolute past."[178] It is precisely because it is a world vanished in the shadows of an absolute past—a world of "such deep anteriority that it will never become the specific content of memory"—that the lost homeland is not a threat to the lived homeland of the here-and-now.[179] On the contrary, returning to haunt the present retrospectively as irrevocable loss, Lemuria paradoxically functions to also remind modern Tamil speakers of what they do possess and hold in the here-and-now—their lived homeland of Tamilnadu, even India. This, then, is the logic of the fascination with this lost world amongst Tamil place-makers, a logic that is rooted in their labors of loss that are passionate, nostalgic, and bittersweet, "bitter because lost, all the more sweet for being lost."[180] Or, as Casey writes, "In being nostalgic, we are moved by a past world that is the more potent for being absent or 'vanished.'"[181] As a lost home-place that is both discursively definite but also fabulously unattainable, Kumarināṭu's potency for Tamil place-makers derives from the fact that their labors of loss have transformed it into a site where they witness the "baffling combination of the sweet and the bitter, the personal and the impersonal, distance and proximity, presence and absence."[182] This, then, too, is the allure of Lemuria for Tamil's devotees, as a place-world that is there—but not really—to remind them of the pristine origin of their beloved language and land in a moment of absolute plenitude and perfection that is now forever lost, and hence safe and secure from the ravages of time—and history.

Chapter 5

Flooding History

Geographies of Loss

It is in the history of the seas that we discover the history of the continents.[1]

OCEAN FANTASIES, OCEAN FEARS[2]

The many labors of loss around Lemuria are fundamentally place-making acts in which the ocean is accorded a creative and destructive role. The ocean is a source of both fantasy and fear to Lemuria's place-makers, although the density of these varies across the different labors of loss I have been considering here. It is also the principal (and in some cases, the only) agent in their labors, and even if the potency of its agency varies, it is the ocean's work that ultimately causes disappearance and loss.

The centrality accorded to the ocean is perhaps not surprising when we recall that Lemuria first surfaced in European place-making when the science of oceanography was virtually nonexistent. In the later half of the nineteenth century, the oceans that cover much of the earth's surface were still the last great frontier for metropolitan men of science, and there was much speculation about what went on in their unfathomed depths. The laying down of submarine telegraphic cables beginning in the mid-nineteenth century, and *HMS Challenger*'s survey of the Atlantic floor between 1872 and 1876, revealed a fascinating underwater world of submerged ridges and hidden valleys whose very existence seemed to scientifically confirm lost lands like Plato's Atlantis.[3] It is no coincidence that Ignatius Donnelly's 1882 bestseller on Atlantis was published a few years later. In a chapter evocatively entitled "The Testimony of the Sea," he wrote of "the revelation" of a great underwater elevation and concluded that "here, then, we have the backbone of the ancient continent which once occupied the whole of the Atlantic Ocean. . . ."[4] Soon on his heels, Blavatsky boasted in her Mahatma letters that the *Challenger*'s findings had only confirmed what ancient legends had known all along of drowned continents which were catastrophically destroyed by Earth's waters. "Science has finally accepted . . . and thus vindicated the truth of one more 'fable.'"[5] And science, indeed, made some room for Atlantis, for even a leading journal like *Nature* associated the remarkable discoveries of the *Challenger* with the lost continent,[6] and Suess's masterly synthesis *Das Antlitz der Erde* (1885) as well included an extensive discussion.[7]

Ultimately, however, the progressive taming of the world's ocean floors from the 1930s have put to rest any lingering doubts about drowned continents in the professional scientific community in the metropole. In the words of Kenneth J. Hsu, "by 1950, only a few aging biologists . . . continued to talk about 'landbridges' or sunken continents."[8] Yet among Euro-American occultists, novelists specializing in the "lost races" fantasy genre, and Tamil's devotees, the ocean and its mysterious depths continue to be the source of enchanted fantasies. As Lewis Spence wrote breathlessly at the beginning of his *The Problem of Lost Lemuria* in 1933: "No episode, perhaps, in the endless narrative of human romance exercises a spell so enthralling as that which tells of lands ancient and cultured submerged by catastrophe in the deep gulfs of ocean. The sentiments aroused by the glowing fictions of the East, the glamour cast by the chronicles of magic and the supernatural, pale before the curiosity which the mere mention of sunken Atlantis or Lemuria invariably excites."[9]

And on the other side of the world from Spence, the underwater mapping of the Indian Ocean by an international team of scientists between 1959 and 1965 only further encouraged Tamil place-makers in their quest for their precious Kumarikkaṇṭam.[10] In this regard as in others, modern science has been both the doing and the undoing of labors of loss around Lemuria.

For those paleo-scientists interested in submerged continents before continental drift theory became the new orthodoxy from the 1960s, the ocean is an eternal but impersonal force of nature whose unceasing work causes transformations in the relation between land and water on Earth's surface. They would have agreed with Ernst Haeckel that "during the course of many millions of years, ever since organic life existed on the earth, land and water have perpetually struggled for supremacy."[11] Some interpreted this struggle catastrophically, providing graphic descriptions of the turbulent oceans, the earth movements that originated in their depths, and the tidal floods that caused subsidence of vast stretches of land and the disappearance of whole species. Others were more partial to gradualist explanations which favored routine but slow mutations of land and sea. Given, however, the disenchanted tenor of the paleo-scientist's labors of loss, there is rarely anguish expressed over the ocean's role in the disappearance of land. This is accepted as one of the indelible facts of nature, although an occasional tremulous voice may be heard now and again.[12]

Like the paleo-scientist, Euro-American occultists as well, especially those influenced by Theosophy, rarely express grief over the loss of Lemuria to the ocean, partly because creation and destruction are part of an endless cycle in which mankind is entangled. Occult labors of loss, however, are invariably catastrophic in their image of the ocean, whose unpredictable work results in the destruction of whole civilizations. Even though the destruction of the Theosophists' Lemuria began with volcanic eruptions, the chasms opening up in the ocean floor deliver the final coup de grâce. In post-Theosophical occultism as well, Lemuria arises out of the sea and returns to the sea, only to reemerge in the future from its submarine slumber at the dawn of the New Age. Accordingly, occultism's labors of loss are shot through with invocations of the terrible convulsions caused by the fiery ocean, the havoc it visited upon the hapless Lemuria, and the tragic submergence of a once flourishing land to its mysterious depths. A 1938 advertisement for Cervé's 1931 book on Lemuria even carried a striking visual of waves pounding away at the shores where men and women are seen in flight as the tall neoclassical edifices of the stricken land start to crumble. The visual is accompanied by the following blurb:

> *Beneath Pacific Sank . . . Lemuria, the Mystery Continent!*
>
> In the depths of the Pacific, shrouded in darkness, lies a vast continent. Where once great edifices reached skyward and multitudes went their way is now naught but the ceaseless motion of the sea. Centuries before the early men of Europe or Africa found the glorious spark of fire or shaped stones into crude implements, the Lemurians had attained an exalted culture. They had wrested from nature her proudest secrets. Then nature reclaimed her power. With a tremendous convulsion she plunged the civilization of demi-gods beneath the leveling waters. Again she reigned supreme, the victor over man's greatest efforts.[13]

Although the ocean's catastrophic deeds are not necessarily celebrated in occult place-making, there is a sense that these produce the generative loss which is responsible for the creation of secret knowledges and lost wisdom that occultists then discover and reveal to a disenchanted world through their various esotericist technologies. Therefore, catastrophic though it might be, the ocean—mysterious, alluring, secretive—is enormously productive for occult labors of loss, which are utterly captivated by its hidden depths. In the words of the occultists of the Lemurian Fellowship, "In the endless history of human experience, nothing exercises a spell half so enthralling as that concerning ancient lands and cultures submerged beneath the placid waters of some vast ocean."[14] In a disenchanted world where virtually everything had been bared to the gaze of science, the ocean's mysteries could still hold one in thrall.

However, it is in Tamil India that the ocean takes on a life of its own, as it emerges as the all-powerful and all-consuming villain in a series of catastrophic encounters between nature and (Tamil) culture, between water and (the Tamil) man. Thus, phrases like the "merciless Indian Ocean," "the hungry Indian Ocean," the "relentless, land-grabbing sea," or the "enraged sea" punctuate the lamentations of Tamil place-makers. The 1981 film produced by the Tamilnadu government provided a dramatic visual reminder when it showed the relentless ocean, not once but twice, plundering the hapless ancestral homeland, destroying property, and seizing lives.

It is a measure of the intense personal relationship that Tamil place-makers establish with "the cruel sea" that they repeatedly address it, chastise it for its actions, and take it to task for its "vengeance." I translate three such place-making reveries,[15] which are occasioned by travel to land's end at Kanyakumari, where the devotee is confronted with the vista of the ocean—rather than of his beloved former homeland—stretching endlessly to meet his gaze on all sides:

> The mind grieves when it thinks of Cape Kumari for it brings back memories of the seizure of southern Tamiḻnāṭu by the ocean, and the consequent loss. . . . Both creation and destruction are part of God's play. Even if we think in this manner of the loss of the southern land, the mind finds no solace even after thousands of years have passed.

So wrote C. Mutthuvirasami Naidu when he visited the Cape in 1955.[16] As he looked out at the vast ocean that stretched before him, the antediluvian history of the Tamil people flashed through his mind as a Herculean struggle between ocean and land, between nature and culture:

> I looked out at the boundless expanse of the ocean, ceaselessly and tirelessly turning and rolling, rising and falling, again and again. The thought sprang to my mind as I stood gazing at this sight that in this very manner, the ocean rose and destroyed Kumarikkaṇṭam several hundreds of years ago. When the southern ocean boiled over and seized the land, fine men, graceful women, and beautiful people, shapely colorful birds, animals and other living beings, all these suffered and perished. Temples, mansions, and palaces, renowned Tamil libraries and the rare arts, colleges and schools, assemblies and meeting houses, market places and port facilities, homes, gardens, and playing fields, all disappeared into the ocean. When I saw the distress all this caused, my heart trembled and drowned in an ocean of suffering.[17]

So overwhelmed was Mutthuvirasami by such memories that even after he retired for the night, "the cries of distress of Tamiḻnāṭu as it sank into the ocean" filled his thoughts. It was almost as if the ceaseless rolling waves continually stoked the embers of distress in his mind: "Alas! What can we say about the suffering heaped upon the ancient Tamil land, which gave birth to the arts and to world civilization? What can one say about the cruelty of the ocean that caused all this? What can we say about the state of the Tamil-speaking land that today stands shrunken and emaciated. . . . How much distress and loss has all this caused? Can we even measure this?"[18]

Mutthuvirasami Naidu was not alone in this regard. Ten years earlier, the Tamil teacher and scholar A.M. Paramasivanandam (b. 1914) had also cast similar aspersions on "the cruel sea" during his visit to the Cape with a friend. After taking in the sights, they bathed in the ocean and then sat on the shore. As they sat there, Paramasivanandam was overwhelmed with "memories of Kumari," and he demanded of the ocean, "Where is the Lemuria continent that extended in the east to Australia and in the west to Madagascar?"[19] In contrast to Mutthuvirasami, his memories took an explicitly geographical turn, and he names the rivers, mountains, cities, and regions declared to have constituted the "49 territories" of the drowned antediluvian Tamil realm:

> O Indian Ocean! Where did you conceal our Tamiḻnāṭu, our ripe old land? Why did you plunder the 49 Tamil territories thousands of years ago? Where is that fine ancient river Pahṟuḷi? Where is that golden stream called Kumari? . . . Where is the incomparable Mount Kumari filled with beauty? Where is Teṉmaturai famed for the first Tamil academy presided over by the Lords Siva and Murugan? Where is the lofty Kapāṭapuram, seat of the second academy graced by Tolkappiyar, Agastya, and many others? . . . Where did these go? Where did these go?[20]

Like Mutthuvirasami, he, too, meditated on the consequences for the present of the momentous submergence of Tamil territory, asking of it: "If today our ancient Tamil land and ancient Tamil learning had survived, we would have ruled the world from Kumari to the Himalayas! We have instead become the laughing stock of others! Is this not your doing? . . . If it had not been for your actions, we would have ruled the world! Why were you angry with Tamiḻnāṭu and the Tamil language and the Tamil people?"[21]

The last of these reveries comes from the pen of R.P. Sethu Pillai (1896–1961), reputed essayist and Professor of Tamil at the University of Madras. In Sethu Pillai's place-making, it is the venerable poet-sage Ilango Adigal, author of the classic narrative poem *Cilappatikāram* who interrogates "the cruel sea" on a visit to the Cape.[22] As Ilango looks out at the sea, his face, which had been "aglow with the light of Tamil," clouds over with sorrow, and his eyes fill up with tears:

> Oh! Kumari Sea! When I see you, my heart surges with distress. Your breeze fans the hot embers of my body! Alas! O wave-filled ocean! What did our dear precious land do to you?"[23]

Speaking through Ilango, Sethu Pillai charges "the mischievous ocean" with destroying not just the antediluvian cities and rivers of the Tamil realm, but also for bringing distress to the famed Pandyan kings who magnanimously ruled over that ancient continent, nurturing the Tamil language and literature:

> O mischievous ocean! You consumed so many towns belonging to the Pandyan king! You drank up so many rivers! You swallowed so many mountains! . . . Our Pandyan shed tears when he learned that you had destroyed [the Pahr̲uḷi] river. Even today I can hear the sound of his crying in your breeze! Is this your only cruelty? You also swallowed the river Kumari. . . . Alas! You ate up our land! You drank up our rivers! You consumed our mountains![24]

Like Mutthuvirasami and Paramasivanandam, Sethu Pillai, too, lamented that because of the ocean's cruelty, "the Tamil land has shrunk. My heart grieves . . ."[25]

In striking contrast, therefore, to the Euro-American paleo-scientist or occultist, the Tamil place-maker's attitude to the ocean is intensely personal, even intimate, if angry. The ocean is not a mere fact of nature waiting to be tamed, or a mysterious occult force, but a wrathful and deceitful entity which decided to take on the hallowed Tamil land for no rhyme or reason. The battle lines are clearly drawn in the Tamil country where labors of loss are also labors of grief, in which the ocean is the principal enemy of the Tamil people and of their drowned ancestral homeland.

FLOODING LEMURIA

The principal term used to designate the catastrophic and destructive agency of the ocean in Tamil devotion's labors of loss is *kaṭalkōḷ*, a word that has attained the same status among its place-makers that was accorded to "the Deluge" or "the Flood" in the discourses of the Judeo-Christian West well into the early decades of the nineteenth century.[26] Literally, "seizure by the sea," *kaṭalkōḷ* is frequently glossed in English as "flood," "ocean swell," and, significantly, as "deluge," or even "Deluge." All of Tamil devotion's ocean fears and fantasies turn around this word, which is used, over and again, to describe the actions of the "cruel sea" in its labors of loss. It is because of *kaṭalkōḷ* that Tamil speakers irrevocably lost their patrimony as embodied in their antediluvian words and works as these had been nurtured in their ancestral homeland. Instead of sustaining Tamil homes and hearths today, their patrimony lies consigned to a "watery grave" at the bottom of the Indian Ocean.[27]

But just as consequential for the politics of the present, because of *kaṭalkōḷ* and its disastrous consequences, today's Tamil speakers had also lost the material evidence to convince others that Tamil is an antediluvian language, the world's most ancient at that. Not surprisingly, as early as 1901 Savariroyan lamented, "So completely has the Deluge done its destructive work that we hardly come across in South India such old manuscripts and libraries as confront the gaze of antiquarians and explorers in Neneveh and other ancient Chaldean towns."[28] Most tragically of all, vast stretches of inhabited Tamil territory *(nāṭu)* resonating to the sound of Tamil were forever submerged, leaving Tamil speakers confined today to a small piece of land on the Indian mainland. No wonder the consequences of *kaṭalkōḷ* are "catastrophic," a characterization that appears in its English guise as early as the 1890s and continues to reverberate through the course of the next century, as the Tamil place-maker labors in grief over the loss of his antediluvian homeland and his ancestors.[29]

As the litterateur Nallasami Pillai noted many decades ago, explicit discussion of *kaṭalkōḷ* appears for the first time in the learned commentary on Aphorism 1 in an ancient grammar of love poetry called *Iṟaiyaṉār Akapporuḷ* (also known as *Kaḷaviyal* and believed to have been authored by the lord Siva himself).[30] Generally attributed to Nakkirar, this commentary (dated unstably to the later centuries of the 1st millennium C.E.) offers the first tentative sketch of an antediluvian land in which Tamil had flourished for 8,140 years under the patronage of Pandya kings in academies *(caṅkam)* established first in the city of Teṉmaturai, and then in Kapāṭapuram, both of which were subsequently "seized by the ocean." As a consequence, all the literary creations of the 4,449 poets of the first academy (which flourished for 4,440 years), and the 3,700 bards of the second (which lasted another 3,700 years) were washed away, before a third was established in postdiluvial (or "northern") Madurai of the present.[31] Nakkirar's commentary, however, does not directly allude to the loss of land accompanying the loss of these literary academies and their learned productions. This followed a few centuries later when Adiyarkunallar wrote his commentary on the *Cilappatikāram.*[32] Adiyarkunallar's commentary supplements Nakkirar's details of antediluvial literary losses with territorial losses as well, as it notes that a stretch of land extending to a distance of 700 *kāvatam* and divided into forty-nine territories *(nāṭu)* south of Cape Kumari was swept away by the ocean.[33] Two other ancient anthologies refer to the catastrophic consequences of the ocean's wrath: scattered verses in the *Puṟanāṉūṟu* and *Kalittokai* hint that "the cruel sea" invaded the Pandyan domain and seized land, and that when this happened, the Pandya king compensated his loss by seizing the equivalent amount of territory from his neighbors, the Cera and the Chola.[34] And in medieval commentaries on ancient texts such as the *Tolkāppiyam* by scholars like Ilampuranar, Naccinarkiniyar, and Perasiryar there are stray allusions to an antediluvian land south of Cape Kumari that was subsequently seized by *kaṭalkōḷ.*[35] Modern Tamil labors of loss around Kumarikkaṇṭam are anchored in this limited but telling corpus of antique words and verses.

These ancient texts and their medieval commentaries were published for the first time in the closing decades of the nineteenth century, after being lost for more than a hundred years to the living memory of the Tamil reading public.[36] Soon after, the labors of loss that I have been detailing in these pages commence in the Tamil country. In the process, they repeatedly turn to these archaic and oblique allusions to the ocean's mischief. But they do so with a crucial difference that points to the very modernity of these labors of loss. Nowhere do the ancient poems or their learned commentaries mention that the land seized by the ocean was a whole continent *(kaṇṭam).* Nor do they mention the name of the land, let alone bestow upon it the scientific toponym of Lemuria, or even Kumarikkaṇṭam or Kumarināṭu. Most crucially, there is no attempt to link the loss of land or of Tamil books produced in the antediluvial academies to the history of the Tamil people as a community or an emergent nation. Yet, within a few years of the publication of these ancient works for the first time, in the closing years of the nineteenth century, all this is asserted, as Tamil's devotees conjure up a new place-world out of these archaic allusions to lost academies, lost books, and lost territory, facilitated by the paleo-scientist's speculations about Lemuria which had begun to circulate by then in the Madras Presidency. So, here is Savariroyan in 1901 confidently invoking "tradition" to advance some really novel claims:

> The old tradition preserved to this day in some of the classical works such as *Cilappatikāram* and *I̱raiyaṉār Akapporuḷ* . . . locates the beginning of the cultivation of Tamil in the hoary past. . . . The tradition asserts (a) that there was a great continent contiguous with South India covering the large portion of the Indian Ocean to the South of Cape Comorin and it was the seat of a civilized nation and of a powerful dynasty for many centuries from very remote times; (b) that the capital of the dynasty was the seat of an assembly of learned men first at South Mathurai [Teṉmaturai], second at Kapadapuram or Alavai and the assembly of literati at South Mathurai is known as the first Sangam and that at Kapadapuram as the second; and (c) that there occurred then a great inundation which washed away the vast extent of land stretching from Cape Comorin southwards with all the literary productions of the time. The fact that a vast land existed south of Cape Comorin and was submerged by the flood receives great support from the modern sciences, Geology and Natural History, which prove the existence of a land south of India and its disappearance beyond the pale of doubt.[37]

Savariroyan's assertions (a) and (c) were only possible because of metropolitan science's labors of loss over Lemuria. But also crucial here is the fact that "tradition" concurs with "the modern sciences, Geology and Natural History," on the matter of the inundation of the antediluvial Tamil land and its subsequent loss. It is this concurrence that allows Tamil place-makers to move seamlessly and effortlessly between literary allusions to *kaṭalkōḷ* and loss of territory on one hand, and scientific findings regarding land submergence on the other.

Numerous sources feed Tamil devotion's ocean fears and fantasies and its dread of *kaṭalkōḷ*. In addition to the ancient poetic anthologies whose occasional verse carried memories of the sea's wrath, religious legends from the 1st millennium C.E. on speak of frequent floods that threatened old cities such as Madurai, which were then subsequently saved by divine intervention.[38] Oral traditions handed down the generations by fisherfolk lament the loss of life and limb to the cruel sea as they recall with wonder stories of submerged palaces and temples that they had heard.[39] In the everyday life of the Tamil country, rumors about the disappearance of built structures in coastal towns like Mahabalipuram actively intermingle with historical accounts of the vanishing of ancient ports such as Korkai and Poompuhar. And then there are the graphic eyewitness reports of those who lived through the fierce cyclones that periodically hit the Coromandel coast, destroying both life and property as well as washing away land, the most talked-about in recent memory being the inundation and disappearance of the temple town of Dhanuskodi in December 1964. From the early years of the nineteenth century, the colonial state assiduously catalogued all such oceanic activities in its district reports and manuals, its gazetteers and scholarly tomes, and so, too, have the postcolonial governments of the region.[40] Not surprisingly, in such a context, metropolitan science's labors of loss around submerged lands find ready accommodation, and indeed, have been welcomed, for they are deemed to offer material proof for what Tamil speakers had known all along—from the time of their antediluvial ancestors—about the ocean's fury. As one of them insisted, "No one can reasonably be a skeptic as regards the flood in the face of such overwhelming evidence."[41]

The Tamil place-maker may dread *kaṭalkōḷ*, but it is also discursively critical, indeed necessary, for his labors of grief around the loss of his ancestral homeland. For *kaṭalkōḷ* is not just a geological event or a periodic historical happening, but an explanatory device that enables Tamil labors of loss to connect the destructive actions of the ocean with the history of Tamil speakers as a singular community, and to the fortunes of a territory that is unambiguously identified as Tamil. The consequences of this nexus are many. It provides an opportunity for Tamil place-makers to imagine an antediluvial past for their beloved land and its inhabitants and empowers them to tap into all the symbolic potency—innocence, purity, authenticity, and singularity—associated with prelapsarian virtue. *Kaṭalkōḷ* also enables the writing of the history of the Tamil people and their territory in catastrophic terms, the tragedy of the catastrophe heightened by the fact that it was the uncontrollable forces of nature, as these manifested themselves in the ocean's fury, that caused the irrecoverable disappearance of the ancestral Tamil homeland. Strikingly, Tamil labors of loss around Lemuria do not single out the usual communities, such as Aryans, Brahmans, North Indians, Sanskrit and Hindi speakers, Muslims, or the British, who appear in other discursive realms as the principal enemies of the Tamil people.[42] Instead, forces far beyond the control of mere mortals had wrought this catastrophic destruction. From this perspective, everything that had affected Tamil speakers and their territory and possessions in postlapsarian times pales in significance when compared with the catastrophic loss of the ancestral antediluvial homeland and its cultural productions. David Shulman has observed that traditionally in the Tamil country both the creative and violent aspects of the flood are typically underscored.[43] Although Tamil place-makers find little that is redeeming in the *kaṭalkōḷ* that catastrophically destroyed the Tamil prelapsarium, even they can see that it provides the occasion for remembering Tamil heroism as embodied in the Pandya kings who battled the ocean and who valiantly salvaged their language from nature's fury, by establishing *caṅkam* after *caṅkam* in their new capitals, by providing succor to the poets and scholars who were members of these academies, and by nurturing works of Tamil literature. As in many pre-modern flood stories that circulated in the Tamil country, in Tamil devotion's accounts of *kaṭalkōḷ* as well, "Order, in the person of the king, replaces the inchoate powers of the ocean, and the flood provides the background to the dynastic foundation, or in other words, to a renewed creation."[44] So, remembering *kaṭalkōḷ* allows the place-maker to remind his fellow Tamil speakers of all that their language, their community, and their land had endured, and gives him an occasion to urge them to protect what remained after the catastrophic losses of the prelapsarian past. Notably, kings, not gods, as they do in other flood narratives, ultimately save the day. Divine intervention has no place in these secular labors of loss of the Tamil country.

Of course, most critically, *kaṭalkōḷ* allows Tamil devotion to write the history of the Tamil people and of their territory as a history of catastrophic and irrevocable loss. Loss, as I have insisted, is sustaining and empowering, even satisfying, in the Tamil country, and *kaṭalkōḷ* its enabling cause. So much so that K.P. Aravaanan ingeniously suggests that the very name for the Tamil speaker, "Tamiḻaṉ," derives from the Tamil word *amiḻ*, "to submerge." In his reckoning, Tamilians are those people who survived submergence by the sea: they were named as such by their ancestors so that they might remember this originary catastrophe.[45] As I earlier noted, when the principle of isostasy was discovered in the middle of the nineteenth century, it soon posed a major problem for advocates in the metropole of sunken landbridges and submerged continents. By the early years of the twentieth century, it had been accepted as geological orthodoxy—perhaps the most important reason that catastrophically drowned continents soon fell out of favor among most paleo-scientists. Yet, not surprisingly, up until today, Tamil labors of loss refuse to accept the prevalent geological theory that continental landmasses are in isostatic equilibrium with ocean basins and that it would be impossible for them to drown and vanish so catastrophically. This refusal is crucial and strategic, given the importance of the cataclysmic submergence of land for the politics of loss that sustains Tamil devotion.[46]

Kaṭalkōḷ, therefore, is put to a lot of hard work in Tamil labors of loss, operating as it does as event, explanatory device, and strategy. It allows Tamil place-makers to establish that their land—at least those parts of it that survived the diluvial catastrophe—is chosen and blessed. It is also invoked to account for why empirical proof of the primordiality of Tamil could not be proffered: its antediluvian literary productions had literally been swept away in cataclysmic floods. As an explanatory device, it also establishes Tamil as the source of all other languages of the world, for was it not the tragic dispersal of its speakers from their original homeland that led to the peopling of the earth? *Kaṭalkōḷ* also helps to reconcile the new revelations of the modern sciences with the old truths of ancient Tamil literature. Not least, it is invoked to claim as part of the Tamil country, if only discursively, all manner of landmasses from the earth's deep past that the paleo-sciences were making visible.

Given all this investment in the catastrophic work of the ocean, most of the labors of loss around Lemuria in Tamil-speaking India are about *kaṭalkōḷ.* Some point to meteorite showers,[47] to volcanic activity,[48] to earthquakes,[49] even to the fall of a comet[50] as having occasioned the ocean to flood and destroy the antediluvian Tamil land. By and large, though, its place-makers are strangely sanguine about what provoked *kaṭalkōḷ,* almost as if the ocean has a destructive mind of its own. Studies of flood stories from across the world note that, generally, the Deluge is presented as an act of divine retribution in punishment for moral lapses among humans.[51] Tamil labors of loss are singularly secular in this regard as well, for the ocean apparently needs no reason to rise up in anger against the hapless Tamil people and to wreak havoc on their prize possessions. The ocean's unpredictability only heightens the tragedy of the loss visited upon the Tamil country.

Tamil labors of loss may be centrally oriented around *kaṭalkōḷ,* but there is surprisingly little consensus over the number of times the ocean "boiled" over and "devoured" the precious Tamil homeland. Some suggest that this might have happened just once.[52] Others insist that the Pandyan kings were forced to shift their capital at least three times as a result of the three floods that hit their domain, first from Teṉmaturai to Kapāṭapuram, and then briefly to Maṇalūr, and finally to postdiluvial (northern) Madurai.[53] A large number concur with the commentator Nakkirar's assessment of two catastrophic deluges. But there are others who refuse to enumerate at all, suggesting instead that an indeterminate number of inundations overcame the ancestral Tamil homeland and caused it to drown over time.

As I noted earlier, there is a similar lack of consensus around the extent of the former homeland lost to *kaṭalkōḷ.* But most of all, there is no agreement on when these floods happened. For Kumarikaṇṭam's place-makers struggle to translate the deep geological time of the paleo-scientist into the secular universal chronologies of disciplinary history, even while trying to reconcile both with a contrary Tamil devotional calendar (with its eccentric calibrations of a specifically Tamil past reckoned in terms of the birth of the poet-sage Tiruvalluvar) and with an archaic mode of temporal reckoning built around diluvial breakpoints called *ūḻi* (Skt. *praḷaya*) which bring an end to periodic aeons (Tam. *yukam*; Skt. *yuga*).[54] One of the earliest attempts to assign a specific date was Savariroyan Pillai's in 1901, when he suggested that the *kaṭalkōḷ* happened in the fifteenth century B.C.E.[55] Soon after, others hazarded dates ranging from as early as 30,000 B.C.E. to as recently as the third century B.C.[56] Several turn to the English Theosophist Scott-Elliot's occult chronologies, in which floods periodically destroy large landmasses between 800,000 years ago and 9,654 B.C.[57] Nineteenth-century Orientalist attempts to date the Puranic flood, as well as the floods of the Buddhist canon, are also eclectically drawn upon, as are metropolitan attempts to date the Noachian Deluge.[58] However, as with attempts to enumerate the number of floods or to measure the land lost to *kaṭalkōḷ*, several refuse to participate in the chronological exercise of dating, noting its futility given that even "the records of the Geological Survey of India do not help us in fixing the exact times of any of the epochal erosions, submergences, upheavals or the like; paleontological or stratigraphical evidence can only give us approximations of *periods, ages,* and horizons, and not actual centuries or years."[59] So, they resort to fuzzy temporalizing with phrases like "once upon a time," "long long ago," "in olden times," or "several thousands of years ago."[60]

What, then, should we make of these epistemological struggles around *kaṭalkōḷ*? Whence the imperative to enumerate the number of times the ocean struck, to measure the extent of the damage it caused, and most pressing of all, to temporalize its destructive bouts? I suggest that if occult labors of loss battle it out in the shade of the material sciences, Tamil labors of loss struggle in the shadow of professional history, which had emerged by the early years of the twentieth century as the dominant knowledge-form through which the past is reckoned and written about among those who deemed themselves modern. This is not to say that academic history is a popular pedagogical subject, or that a modern historical consciousness—or historicism more generally—was disseminated widely among colonial Indians.[61] Far from it. But it certainly means that history, by virtue of being a modern metropolitan "science," becomes the ground on which all ways of remembering the past, ranging from the non-historical and the a-historical to the historical, come to be progressively adjudicated for their truth-value, as does the very idea of modern citizenship. It also means that a novel specialist—"the historian," formally trained in the protocols of the discipline—emerges in this society that had hitherto been rather indifferent to its truths, sharing space with others who also took to the business of (re)thinking the past through the new device of history even while living in and with other modes of reckoning.[62] In Ashis Nandy's insightful characterization:

> Historical consciousness now owns the globe. Even in societies known as ahistorical, timeless, or eternal—India for example—the politically powerful now live in and with history. Ahistoricity survives at the peripheries and interstices of such societies. Though millions of people continue to stay outside history, millions have . . . dutifully migrated to the empire of history to become its loyal subjects. The historical worldview is now triumphant globally; the ahistoricals have become the dissenting minority.[63]

Occult labors of loss, I earlier proposed, had managed to clear for themselves some moments of freedom from the shackles of this empire of history, but Tamil place-makers are not so successful, as again and again the free play of an unabashed fabulous imagination—operating outside the usual range of facts—is compromised by the historicist demand for documented certitude, causal consistency, and especially for chronological precision and accuracy. As many a professional historian in the Tamil country intoned, "The first essential to history is chronology. . . . There can be no history without chronology."[64] This was especially true when it came to matters dealing with ancient India, whose chronology "is and will continue to be in an unsettled condition, and so afford ample room for patriotic megalomania and prejudiced micromania."[65] As the litterateur (and Tamil devotee) P. Sundaram Pillai (1855–97) observed in 1895 on the very eve of the commencement of the Tamil labors of loss around Lemuria, "We have not, in fact, as yet, a single important date in the ancient history of the Dravidians, ascertained and placed beyond the pale of controversy. It is no wonder then that in the absence of such a sheet anchor, individual opinions drift, at pleasure, from the 14th c. B.C. to the 14th c. A.D."[66] Given this preoccupation with secular chronology (described as "the eye of History"[67]), the pressure on Tamil labors of loss to date a momentous event like *kaṭalkōḷ* is enormous, for only then would this all-important catastrophic event be taken seriously, especially by professional historians.

But there is more to all this than contending with the historicist demand for chronological certitude and consistency. By the time the labors of loss around Lemuria commenced in the later half of the nineteenth century in Europe, the Flood as a global geological event had become questionable in professional scientific circles, and from the 1840s credible geologists or natural historians were disinclined to draw upon it to theorize the earth's deep past.[68] Not surprisingly, the paleo-scientist never invokes the universal Flood to account for submergence and subsidence in his labors of loss around Lemuria. It is another measure of occultism's contentious relationship with metropolitan science that leads its practitioners to turn to diluvial events in their own catastrophic labors of loss around Lemuria, but even occult place-making does not accord them the centrality that is given them in Tamil devotion.[69] As an explanatory device for global change as well, the Flood had fallen on bad times among scientists in the metropole by the late nineteenth century. In contrast, it had been, from the Renaissance into the Enlightenment, the central moment in the drama of human history, to whose unveiling some of the best intellectual minds of Europe had dedicated themselves. But the professionalization of the various sciences in the nineteenth century had pushed the Deluge to the margins of serious thought, where, dismissed as "myth," "legend," or "fable," it languished as yet another example of the magical and miracle-mongering mentality of the pre-moderns.[70]

The Flood, therefore, as a geological or historical event, or even as an explanatory device, had little credibility in disenchanted knowledge-practices, history included, by the end of the nineteenth century.[71] Yet, *kaṭalkōḷ* remained the principle engine that sustained the labors of loss that are so important to Tamil devotion's place-making which commenced around this time. For the Tamil place-maker, it is crucial to establish its historicity, as it is to demonstrate that it had actually happened, at least once upon a time, if not more than once. The very crux of the Tamil project around Lemuria is to demonstrate that a Tamil antediluvium had really existed, and more importantly, had mattered. But how can the place-maker go about the business of reconstructing an *ante*-diluvial past in a post-Enlightenment age that is so vigorously *anti*-diluvial? How can he use the protocols and procedures of history to pursue an antediluvial project, especially in the face of that discipline's very hostility to such an enterprise? What are some of the gains of doing so, and what are some of the losses incurred in allowing himself to be commandeered by history?

It is to these questions that I now turn, as I explore Tamil devotion's complicated and tendentious relationship to the master discipline of history in twentieth-century India.[72] To anticipate my principal argument, I suggest that Tamil labors of loss are compromised precisely because they can neither give full rein to an untrammeled fabulous imagination befitting (and, indeed, demanded by) the task of fantasizing about a prelapsarian time free of the shackles of history, nor can they afford to completely cater to the disenchanted expectations of that demanding discipline with its dogmatic adherence to documenting what actually happened. The result is that these labors of loss are fatally suspended between pure fantasy and disciplinary history, satisfying the demands of neither, vulnerable to ridicule by both.[73]

FLOODED BY HISTORY

I begin developing this argument by proposing that metropolitan labors of loss around a drowned Indian Ocean continent found such a congenial home in the Tamil country at the end of the nineteenth century because of the failure of disciplinary history to seize the intellectual or popular imagination. This, in itself, is perhaps not surprising. Although history was increasingly (albeit routinely) taught in schools established by Christian missionaries, by reform-minded natives, and by the colonial state beginning in the 1830s, and in the Presidency's few colleges from the 1840s, it would be hard to conclude that a modern historical consciousness had disseminated widely and broadly through the region, given that an overwhelming majority had not secured a formal education at the beginning of the twentieth century.[74] And while the reach of modern historical pedagogy expanded with the growth of literacy throughout the course of the twentieth century, the very ease with which Tamil devotional notions of a drowned Tamil prelapsarium have penetrated even professional historical circles suggests that the protocols and procedures of a disciplinary history, with its drive for empirical facts, chronological certitude, causal reasoning, and "scrupulous accounting"[75]—what I characterize here as historicism—have coexisted with fabulous and enchanted modes of reckoning and remembering the past. This was so much the case that in 1971, when the Government of Tamilnadu established a formal committee consisting of professional scholars, including historians, "to write the history of Tamiḻakam from ancient times to the present," it was observed by R. Neduncel̲iyan (the state education minister) in the State Legislative Assembly that "when we say history, we mean from the beginning, that is, from the time of Lemuria that was seized by the ocean."[76] Given this, it is perhaps not surprising that as recently as 1996, N. Subrahmanian (b. 1916) lamented that a disciplinary historian like himself is under "heavy pressure while pursuing his profession," as he battles the forces of "anti-history" most characteristically represented by the "myths" of Lemuria and of *kaṭalkōḷ*. "A social tradition has helped stabilize [these] myths in the minds of the Tamils. The hold of these myths on the native intellect is so tenacious that it has not loosened its grip even when the intellect is 'improved' by modern education."[77]

A century before Subrahmanian, almost identical lamentations might be heard, not just from the few professional historians in the Presidency, but also from students of literature trained in the new "critical methods" of the metropole, devoted to Tamil though they might be. So, in 1895, Sundaram Pillai complained that his fellow scholars had failed "to imbibe the historical spirit of modern times, and [did] not stir themselves to help forward the researches made regarding their own antiquities," with the result that all manner of "absurdities" assume the status of "axiomatic truths."[78] Similarly, in one of the earliest criticisms of the "myth" of *kaṭalkōḷ*, M. Seshagiri Sastri, Professor of Sanskrit and Comparative Philology at the University of Madras, railed against the fact that among the majority of the Tamil pandits and others studying Tamil literature there is not much difference between a real history on the one hand and traditions, myths and legends on the other, and Tamil poems are studied and taught with a ready credulousness which has been handed down from generation to generation; and the conservatism imbibed at the feet of the Tamil teachers forms a stronghold too impregnable to all the cannons [*sic*] of critical and comparative study. Any person who breaks through these ancient barriers of real knowledge has to meet with much opposition, and the war against untruths, falsehoods, absurdities and inconsistencies terminates with little or no success. But when the Western education spreads more and more and turns out scholars who make original researches, and the combined results of the literary researches of these scholars are made accessible to the reading public, we shall hope that false knowledge will disappear before real knowledge.[79]

This hope that "Western" and "modern" education would dispel "the untruths, falsehoods, absurdities, and inconsistencies" associated with the "legend" of *kaṭalkōḷ* is ironic, given that it was through the circulation of metropolitan labors of loss over Lemuria that the antediluvial memories of ancient Tamil literature received a new lease on life in the late nineteenth century and found a novel perch on Kumarikkaṇṭam. Without the modern West and its labors of loss, I would reiterate, Tamil labors of loss over a vanished prelapsarium would not have flourished.

All the same, the dissemination of the protocols of a "Western" education, with its emphasis on "critical and comparative study," led those Tamil intellectuals who prided themselves on their rationality to diligently distance themselves from a body of knowledge and memory that they had hitherto inhabited unconsciously and spontaneously, to name it as "tradition," and to subject it to the apparently disinterested gaze of a disenchanted historicism.[80] The goal ought to be, as one of them put it, to pass from "the dubious field of blind faith and tradition to the domain of reason and history."[81] Here is another, K.N. Sivaraja Pillai (1879–1941), historicist-author of *The Chronology of the Early Tamils,* declaring in 1932 that "venerable as the Sangam tradition is in the Tamil land, first put into shape by the commentator on [*Iṟaiyaṉār Akapporuḷ*] and then sedulously propagated by later commentators, we have to examine it closely and satisfy ourselves first about its authenticity and secondly about its evidentiary value for purposes of history."[82] Why, in Sivaraja Pillai's assessment, was it necessary to do this? Because it is "our bounden duty" to ask questions of tradition:

The main purpose of tradition itself is to supply us with the means of asking questions, of testing and inquiring into things[.] . . . If we misuse it and take it as a collection of cut and dried statements to be accepted without further inquiry, we are not only injuring ourselves here but by refusing to do our part towards the building up of the fabric which shall be inherited by our children, we are tending to cut ourselves off and our race from the human line.[83]

When tradition becomes an object of critical reflection, instead of an entity to be seamlessly inhabited, and when it is seen as necessary for the fostering of citizenship, we witness the arrival of a modern historicist consciousness. This happens when the past appears to us as radically other, when we realize that, as Pierre Nora tells us, "the past is a world from which we are fundamentally cut off. We discover the truth about our memory when we discover how alienated from it we are."[84] Yet, the spirit in which Sivaraja Pillai approaches "tradition" also clearly indicates that such a consciousness was still wanting in southern India, even by the middle decades of the twentieth century, and that this encouraged the "untruths, falsehoods, absurdities, and inconsistencies" in the "traditional" accounts of *kaṭalkōḷ* to find fertile ground among Tamil speakers and scholars fond of "tedious parrot-like repetitions of fictions and facts."[85]

For the newly hatched historicist (some explicitly devoted to Tamil, others not necessarily so), there was much that was suspect in the tradition regarding *kaṭalkōḷ* that Tamil place-makers so readily embraced. For one, the fact that the tradition was largely enshrined in poetry rather than transmitted through prose made it less than credible. Poetry, it is maintained by many a historicist, is incapable of speaking to truth, and the ancient Tamil poet, in particular, is frequently indicted for his "prolific imagination," his tendency to "exaggerate," and for his fondness for "fancy." Seshagiri Sastri's early critique rested on the claim that the memories of the ante-diluvial *caṅkams* were "a mere fiction originated by the prolific imagination of Tamil poets."[86] Similarly, historian S. Krishnaswami Aiyangar (1871–1947) charged that in the poet's "active imagination," and in "the idle tongue of tradition," truth is "generally overgrown and interwoven with fable and legend."[87] The Tamil historicist is not alone in his suspicion of poetry as a transmitter of historical truth. This is a suspicion that he inherits from his colonial masters in many of whose Utilitarian narratives the disparagement of Oriental poetry is coupled with the argument that it is the dependence on poetic forms that had made Indian society indifferent to genuine historical consciousness.[88] History as a modern historicist discipline in the subcontinent (as indeed elsewhere) has been resolutely prose-centric, and correspondingly, knowledges transmitted through poetry have been rendered suspect.

Similar suspicion is voiced about the ability of orally transmitted commentaries such as Nakkirar's and Adiyarkunallar's—the very fount of Tamil labors of loss around Kumarikkaṇṭam—to speak to truth. Thus, the venerable doyen of Tamil historical studies (and Professor of Indian History at the University of Madras from 1929 until his retirement in 1947) K.A. Nilakanta Sastri (1892–1975) conceded that while such commentaries on the ancient poems have "considerable merit," they "cannot possess the same indelible format that, for instance, rock epigraphs and copper plate grants have," since they had been subjected to all manner of "alterations, modifications, and aberrations successively at the hands of the compilers, scribes and even the scholars who learnt them by rote as well as from cudjan [*sic*] leaves."[89] The difficulty of determining their date of composition, the cloud of doubt that hung around their authorship, the fear of latter-day interpolations corrupting the authenticity of original truths—these made such commentaries vulnerable to the historicist's disavowal. But as importantly, orality itself suffered a reversal of fortune with the growing consolidation of print culture in colonial modernity. The printed word now assumes the aura of truth, transparency, and fidelity that was once accorded to orality, and in turn, oral knowledges are relegated to the status of legends and myths, or worse, rumors, that no historicist could or should take seriously.

But perhaps the most important reason that ancient memories of *kaṭalkōḷ* become highly suspect to the new historicist imagination is because they are deemed "Puranic," and hence archaic, absurd, and fabulous. Hostility to the *Purāṇas*—the foundational Sanskrit texts of medieval Hinduism—had accumulated through the course of the nineteenth century, both within and outside southern India.[90] As late as the 1890s, almost half a century after Macualay and Co. had tried to banish the hold of the *Purāṇas* on the Hindu mind, one M. Rangachariar continued to lament:

> These [Puranic] tales took ready hold of the people's minds, and are even now the basis of current popular, unscientific, cosmogenetic, and chronological beliefs. How much of the false science, false philosophy, and false history now flourishing in all the strata of Hindu society, owes its origin to the fancy of the poet and the rule of three of the astronomer, it is indeed impossible to tell.[91]

The critique of Puranic consciousness led many Tamil historicists to establish an important distinction between "*katai*" (lit. story, but also, by extension, legend and myth, "*kaṭṭukatai*"), and "*carittiram,*" "history." Whereas the former is prone to falsehood *(poy)*, the latter speaks to truth *(uṇmai)*. Correspondingly, the former comes to be associated with "fancy," the latter with hard "facts." As early as 1892, when Nakkirar's commentary on the three *caṅkams* and *kaṭalkōḷ* was just beginning to gain visibility among Tamil intellectuals, T. Saravana Mutthu noted that it would take a "Herculean effort" to determine what was true and what was false about it, for "the truth *(uṇmai)* is lost in stories *(katai)* such as these."[92] And a few years later, Seshagiri Sastri acerbically insisted, "With reference to the first two Sangams, I may say that the account is too mythical and fabulous to be entitled to any credit, and I do not think that any scholar who has studied the histories of the different countries of the world will be bold enough to admit such tales within the pale of real history."[93]

For the committed historicist, therefore, the memories of an antediluvial land and literature enshrined in Tamil's poetic "tradition" amounted to "fabulous" *katai* rather than "real" *carittiram.* Even a sympathetic critic like T.G.

Aravamuthan was compelled to note, "Traditions so long and so deeply rooted have naturally attained to the status of articles of faith and are not to be lightly brushed aside. Still, *legends are only legends and can be no substitutes for history.*"[94]

But it was not just the formal apparatus in which the tradition regarding *kaṭalkōḷ* came embedded—as orally-transmitted "Puranic" poetry and commentaries—that invited the realism-soaked ire of the historicist, but its very content. There was, first of all, the problem of gods mixing it up with humans in the antediluvial academies. Saravana Mutthu offered the following ridicule as early as 1892, in a voice of disbelief that one hears even in another century: "They say that each of these *caṅkams* lasted for several thousand years, that Siva and Murugan and other gods presided over them, that there was a divine testing seat, etc."[95] More than half a century later, the learned Professor S. Vaiyapuri Pillai similarly declared, "Gods are also said to have participated in the celebrations of the first Sangams. We may leave such fables alone and seek for historical truth elsewhere."[96] The belief transmitted over the centuries that the forty-nine members of the third *caṅkam* were "incarnations of the forty-nine letters of the Sanskrit alphabet which represented the Goddess of Speech," is dismissed as "legendary" and "miraculous."[97] Like all good historicists, these intellectuals are deeply suspicious of any consciousness or memory which admits the agency of gods, spirits, or other supernatural beings, for as Dipesh Chakrabarty observes, "The secular code of historical and humanist time . . . is a time bereft [of] gods and spirits."[98] With the rise of disciplinary history, in Pierre Nora's evocative reading, "A world that had once contained our ancestors has become a world in which our relation to what made us is merely contingent."[99] Not surprisingly, the very existence of *caṅkams* presided over by divinities such as Siva and Murugan becomes vulnerable to historicist ridicule.

Our new-fangled historicists, committed as they are to an empiricist worldview with its indelible faith in material evidence, are also deeply troubled about the fact that there is no proof of any literary work anterior to the anthologies that were discovered and printed at the end of the nineteenth century, and whose composition is dated to a period ranging from the third century B.C.E. to the third century C.E. Where, they ask, are the texts putatively composed by the numerous poets of the antediluvial academies? Some among them, who are also Tamil devotees, such as P.T. Srinivasa Iyengar (1863–1931), concede that there must have been an "extensive" poetry prior to the surviving specimens, given the latter's full-blown literary form. Yet, Srinivasa Iyengar's question, "Why did this ancient literature come to be lost?" does not take him down the fabulous route that Tamil place-makers routinely adopt, namely, to suggest that this was swept away in catastrophic *kaṭalkōḷ*. Instead, prosaically and disenchantedly, he concludes that it is their very orality that accounts for their demise.[100]

The historicist critique is at its positivist best when it takes on the "incredible" number of regnal years assigned to the antediluvial Pandyan kings who presided over the *caṅkams* which lasted for thousands of years each, producing innumerable works authored by countless poets. Armed with his secular chronologies and with his dynastic lists, the most obvious gifts of the new science of history, the historicist summons up the voice of disenchanted reason and asks if it would be possible for any single institution to last for a period of 9,990 years? "It is absurd to say that 59 kings reigned for 3, 700 years. . . . [Agastya] is made to endure for 8,410 years, the period of the first two Sangams put together. Nor can we believe that 49 members of the last Sangam lived for 1,850 years, though 49 kings died during the period. Therefore, not one of the figures in this account can be believed."[101] The absurdity of these numbers is only further heightened by their obvious artifice: "The number of years allotted to the different Sangams is incredible, not only on account of the length assigned to the three periods, but also on account of their symmetry. The length of the period for each Sangam is a multiple of 37, and the total duration is $37 \times (120 + 100 + 50)$."[102] Similarly, Sivaraja Pillai noted that "the period of duration for the three Academies put together, *viz.*, 9,990 years, if distributed among the 197 Pandiya Kings will be found to give us an average of fifty and odd years per generation—certainly an impossible figure in the history of man."[103] Confronted with such absurdities and impossibilities, the historicist responds, borrowing from James Mill, that "a chronology, involving such immeasurably long periods of time, is a sure sign of savagery on the part of the people who adopt it."[104]

But the historicist saves his most biting critique for the fact that the place-maker has dared to legitimize the traditional poetic and literary account of *caṅkams* and *kaṭalkōḷ* by resorting to the modern geological truth of Lemuria. Science has been sullied by being dragged into the Puranic mire. Here is Nilakanta Sastri at his disenchanted historicist best in 1956:

> Somehow it has happened that in discussions of early South Indian Chronology there has been prevalent a fairly widespread error of using geological arguments in historical discussions. Now this has to be said with some emphasis, because the talk of Lemuria, of Tamil having been spread all over the area of the Indian Ocean before the ocean submerged the land, and of its being the oldest language of the world—this talk has been the pastime of some persons for too long. It is time that some one stood up and said, "It is all bosh!" Human life on earth in any form that concerns us as students of history had its first beginnings not more than thirty or thirty-five thousand years ago at the highest. But geological changes relate to conditions of earth before any life (not only human life) came into existence. Submergence of continents and emergence of oceans are not occurrences of every day, and

> the last great change of this character is put by geologists some millions of years ago. What has this got to do with the history of humanity which stretches back at most to about five, six, or ten or twenty thousand years from now? For that length of time would take us back to the old stone age, an age when men were hardly different from animals, when they had no language, no speech and no culture, and were still living in the food-gathering stage.[105]

It is notable that the Tamil historicist continues to accept the fact established by nineteenth-century science that there had once been a landmass called Lemuria which had drowned in the Indian Ocean, even though by the middle of the twentieth century that very science had disavowed its earlier labors of loss. Nevertheless, for the historicist, it continues to be Truth. But to confuse Lemuria's disappearance with the submergence of "mythical" cities like Te<u>n</u>maturai and Kapāṭapuram through *kaṭalkōḷ* was to mix "fact" with "fancy": "Surely the quite respectable antiquity of Tamil literature—the oldest in any living language of the modern world,—can be adequately assured otherwise than by pushing it down the dizzy depths of the prehistoric pluvials to almost the date of man's debut on the subcontinent, if not the planet."[106] In summation, then, modern historicism in south India, as in modern Europe, appears to have little patience with "prehistoric pluvials."

I write "appears" here deliberately, because historicism's career at its colonial address is not as straightforward as Clio might have wished. In other words, the dramatic and violent rupture that Pierre Nora sees between traditional memories and professional history, with the consolidation of the modern in Europe, is not so self-evident in India. For Nora, "memory installs remembrance within the sacred; history, always prosaic, releases it again."[107] Yet in places like the Tamil country, the release is not always complete, the installation not utterly undone. When historicism first encountered, in the 1890s, the "absurd" tradition of *kaṭalkōḷ* and the three *caṅkam*s, "which is implicitly believed by everyone . . . without any inquiry whatever," the hope was that good research would help to "sift the truth from legendary and fabulous accounts."[108] Reason, it appeared, would turn its searing light on tradition, liberate it from wild fable and fantasy, and deliver it to the always-reliable history. Yet, as the new century dawned and wore on, and Tamil devotion's preoccupation with Lemuria flooded the Tamil country with its labors of loss, historicism began to lose ground to the "fabulous" tradition's "untruths, falsehoods, absurdities, and inconsistencies." Increasingly, its critique was merely peppered with charges such as "incredible," "stupendous," "marvelous," "unbelievable." Faced with the tide of Tamil labors of loss over Kumarināṭu, and having to contend with its polar Other in the fantastic, the historicist challenge was reduced to such semantic displays in the name of reason.

But even more dangerously (dangerously, that is, to the cause of a positivist disenchanted discipline), historicism itself did not remain unscathed and pure in its encounter with the "fabulous" tradition of *kaṭalkōḷ*, especially as this tradition came to be rejuvenated through Tamil labors of loss. We detect traces of a compromise early enough in the encounter, when the historicist begins to admit—incredibly enough—that it was not necessary to discard "tradition of a reliable character."[109] As time goes on, the historicist goes on to suggest, "the entire tradition concerning the Academy does not seem to have been a fiction, for in the first place, traditions do not arise normally without any basis."[110] And before too long, the historicist is involved in a mining operation, sifting "fact" from "fancy," sternly applying the positivist canons of historical criticism to tradition, and rationalizing the fabulous and the fantastic, for "even in the traditions handed down to us, much distorted though they are, there are certain critical facts and characters standing clearly marked out from the rubbish outgrowths. It will not, therefore, be without interest to attempt to place these facts in the light in which they appear on an unbiased and impartial inquiry."[111]

So, the historicist begins to persuade himself—and his reasonable readers—that the periodic sea squalls and the powerful erosions along the Tamil coast may well account for the belief in *kaṭalkōḷ* that swept away a whole civilization. Such marine events might have also led the Pandyan kings to shift their capital cities, and as these capitals moved, so did the Academy that flourished there, leading to the creation of the "incredible" memories about the three successive *caṅkam*s.[112] The historicist is still hard-pressed to come up with reasonable explanations for the long duration of the *caṅkam*s, for the numerous kings who acted as their patrons, or for the extraordinary number of poets and bards who produced countless literary gems. But it is bravely conceded that "certain kings and poets, mentioned in the traditional accounts, figure in more than one classic of the Sangam Age, which fact strengthens the historicity of these personalities."[113]

It is possible to read the historicist compromise as marking the eventual victory of tradition, since the latter's imagination, deemed "absurd," "incredible," and "fantastic" by its detractors, appears to have infected the stern science of history as it came to be practiced in Tamil India. Such a compromise is also not surprising, given that many a historicist also moonlighted as a devotee of Tamil, to whose enchanted demands he succumbed again and again. At the same time, as I have just detailed, as historicism compromises with the fabulous tradition of *kaṭalkōḷ*, it translates it into realist terms, making it historically plausible, possible, and palatable, denying it its moments of freedom from history by bringing it *within* the usual range of facts, curtailing its power to circulate as pure fantasy uninhibited by the imperative to "touch solid earth."[114]

Historicism's refusal, even inability and inadequacy, to speak the language of the fantastic—to deal with the fabulous qua fabulous—is telling but not surprising. History's power in the (post)colony emerges from the paternalistic pedagogic role it assumes for itself as it attempts, in the name of a superior reason, to wean away the irrational native from "incredible," "stupendous," and "fantastic" ways of reckoning and remembering his past through (misguided) tradition. Backed by the historicist demands of the colonial (and postcolonial) state, history takes upon itself the burden of bringing the ahistorical into the historical, for in this lies the virtue of being modern, reasonable, civilized.[115] Given this role it assumes for itself, historicism in Tamil India cannot afford to speak the language of fantasy. But nor can it afford to ignore traditional memories, especially in the face of Tamil devotion's labors of loss which both resurrected them and added to them the seductive veneer of modern science. The result is the compromise of historicizing, rather than totally discarding, tradition. Yet, this is a historicization that is necessarily incomplete and inadequate, for the fantastic, bolstered as it is in the Tamil country by the labors of loss of Tamil devotion, quietly and subversively forestalls the empire of disenchanted history. In this process, it manages to lurk at history's edges, ever threatening to undo historicism's work in the Tamil life-world.

FLOODING HISTORY

Compromised and vulnerable though it might be, for much of the twentieth century history remained a force to contend with in Tamil-speaking India, backed as it is by the power of metropolitan knowledge formations, the bureaucratic agencies of the modern state, and the persuasions of universal reason. This was especially so for the duration of the colonial period, when Indians, especially the Hindu among them, were caricatured as ahistorical, hence savage, people, destined to be left behind in the march of (linear) time. So much so that historian Krishnaswami Aiyangar took it upon himself to refute the English essayist Thomas Carlyle's observation that "happy is the nation whose annals are a blank" by retorting, "If we can derive comfort from this seeming blankness, we shall perhaps be in a delusion."[116] Indeed, in their dedication to the pursuit of history, Tamil historicists would have agreed with Bankim Chandra Chatterjee's sentiment from across the subcontinent that "Bengal must have her own history. Otherwise there is no hope for Bengal."[117] And from the closing years of the nineteenth century into the next, numerous histories of the Tamil country were written, in both English and Tamil, by professional historians and others, as the "Tamil" past came to be periodized, secularized, linearized, and inserted into universal history. In other words, heterogeneous pasts were disciplined to conform to the protocols of what Prasenjit Duara has recently characterized as the Enlightenment mode of history.[118]

Given the disparagement of *kaṭalkōḷ* in the course of this disciplining, however incomplete, how did Tamil place-makers respond to the harsh gaze of historicism and to the assault on the "fabulous" and "fantastic" antediluvial tradition that feeds their labors of loss? We get an early glimpse of this in a 1901 exchange in the literary journal *Siddhanta Deepika or Light of Truth* between Jules Vinson, learned Professor of Tamil and Hindustani at the Living Oriental Languages Institute, Paris, and D. Savariroyan, whom I have singled out as among the first of Tamil's devotees to connect the premodern Tamil memories of an antediluvium to metropolitan labors of loss around a submerged Lemuria. In a letter to the journal, Vinson repeated the oft-heard colonial and Orientalist assertion that Tamil literature's antiquity pales in comparison with Sanskrit, dating its beginning to the fifth or sixth centuries C.E.[119] Savariroyan protested this confident assertion by pointing to the *Tolkāppiyam,* which in his reckoning was an antediluvial work and hence "anterior to the advent of the Aryans in [India]":

> The author of this work . . . lived before the inundation which swamped the stretch of land that existed to the South of the modern Cape Comorin. . . . From the references in Tholkappiyam itself we are led to surmise that there were many literary works even before it, which perished at the cruel hands of Time and from the big *Deluge.*[120]

Since the Deluge itself happened no earlier than the fifteenth century B.C.E., Tamil literature is certainly older than this. "I believe that, from the above, Prof. Vinson will find that he was very hasty in his conclusions about fixing the date of the Tamil literature. . . . We have indisputably shown that the literature of the Tamilians dates from a period far beyond the 15th century B.C.," he concluded.[121]

It is a measure of (undevoted) historicism's anxiety regarding such labors of loss that Vinson found it necessary to respond to Savariroyan. Noting that there is no empirical evidence for an antediluvial Tamil corpus which had been lost to the ocean, he also insisted in good historicist fashion that it was "most improbable" that such works ever existed.[122] Savariroyan was quick to take on this historicist parry, and his response offers glimpses of the ways in which some Tamil place-makers labor to liberate themselves from the shackles of history. First, Savariroyan invokes and then defends the tradition regarding a Tamil antediluvium: "Our Professor says that it is impossible to believe that all the works preceded [*sic*] Tolkappiam are lost, but he has not stated any reason for his not believing the tradition accorded in the ancient classical Tamil works. . . . The old tradition preserved to this day in some of the classical works as *Cilappatikāram* and *I̱raiyaṉār Akapporuḷ* . . . locates the beginning of the cultivation of Tamil in the hoary past."[123]

Savariroyon is not alone in defending the tradition regarding a Tamil antediluvium against historicism's disavowal. As early as 1897, another editorial published in *Siddhanta Deepika* had noted, "the [traditional] references to one or two deluges are too numerous to be untrue. . . . At any rate, we cannot be too dogmatic in these matters and some of the tests applied by the people of the new school of criticism are in themselves too artificial."[124] Similarly, a few years later, another essayist in the same journal listed the ancient authorities who spoke of the invasion of the Pandiyan kingdom by the ocean, and insisted that "no one can reasonably be a skeptic as regards the flood in the face of such an overwhelming evidence."[125] Indeed, as recently as 2001, in response to a vigorous historicist questioning of the value of the tradition regarding antediluvial events in the Tamil past, one place-maker (with the revealing pen name Kumari Maintan, Son of Kumari) defends its veracity, noting that "without evidence the commentators cannot just make up stuff."[126]

For the place-maker, therefore, the tradition regarding a Tamil antediluvium is not *katai* (a fictive story), as it might be for the (undevoted) historicist; it is instead *carittiram* (history) and hence necessarily speaks to truth.[127] Purnalingam Pillai insisted, therefore, that "tradition is overgrown truth, as *Purana* is masked history."[128] Not for him the casual dismissal of tradition as falsehood. In fact, this particular tradition is hallowed precisely because it is home-grown, transmitted from generation to generation, and a better "custodian" of a people's memories than history's acclaimed archive. So thundered Somasundara Bharati in 1935:

> The story of deluges that successively swallowed up large slices of land in the south of the Indian continent rests not merely on legends and folklore. It is enshrined and embalmed in standard old classic poetry of the Sangam poets, and has been handed down to posterity by all the ancient and medieval classic text-writers and commentators of great fame and undoubted authority in an unbroken series of standard works that ever commanded the uniform respect and universal homage of all the students of Tamil literature.[129]

This is so much so that the venerable Tamil pandit M. Raghava Aiyangar (1878–1960) dismissed as "moderns" *(navīṉarkaḷ),* and therefore, implicitly, upstarts, those who are dismissive of tradition.[130] So does the place-maker neatly turn the tables on the historicist in his recuperation of this particular tradition.

All the same, this defense of the tradition regarding the Tamil antediluvium is not generally conducted in traditional terms. Instead, the place-maker typically resorts to modernity's most prized possession, its much-vaunted natural sciences, which even in the metropole were accorded far more prestige than the historicist's discipline. To return to Savariroyan's response to Vinson, having invoked the many allusions in Tamil literature to *kaṭalkōḷ,* he then notes that according to this old tradition, "a vast land existed south of Cape Comorin and was submerged by the flood." This fact "receives great support from the modern sciences, Geology and Natural History, which prove the existence of a land south of India and its disappearance beyond the pale of doubt." In fact, "centuries before the birth of Natural History and Geology, this old tradition was recorded in the Ancient Tamilian classics. Therefore the Professor is obliged to give credit by all means to this tradition, corroborated by modern sciences and discoveries."[131]

Savariroyan was not alone in thus enlisting the assistance of the modern natural sciences to counter the power of academic history and to launch a defense of the tradition regarding the Tamil antediluvium. This, in fact, is the most popular strategy that Tamil devotion's place-making adopts through the course of the twentieth century, and indeed the next.[132] For Tamil labors of loss, this tradition is worth recovering, saving, invoking, and deploying, precisely because it is scientific.[133] At the same time, unlike the modern sciences, which are, after all, newfangled knowledges, tradition is simultaneously old, and hence hallowed. Some enthusiasts were even confident that their tradition could be used to correct the claims of the material sciences, anticipating as it had the latter's conclusions by several centuries. Hence, A. Mutthuthambi Pillai (1858–1917) offered in 1906 a passionate defense of the traditional authorities on *kaṭalkōḷ*, and then noted that while there are skeptics who are quite willing to believe European scholars when they claim that there were deluges in the past, they are not as willing to accept the evidence of the Tamil *Purāṇa*s. Why are Sclater and other Europeans believed when our own tradition is disavowed, he asked rhetorically. Further, "that continent has been named 'Lemuria' by these European scholars. That cannot be the name of this continent in its own time. If [the Europeans] had known our *Purāṇa*s, they would have named this continent 'Lanka.' Their 'Lemuria' is the land referred to as 'Lanka' in our *Purāṇa*s."[134]

But there is still the matter of Vinson's principal contention that the much-vaunted antediluvial works that Tamil labors of loss invoked were simply not there. And here, Savariroyan's response resorts to the second strategy that Tamil place-makers typically adopt to cope with the terror of history, for historicism's empiricist demands are countered by invoking the flood itself as having destroyed the evidence. If *kaṭalkōḷ* had not happened, then, surely, Tamil place-makers would have no difficulty at all in proving the existence of their antediluvial patrimony:

> That the Tamilians even in those days possessed an extensive literature will strike every one who goes through any extant old commentary of any one of the Tamil classical works . . . . [They] all go to impress strongly, when compared with the meagre portion that is left to us, the possibility of a vast store of ancient literature displaying considerable erudition and the sense of the loss that Tamil has sustained by a great catastrophe. The lost works of which there seem to have been quite an ocean pass in view before us and remind us

> of the ancient grandeur and wealth of Tamil. This fact also cannot but be admitted by our Professor.[135]

Historicism's metaphysics of presence is thus countered by Tamil place-making's metaphysics of loss. Hence also the importance of diluvialism to Tamil labors of loss. *Kaṭalkōḷ* produces loss, accounts for it, and clears the ground for its invocation.

All the same, it is telling that instead of merely asserting—as the traditional commentators simply did—that "innumerable" books had drowned in the ocean, the place-maker feels compelled to explain the loss, to account for it, and to do so by invoking the findings of modern science. In other words, to rationalize—and to historicize. Tamil place-making may insist that *kaṭalkōḷ* did really happen, but its labors of loss are not able to entirely free themselves from history's torment. On the contrary, the place-maker repeatedly turns to historicization in order to rehabilitate *kaṭalkōḷ* and loss. We see this as early as 1903 in Suryanarayana Sastri's spirited defense of Nakkirar's ancient commentary on the *Iṟaiyaṉār Akapporuḷ*. After insisting that "there is no doubt" at all regarding the commentator's narration of *kaṭalkōḷ* and of the loss of the *caṅkam*s and their literary productions, even Suryanarayana Sastri is skeptical when he considers the long regnal years assigned to the antediluvial Pandyan kings. "This is rather open to doubt. It is improbable that they continually reigned for such long periods."[136] This is indeed the disenchanted voice of history halting the ability of traditional memories to circulate as pure fantasy and diluting the fabulous reach of Tamil devotion's labors of loss. Similarly, Suryanarayan's colleague Purnalingam Pillai in his *A Primer of Tamil Literature*—one of the first books in English to subject Tamil's literary past to historicization—reminds its readers of the traditional account of the three *caṅkam*s and their seizure by *kaṭalkōḷ*, and then notes:

> But now it is challenged by critical scholars, both Indian and European, on the ground that it is full of improbabilities and inconsistencies and draws too much on the marvelous as it gives an incredible longevity to each poet and prince who had anything to do with the Sangams. They believe that these Academies must have been the figments of some poetic imagination. . . . *But the question of their existence cannot be easily decided until the researches of the archaeological society in South India bring to light facts and materials enough to explain away the apparent improbabilities and contradictions.* Till then the commentators' accounts will bear sway and must be accepted *cum grano salis.*[137]

Here, Purnalingam Pillai's defense of the tradition regarding the antediluvium is compromised by his concession to the historicist demand for empirical evidence that would "explain away the apparent improbabilities and contradictions." Until that happens, he is willing to go along with this particular tradition as is—but only until empirical evidence vindicates its claims, as he is convinced it will. However, this very deferment is predicated on an ultimate surrender to historicism. So, here, too, the fabulous reach of Tamil labors of loss is ruptured by the disenchanted demands of history.

Similarly, others resort to the comparative method—another historicist favorite—to defend traditional memories of *kaṭalkōḷ*. It is repeatedly noted that the Sanskrit scriptures, too, speak of a universal Deluge, as do the Hebrew "records," some even noting that Manu or Noah were none other than antediluvian Tamilians who had lived on Kumarikkaṇṭam.[138] Indeed, it is a telling sign of the global and Orientalist politics of knowledge that even the skeptical historicist occasionally concedes the fact of a catastrophic deluge in the subcontinent's past, merely because Sanskrit literature attested to its occurrence.[139]

But nothing betrays the undertow of the disenchanted voice of history more clearly than the enormous trouble to which Tamil labors of loss subject themselves in order to date the three (and sometimes more) *caṅkams*, the two (or more) *kaṭalkōḷ*, and the regnal years of the various antediluvian Pandyan kings. These temporalizing labors create a distinct antediluvian beginning for the Tamil past, whose relation to the successive periods of the history of a territory that is clearly identified as "Tamil," and whose relation to a community of speakers identified as "Tamilian," is also delineated. All of this is then inserted through the device of secular chronology into universal history.[140] In insisting that the geo-body of the Tamil nation had a past that originated in an antediluvial moment, Tamil labors of loss insert *kaṭalkōḷ* as a catastrophic break-point. Disciplinary history stands flooded, in a manner of speaking, as its normative starting point of a prehistoric moment now makes room for an antediluvial beginning, even in school and college lessons. Nonetheless, all of this only translates and normalizes the antediluvial moment, inserting it into a universal history in which its relation to other pasts and other times is specified. It loses the fabulous singularity it had enjoyed when it had resided in the poetry, commentaries, and memories of ancient Tamil literature, free of modern Tamil labors of loss around Lemuria.

So, just as "[the] threat of fantasy . . . lies at the heart of the modern historical imagination" as it unfolds in the Tamil country,[141] the torment of history as well disenchants Tamil labors of loss, constantly compelling it to compromise with historicist imperatives to show empirical evidence, indulge in causal reasoning, demonstrate chronological consistency, and erase incongruities. It is not surprising, as I have observed, that history refuses to speak the language of fantasy, but, remarkably, even Tamil labors of loss refuse to do so when it comes to the imagination of the Tamil prelapsarium. One would have thought that detailing the life and times of a prelapsarian land and people would have allowed the marvelous play of fantasy, free of the shackles of universal chronologies, the secular reckoning of clock time, the demands of empiricism, and so on. But this does not happen. Every now and then, especially in the postcolonial period when the imperative to prove that "we too have history" has not been so pressing, the occasional place-maker gives free rein to the fantastic. So, M. Raghava Aiyangar refused to give in entirely to the demand to rationalize the presence of supernatural beings in the first and second *caṅkams*, as some of his fellow-devotees were wont to do in response to historicism's disavowals.[142] Similarly, some postcolonial labors of loss have encouraged occasional flights of untrammeled fantasy that clear a moment of freedom from the empire of history.[143] But such examples are episodic, and their very sparseness points to the difficulties faced by Tamil labors of loss over Lemuria in giving in completely to undiluted fantasy in the face of the hegemony of history. The magic of make-believe is undermined by the historicist imperative to demonstrate what actually happened. So much so that instead of appearing marvelous and wondrous, the Tamil antediluvium seems shabby, incongruous, even impoverished as it is compromised by catering to the demands of historical realism.

Tamil devotion's refusal to give full rein to the enchanted magic of fantasy may be understood when we remember its principal function in the Tamil life-world. If history's role is paternalistic and pedagogic, Tamil devotion's is memorializing. Today's Tamil speakers had to be necessarily and constantly reminded of the reality of their prelapsarian past because it was then (and there) that they had been pure and sovereign; it was then (and there) that their language and literature had been truly secure, receiving the love, support, and patronage of the state; it was then (and there) that they had indeed been truly Tamil. Given this imperative, the Tamil place-maker cannot afford to imagine the antediluvial moment through the undiluted lens of fantasy, for to do so would take away from its realism and its certitude—that it actually happened—so necessary for Tamilian self-pride and self-respect—indeed, for Tamil modernity and nationalism. The consequence, therefore, of daring to imagine an *ante*-diluvium in a historicist environment which is rigorously *anti*-diluvial, is that Tamil labors of loss are neither empowered by the magic of pure fantasy nor anchored in the realism and certitude of disciplinary history. A mongrel formation, neither pure fantasy nor respectable history, Tamil labors of loss are vulnerable to disavowal and dismissal both as fantasy and as history.

GEOGRAPHIES OF LOSS

Tamil labors of loss stand out among the others certainly because *kaṭalkōḷ* is accorded a centrality that is not the case even in Euro-American occultism with its own substantial investment in a catastrophic antediluvial vision of the earth's past. As strikingly original, however, are the landscapes of loss that Tamil place-makers summon into existence. These are marked by an intensity of detail and a demonstrated intimacy of acquaintance with its topography that is quite lacking in the remote geographies of the metropole's Lemuria. This intimate antediluvial geography is what sustains the Tamil place-maker's insistence that the loss he has endured is not of any territory, but of his beloved ancestral homeland whose irrevocable disappearance has compelled him and his fellow speakers to live in perpetual exile.

While the imperative to demonstrate that the antediluvial Tamil past *really* existed leads Tamil labors of loss to contend with history, the pressure to delineate an antediluvial topography entraps the place-maker in the web of academic geography, the discipline that emerged in later modernity in Western Europe to describe Earth as a disenchanted realm that is realistically knowable with certitude and precision—the real and present world, as it were.[144] But the thralldom to geography does not just arise from the fact that in (colonial) modernity it commanded the language and tools with which to conduct legitimate scientific discourses on space and place. As importantly, as I have insisted, Tamil place-makers are preoccupied with working out the complicated relationship between their lost homeland of Kumarikkaṇṭam and their lived homeland of Tamilnadu/India. By the early years of the twentieth century, the geography of the lived homeland had been well documented by European missionary-scholars, colonial bureaucrats, and educated natives.[145] In contrast, however, the lost homeland was an utterly unknown entity. Yet, Tamil place-making could not afford to let it remain a terra incognita, especially because it was necessary to generate attachment to it in order to produce love and loyalty for the lived homeland, its only surviving remnant. So, Tamil labors of loss translate Kumarikkaṇṭam from an antediluvial terra incognita into a known and familiar landscape, and in so doing, they turn repeatedly to the geography of their lived homeland as the model. This, too, leads them to engage with the modern discipline of geography, with its penchant for describing the earth as a measurable, empirically knowable, disenchanted place. How does one *realistically* describe a fabulous antediluvial landscape that is no longer available to empirical knowledge, if it ever it was? This is the dilemma facing the Tamil place-maker, and leads him to dilute any flights of fantasy in order to cater to disciplinary geography's disenchantment.

The paleo-scientist is, not surprisingly, quite silent about Lemuria's geography, given that it is a remote Mesozoic continent utterly uninhabited by humans, but occultists give free rein to fantasy in this regard. In the English Theosophist Scott-Elliot's fabulous imagination or in the German Rosicrucian Max Heindel's, the Lemurian landscape is fiery and frightening, for its atmosphere "was still very dense. . . . The crust of the earth was just starting to become quite hard and solid in some places, while in others it was still fiery, and between islands of crust was a sea of boiling, seething water. Volcanic outbursts and cataclysms marked this time."[146] The world then thronged with monstrous Saurians, fierce dragons, and gigantic birds with whom Lemurians had to share space. The Lemurian landscape of the Lemurian Fellowship is less storm-swept and cataclysmic, and unique among metropolitan place-makers, twelve "great valleys" of the Pacific Eden are distinguished with fantastic names such as Tama Valley, Chi, Thibi, Upa, Mu, Levi, Xion, Cari, Beni, Opu, Judi, and Hata. The Mukulian civilization's heartland lay to the north of these valleys in the Rhu Hut Plains, where the Lemurian elect lived.[147] By the time the New Age turns its attention to the lost continent, it has been transformed into an ethereal Nowhere, a non-geographical place-world.

Fantastic though these occultscapes might be, they lack, however, the intimacy of Tamil landscapes of loss. This intimacy is largely achieved in Tamil place-making by turning to Tamil's own precolonial literary corpus, which furnishes the localized names and details that help in the recasting of the remote continent of Euro-American labors of loss into a Tamil home-place. To recall, toponyms (apparently) derived from Tamil literature, such as Kumarikkaṇṭam and Kumarināṭu, or based on the everyday geography of the Tamil life-world, such as Tamiḻakam and Tamiḻnāṭu, are bestowed on Sclater's Lemuria, dislodging it from the impersonal world of metropolitan science and relocating it within a very Tamil sensibility. Further, just as ancient Tamiḻakam was divided into five poetic ecoscapes (*tiṇai*) celebrated in its early poetry, Kumarikkaṇṭam, too, had its *marutam* (riverine plains), *neytal* (littoral), *kuṟiñci* (forest lands), *mullai* (pastures), and *pālai* (desert).[148] Such exercises undertaken by Kumarināṭu's place-makers are noteworthy, given that disciplinary geography in colonial India generally disavowed native categories and spatialities as so many Puranic "seas of treacle and seas of butter."[149] The recuperation of precolonial Tamil categories and spatial visions, imagined or otherwise, thus clears some moments of freedom from scientific geography, even as it enables Lemuria to be incorporated into a Tamil horizon of meaning and memory.

But such moments are few and far between, for Tamil labors of loss are generally indebted to disciplinary geography's spatial vision in imagining the drowned homeland's landscape. Accordingly, Kumarināṭu's antediluvial extent is reckoned on the basis of the distribution of present-day landmasses and oceans. And as we will see, maps of the lost continent, too, are inevitably drawn in terms of today's geography, inviting their readers to imagine a world that once *was* in relation to the world that now *is*. A modern geographic and cartographic common sense is essential therefore to the many labors of loss around Kumarikkaṇṭam, orienting them and anchoring them in numerous ways, even while trapping them in geography's realist imperatives. Given that Tamil place-making is centrally preoccupied—unlike the others—with working out the relationship between lived and lost homelands, it is also fixated, much more so than the others, on the borders and boundaries of Kumarikkaṇṭam. The anxiety over borders is crucial to a modern territorial sensibility that emerges with the rise of nation-states, the reach of whose sovereignty is crucially limited to a clearly bounded piece of Earth that they are sworn to police and defend. Geography aids and abets in the consolidation of this sensibility, and not surprisingly, the heyday of the discipline in Europe, as indeed elsewhere, coincides with the consolidation of the national idea.[150]

Metropolitan place-makers of Lemuria, with the exception of some American occultists and fantasists, are not concerned with today's nation-states. Tamil labors of loss are strikingly different in this regard as well, as its place-making repeatedly questions the delineated borders of the emergent nation-state of "India." The reasons for this are not far to seek, for from the 1930s, more and more of Tamil's devotees came increasingly under the influence of the Dravidian movement and its radical imagination of a nation outside the spatial confines of India. Indeed, from the vantage point of an emergent Indian nationalism, Tamil labors of loss are clearly extraterritorial, preoccupied as they are with an ancestral homeland most of which falls outside the delineated borders of India. So, the desire to imagine an alternative to India leads many to spatially dissociate themselves from a territory that is deemed to be contaminated by Aryanism, Brahmanism, Sanskrit, and Hindi, propelling them in turn to locate their Utopia of perfection and plenitude *elsewhere*. But because India is also deemed to be originally and fundamentally Dravidian and Tamil, before the Aryan hordes took it over, and because it is after all the ground for the conduct of practical politics, Tamil place-makers cannot give up on it entirely. Hence, spatially as well, Tamil labors of loss compromise, by locating their imagined elsewhere partly within the borders of contemporary India (whose own political contours changed dramatically over the course of the twentieth century), and partly outside.

But a separatist Dravidian nationalism is not the only force compelling Tamil devotion's extraterritorial place-making. Working sometimes in opposition to Dravidianism's separatist agenda prior to the 1940s, and then increasingly contrary to it from the 1950s, is the pressure of Tamil nationalism, which found progressive accommodation with Indian nationalism and with the latter's territorial imperatives. In fact, for a decade and more from the late 1940s Tamil nationalism's most important territorial concern was to ensure that all those areas of the newly formed Indian nation-state which were deemed to be Tamil-speaking should come under the rule of a Tamil polity, and should become part of the newly formed Tamil state of Madras. In these years, the anxiety is palpable that these regions would be "lost" to Tamil-speakers in the process of accommodating the territorial demands of their neighbors. It is in such a context that I locate the intensification of Tamil labors of loss around Kumarikkaṇṭam as well. In these years, not only do Tamil devotion's lamentations over territorial loss dramatically increase, but its place-making typically resorts to linear catastrophism.[151] So the place-maker begins by detailing the vast extent of Kumarikkaṇṭam once upon a time, which, as a consequence of the various catastrophic deluges, was reduced to Tamiḻakam, celebrated in postdiluvial poetry as extending from India's present coast to coast, and from the Tirupati Hills in the North to the land's end of today. In subsequent centuries, Tamiḻakam suffered further territorial losses, no longer to nature's fury but to Tamil's human enemies. Most tragically, by the late nineteenth century when Tamil labors of loss around Lemuria commence, even Kanyakumari, which had historically served as the boundary of postdiluvial Tamiḻakam, belonged to the neighboring Travancore state. So, from the late 1940s nationalists of various hues in the Tamil country launched a series of powerful protests to ensure that the newly formed state of Madras would conform to the borders of postdiluvial Tamiḻakam, extending from the Tirupati Hills to land's end. Particularly satisfying to all concerned was the final incorporation—not without sacrifice of lives and the loss of other territory—of Kanyakumari into Madras state in December 1956.[152] The redeeming of Kanyakumari was politically important because it had served as the historic southern border of the postdiluvial Tamil realm, as it was claimed again and again by Tamil devotees. Symbolically, of course, its significance lay in the fact that it was a remnant of the doomed Kumarināṭu, and its incorporation into the modern Tamil state meant that this hallowed remnant of the ancestral homeland was now a part of the modern Tamil body politic.[153] Hence, for all its extravagant claims regarding the extent of antediluvial boundaries and prehistoric and historic borders, Tamil place-making is quietly rooted in the logic of territorial pragmatism, and this pragmatism also increases its indebtedness to geography as discipline and discourse.

Above all, Tamil place-making's entanglement with disciplinary geography's realist imperatives is illustrated by its compulsion to fill up the prelapsarian elsewhere of Kumarikkaṇṭam with mountains, rivers, regions, and cities, all of which are described in the manner that professional geographers routinely use to describe the real and present world. So driven are they by the imperative to demonstrate that Kumarikkaṇṭam was an ancestral homeland in which Tamil speakers had once *actually* lived that Tamil place-makers miss the opportunity to present it as a truly fabulous antediluvial landscape, as some Euro-American occultists and novelists do. Instead, they turn to the historical geography of their lived homeland and recast their lost homeland in its image. Just as postdiluvial Tamiḻakam was divided into various *nāṭus*—Chola, Chera, Pandya, and others—Kumarikkaṇṭam, too, was divided into "49 nāṭu," whose names are conveniently furnished by the medieval commentator Adiyarkunallar.[154] These are E̲lu teṅku nāṭu ("Seven coconut lands"); E̲lu Maturai nāṭu ("Seven Madurai nāṭu"); E̲lu mun̲pālai nāṭu ("Seven front sandy areas"); E̲lu pin̲pālai nāṭu ("Seven back sandy tracts"); E̲lu kun̲r̲a nāṭu ("Seven hilly villages"); E̲lu kuṇakārai nāṭu ("Seven eastern littoral hamlets"); and E̲lu kur̲umpan̲ai nāṭu ("Seven dwarf-palm districts").[155] Following Adiyarkunallar, these regions are described as "fertile," "temperate," and "prosperous." Adiyarkunallar himself had only identified the forty-nine *nāṭu* as part of the antediluvial Pandyan kingdom, but in Tamil labors of loss around Kumarikkaṇṭam these are nationalized and frequently referred to as the "49 *Tami̲l* nāṭu."[156] This not only enables Lemuria to be incorporated into a horizon of Tamil meanings and memories, but also anchors the modern polity of Tamilnadu in antediluvial times. Over the years, other antediluvial regions are named and added to this evolving geography of loss. Some, following Adiyarkunallar, name the mountainous regions adjacent to the forty-nine *nāṭu* as Kumari and Kollam, and the southern province to the south as Ten̲pālimukam. Others imagine other regions of the antediluvian landscape, such as Peruvaḷa nāṭu ("the great fertile land"), Pan̲ai nāṭu ("the land of the palmyra"), Oḷi nāṭu ("Land of light"), Pan̲malai nāṭu ("land of many mountains"), and so on. However, other than naming them and invoking their fertility, Tamil labors of loss are quite unforthcoming about these lost territories whose seizure by the "cruel sea" is lamented over and again, in a stereotypic fashion. Here, as in other regards, the refusal to wholeheartedly think an antediluvial landscape as singularly different or fantastic is as much a consequence of the hold of disciplinary geography on these labors of loss as it is a result of the imperative to imagine the lost homeland in the image of the lived homeland.

So, since postdiluvial Tami̲lakam extended from ocean to ocean, antediluvial Kumarikkaṇṭam, too, was surrounded by numerous oceans, variously named Kumarikkaṭal ("the Kumari sea"), Kuṇakaṭal ("the Western sea"), Ten̲kaṭal ("the Southern sea"), and so on. Rivers played an important role in the antediluvial Tamil homeland, as they do in postdiluvial times, and a considerable amount of labor is invested in reckoning the source, the length, and the direction of flow of the two principal rivers of Kumarikkaṇṭam, the Kumari and the Pahr̲uḷi, both of which have vanished today. Opinions vary as to which of these served as the southern boundary of the ancestral homeland and as to which was seized during the first *kaṭalkōḷ*, but there is little doubt about the historicity of these lost rivers.[157] It is widely assumed that the stretch of land, constituting 700 *kāvatam*, that extended between these rivers was the heartland of Kumarikkaṇṭam, also known as Peruvaḷanāṭu. Occasionally, the place-maker waxes poetically about the beauty of these rivers, which are described as "large" and "sprawling." But, here, too, he misses the opportunity to truly think outside realist geography or the hegemony of the lived homeland's topography. So, because disciplinary geography tells the place-maker that rivers in Tamilnadu emerge from the mountains that run along its western borders, not surprisingly, the rivers of the antediluvial Tamil homeland find birth as well in western mountains and flow down to the eastern ocean. The most important mountain of Kumarikkaṇṭam is named—like the river whose source it was—after Kumari, as Kumarikkōṭu, Kumari Malai, or Kumari Ku<u>nr</u>u. This lay along the western flank of the lost homeland, and its height and grandeur leads many to compare it to the mighty Himalayas (which, it is repeatedly emphasized, had then not even existed). Other mountain ranges dotted the antediluvian landscape: Maṇimalai, Pa<u>n</u>malai Aṭukkam, and Makēntiram. Even Meru, the axial mountain of the Sanskritic *Puraṇa*s, finds a place in some labors of loss, Tamil nationalist though they might be.[158]

Since urbanism comes to be prized in both Victorian Europe and colonial India as the highpoint of civilization, Kumarikkaṇṭam, too, is declared an urban civilization, the world's first and most ancient at that. Its urbanity finds focus in the two antediluvial metropolises of Te<u>n</u>maturai and Kapāṭapuram, which served as the successive capitals of the antediluvian Pandyan kingdom under siege to the ocean.[159] A third city—Maṇalūr—also finds a place occasionally in the labors of loss of those who insist that there were three and not two *kaṭalkōḷ* that demolished the ancestral Tamil homeland. Some suggest that Te<u>n</u>maturai stood on the banks of the Pah<u>r</u>uḷi, while others insist that the Kumari flowed adjacent to it. But we learn little else about their cityscapes, although the 1981 government film—following the canons of Tamil cinema—fills them up with extravagant mansions and palaces, replete with towering arches and ornate balconies.

Of course, the fame of these cities also derives from the fact that they hosted the foundational literary academies of Tamil poets. And here the place-maker waxes on, armed as he is with abundant, if stereotypical, knowledge about these academies from premodern commentaries, and providing him, as they do, an occasion to lament the loss of Tamil literature. Thus, the city of Te<u>n</u>maturai played host to the first academy, *mutar̲caṅkam,* which was presided over by the gods Siva and Murugan, and which counted among its 549 notables such celebrities as Agastya and Kubera. The 4,449 poets of this academy, which was in session for 4,440 years, composed "ever so many" poems, including the "lost" works *Mutunārai, Mutukuruku,* and *Kaḷariyāvirai,* with the help of the "lost" grammar *Akattiyam.* They were patronized by eighty-nine generations of Pandyan kings, beginning with Kalcina Valuti and ending with Kadungon, seven of whom themselves wrote poetry. The academy was reestablished at Kapāṭapuram after Te<u>n</u>maturai was washed away by the ocean. This "middle" academy *(iṭaiccaṅkam)* counted among its 3,700 members both Agastya and Tolkappiyan. During the 3,700 years it was in session, numerous poems and treatises were composed, including the *Tolkāppiyam, Māpurāṇam, Icainuṇukkam,* and *Pūtapurāṇam,* the last three of which are lost today. Fifty-nine generations of Pandyan kings, beginning with Venterceliyan, ruled over the city when this academy was in session. This city, too, was washed away by the sea, during the reign of Mudatirumaran. In these labors of loss, Teṉmaturai and Kapāṭapuram are only the earliest among a long list of "cities" lost to the ocean, These included the famed Kāveripūmpaṭṭiṇam (or Pūmpukār), Korkai, Kayāl, and parts of Makāpalipuram. Today's surviving Tamil ports, therefore, are only a fraction of what had flourished, once upon a time.

Therefore, just as much is made of the literary works that are imagined to have survived the ravages of *kaṭalkōḷ*, much is also made of territorial remnants. As I noted earlier, a good deal of ambiguity plagues Tamil place-making regarding how much of India was part of the antediluvial Kumarikkaṇṭam and survived its loss. Many insist it is only India south of the Vindhyas—the so-called Dravidian heartland—that had formed part of the antediluvial homeland and had survived the ocean's ravages. But a few are willing to include all of India south of the Himalayas, which erupted as Kumarikkaṇṭam drowned. For others, only the present-day Tamil speaking regions of the Indian mainland were part of the former homeland.

There is also a good deal of ambiguity about the island of Sri Lanka in this geography of loss. As early as 1906, P. Arunachalam (1853–1924), a member of the Ceylon Civil Service, insisted in a lecture before the governor in the Legislative Council in Colombo that his island had been part of "an Oriental continent which stretched . . . from Madagascar to the Malay Archipelago."[160] His statement echoed metropolitan labors of loss around Ceylon. Since the middle of the nineteenth century, there had been some discussion among European geologists about the island's relation in paleotimes to the Indian mainland, with some proposing that into the Tertiary period and before the rise of the Himalayas, Ceylon had been connected to the Deccan.[161] Colonial historiography had also resurrected ancient Buddhist contentions that Lanka had once been a much larger land that been overcome by numerous catastrophic floods.[162] Tamil place-making inherits these knowledges. In its labors of loss, there is no question that the island had been once part of Kumarikkaṇṭam, and that once the former homeland began to submerge, Lanka separated from what would become the postlapsarian Tamiḻakam on the Indian mainland. As I note later, in Tamil maps from the 1920s as well, Lanka is an intimate part of Kumarikkaṇṭam's antediluvial landscape. Remarkably, however, Lemuria hardly figures in the Tamil nationalist imagination of modern Sri Lanka, and even in the labors of loss of those mainland devotees of Tamil who are Sri Lankan in origin (like Kandiah Pillai, Kanakasabhai Pillai, Mutthuthambi Pillai, and others), the island's relationship to Kumarināṭu is not accorded any special treatment. While it is clearly a surviving remnant of the former homeland, its loss is subordinated to the loss of Kumarināṭu, which is what is principally grieved over. No attempt is made to reclaim the island for the Tamil mainland in the name of the lost homeland.[163]

This is especially striking when we consider the special place accorded to Kanyakumari in this geography of loss and the concerted political effort made in the 1950s to incorporate the Cape into the Tamil polity. For, it is the one territorial survivor about whose significance all are in agreement, hence, the multiple referents for the place-name Kumari in this landscape of loss—a telling reminder of the "large tasks that small words can be made to perform."[164] In Tamil labors of loss, Kumari serves as the name of the entire lost homeland, that of one of its constituent regions, that of one of its mountains (and of its peak), as well as that of a major river that ran through it. Today, of course, it is the name borne by land's end, a hallowed Hindu pilgrimage site for at least two millennia and a popular tourist spot in recent times. For the last two thousand years, it has also been the name bestowed upon the traditional southern boundary of postdiluvial Tamiḻakam, which extended, as poem after poem reiterates, from "the northern Vēṅkaṭam [Tirupati] mountain" to "the southern Kumari." For Tamil place-makers, the very ambiguity of this latter phrase has offered a strategic opportunity to insist that Kumari here refers not to the Cape—the land's end of today—but to the antediluvian river that ran through Kumarināṭu and that served as *its* southern boundary. It is with the loss of this river—and the territory through which it ran—and with the consequent shrinkage of the antediluvial homeland to its postdiluvial extent that the Cape came to serve as the southern limit, it is argued. In the reckoning of some, today's Kanyakumari is the *only* surviving remnant on the Indian mainland of the glorious Tamil homeland lost to the ocean.[165]

For all who participate in these labors of loss, Kanyakumari is the one place that still carries the name of this lost world and that survived the ravages of *kaṭalkōḷ*—hence its capacity to trigger labors of grief over loss. As such, it is a *lieu de mémoire,* in Pierre Nora's terms, standing on the threshold, as it were, between two places, that which is imagined to have been lost and that which is believed to have survived.[166] It is a vestige of the vanished, a boundary stone of a lost time, a place-marker of eternity. It stops time, inhibits forgetting, fixes a state of things, immortalizes death, and materializes the immaterial. It is a site of excess memory.[167] Not surprisingly, the travelers to Kanyakumari whose exhortations to the sea I earlier discussed launched into their place-making reveries on reaching land's end, for since the closing years of the nineteenth century it has come to serve as the reckoning spot for Tamil labors of loss over a drowned homeland in the Indian Ocean.[168] It is a place where the place-maker remembers another place and another time, or at least receives their intimation. It has become, in Bakhtinian terms, a chronotope, a point "in the geography of a community where time and space intersect and fuse" and "where through the agency of historical tales, their intersection is made visible for human contemplation." Through such tales, it has been transformed into a paradigmatic site of nostalgic place-making, a "mnemonic peg" on which to hang fantasies of dispossession and geographies of loss.[169]

TEACHING LOSS

Nothing more clearly demonstrates the Tamil place-maker's contention that his lost homeland *really* did exist, if only in some long vanished prelapsarian past, than his effort to introduce Kumarikkaṇṭam into the curriculum of schools and colleges. Indeed, this also distinguishes Lemuria's presence in the Tamil country from its appearance elsewhere, for nowhere else does Sclater's lost continent become a part of pedagogical processes, and indirectly, therefore, of the technologies of modern citizenship. Not surprisingly, there are no attempts to fantasize outside the usual range of facts, as the place-maker works to convince his young audience of the reality of Kumarikkaṇṭam, whose existence is presented until this day as an indubitable "fact" of the (antediluvial) history and geography of Tamilnadu, of India, and of the world. In these textbook labors of loss, Kumarikkaṇṭam is too important to be taken lightly—or fantastically.

To recall, Blanford's physical geography textbook speculates about a drowned paleo-continent in the Indian Ocean as early as 1873,[170] and from the early years of the twentieth century history schoolbooks which circulated in Madras in English and Tamil occasionally ponder over its status as a Dravidian homeland. But none of these identify Lemuria as a singularly Tamil place, nor are they freighted down with the sense of loss that so distinguishes the spatial preoccupations of Tamil devotion.[171] To encounter these, we have turn to textbooks meant for Tamil-language instruction, in which, from the 1930s, discussions of Lemuria as a lost Tamil homeland begin to proliferate. Indeed, both in schools and colleges in the colonial and postcolonial period, instruction in Tamil and its literature has been the primary site for the institutionalization of Tamil labors of loss around Lemuria. This is mostly because until 1967, when a Tamil nationalist government was voted into power in the state, Tamil pedagogy, beleaguered thought it might have been (especially in the early decades of the twentieth century), was the principal institutional means through which Tamil devotional ideas were disseminated outside a narrow circle of scholars. As the brief biographies of Tamil place-makers scattered through these pages indicate, a majority of them have been instructors in Tamil in schools and colleges. In that capacity, they also frequently wrote textbooks in which their labors of loss around the drowned Kumarināṭu both found expression and were passed on to the young student, especially to those who were largely reliant on Tamil instruction.

Schoolbooks for Tamil instruction were published in the Presidency from at least the 1840s, but for much of the nineteenth century, rarely if ever did even a mention of Tamil's ancient literary archive make it into the pages of these works published by European missionary educational societies, by the colonial Directorate of Public Instruction, and by the occasional local publisher. From the second decade of the twentieth century—as Tamil devotionalism begins to pick up momentum, along with the slow but increasing Tamilization of the pedagogical and political apparatus—this situation changes, as students who took language instruction began to read lessons on the greatness of Tamil and its hoary antiquity.[172] Soon, the narrative of the three *caṅkams*, their patronage by the Pandyan kings, and their role in the fostering of the Tamil language becomes a favorite staple. Progressively, the telling of this tale offers the occasion to mention—if only briefly—the loss of ancient Tamil works to the ocean.[173]

From the 1930s, the *caṅkam* narrative is supplemented by explicit references to the lost continent, and this is a practice that continues to this day.[174] There is no discernible pattern or even predictability as to when or why Lemuria gets mentioned in any specific textbook in any particular year, although in general it is middle-school students who are taught a Tamil lesson in which it puts in an appearance. Sometimes the citation is cursory and vague; at other times, it is extended and detailed in its recycling of the many ideas about the antediluvian homeland in circulation in Tamil labors of loss.[175] Regardless of such variations, in the vast majority of language-instruction lessons, Lemuria or Kumarikkaṇṭam is *always* presented to the schoolchild as a Tamil place, the ancestral Tamil homeland, and the cradle of Tamil literary works. Its very existence is a reminder, the children are told, of the great antiquity of the Tamil language, land, and people—an antiquity that is beyond historical reckoning in the view of some schoolbooks.[176] Occasionally, it is also presented as a prelapsarian land of plenitude and perfection, inhabited by virtuous rulers and creative poets.[177]

Most significantly, whenever Kumarināṭu appears in the schoolroom it is always used to drive home a sense of loss—of the great works of Tamil literature, but more persistently and consistently, of the loss of Tamil territory.[178] Invariably, then, the students are asked to reverence those literary works which have survived such catastrophic loss, as indeed the remaining territory, namely, the lived homeland of Tamilnadu. The linear catastrophism that undergirds Tamil devotion's place-making is clearly apparent in these textbook labors of loss, as in this telling and typical example from 1959 (authored by the renowned journalist and Congress nationalist Kalyanasundaram):

> The boundaries of the Tamiḻnāṭu in which we reside are confined to 13 districts. Some few centuries ago, Tamilnadu included Kerala and one part of the

Telugu country. Some thousands of years ago, Panamparanar, who recited the preface to the *Tolkāppiyam,* described the boundary of Tamiḻnāṭu in the following way: "The good land between the northern Tirupati and the southern Kumari is where Tamil is spoken." Before the time of the *Tolkāppiyam,* the northern boundary of Tamiḻnāṭu touched the Vindhya mountains, and extended even beyond, according to scholars. The southern limit designated as Kumari which we use to refer to the cape today, was a river by that name according to old Tamil literary works. These literary works also tell us that south of that river flowed another river called Pahṟuḷi. The land that these rivers made fertile is called Kumarināṭu. That land was seized by the ocean thousands and thousands of years ago.[179]

There are three things worth underscoring about such textbook labors of loss. First and very importantly, although terms like "Kumarināṭu," "Kumarikkaṇṭam," "Lemuria," or even "a sprawling landmass" are occasionally used, the land that is lost is invariably identified for the children's contemplation as "Tamiḻnāṭu" or "Tamiḻakam." This naming strategy obviously aims to present what might be otherwise an alien (home)land to the child in terms that s/he would be familiar with from everyday usage and from geography lessons, even as it establishes a proprietary bond between the young student and Sclater's Lemuria, which is recast as "the motherland" or "our Tamiḻnāṭu."

Further, alongside specifically naming the lost land as "our Tamiḻnāṭu," the textbooks generally invite the young student to consider its diminution over time, so that what was once "vast" and "sprawling" shrinks over time, as more and more chunks of "Tamil" territory were lost. So, for example, a 1956 lesson asks the child: "The area where Tamil is spoken is called Tamiḻakam. What was the state of Tamiḻakam in former times? What was its state in the middle period? What is its condition now? Is it not necessary that we know about this?" Prosaic and disenchanted though these questions might appear on their face, the response that the lesson offers is couched in the language of exilic loss:

In ancient times, Tamiḻakam was a sprawling landmass. It stretched all the way across the ocean to Australia and Africa. It was called "Kumarikkaṇṭam." The continent was divided into 49 nāṭu. Other features included the Kumari mountain, and the rivers Kumari and Pahṟuḷi. The first and second *caṅkams* also flourished on that continent. Because of the boiling over of the ocean, that continent drowned. . . . After this, the boundaries of Tamiḻakam were the Tirupati hill in the north, Cape Kanyakumari in the south, and the two seas on the east and west. . . . Tamiḻakam which flourished in such an excellent manner has now shrunk in size. Tamil is spoken well today in only 9 districts.[180]

Frequently, maps of postdiluvial Tamiḻakam, often labeled "Tamiḻnāṭu," accompany these discussions, to visually capture the attention of the students and to remind them of what had survived the catastrophic loss of

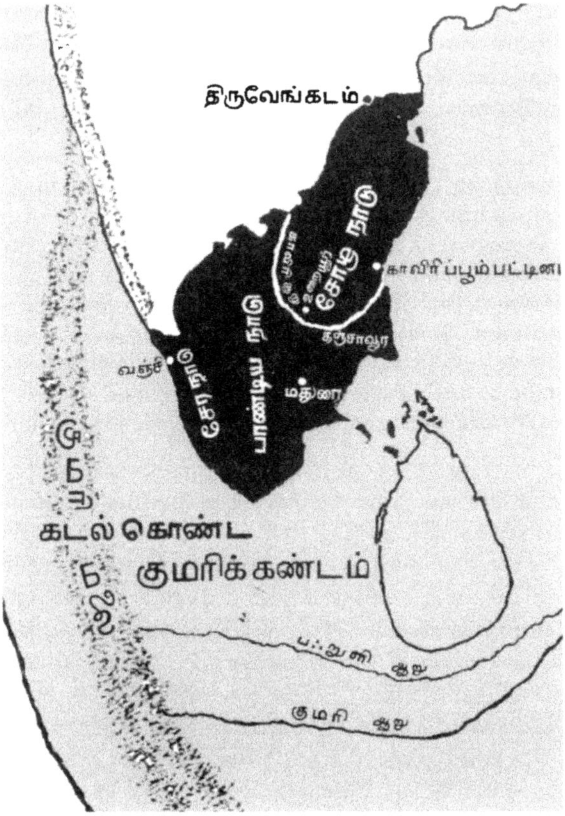

Figure 4. Sivasailam Pillai, Kaṭal Koṇṭa Kumarikkaṇṭam [Ocean-seized Kumarikkaṇṭam]. This map, in Tamil, shows the Indian peninsula (the blackened part represents post-diluvial Tamiḻakam), south of which stretches antediluvial Kumarikkaṇṭam (shaded in gray). From Sivasailam Pillai, *Putumuṟaik Kaḻakat Tamiḻppāṭam (Mutaṟ Puttakam: Mutar Pārattiṟku)* [New Kazhagam Tamil reader for class six]. Reproduced with permission of the South India Saiva Siddhanta Works Publishing Society. Redrawn by Rich Freeman.

land.[181] Curiously, although maps of Kumarināṭu frequently accompany the publications on Lemuria of the Tamil devotee, few appear in schoolbooks. An exception is a 1951 map which shows peninsular India, south of which stretches the shadowy outline of a territory labeled "ocean-seized Kumarikkaṇṭam" (see Fig. 4). The mountain range called Kumari sprawls across its western edge, while the rivers Pahṟuḷi and Kumari flow westward and disappear into the righthand border of the map. North of the territory "seized by the ocean" is the postdiluvial region extending from the Tirupati mountain to the Cape, the Tamiḻakam of the early Tamil anthologies. The text accompanying this map, written by one Sivasailam Pillai, tells the student-reader:

> Our land is Tamiḻnāṭu. . . . Tamiḻnāṭu is a very ancient land. In ancient times this was a vast sprawling land. Formerly, the ocean that is to the south of today's Cape Kumari was a vast landmass that was attached to Tamiḻnāṭu. The Western Ghats extended to great length into this ancient land. That section of the mountains was called "Kumari." Moreover, two rivers called Kumari and Pahṟuḷi flowed there. Because of these, this territory was very fertile. Several thousand years ago, this prosperous landmass drowned in the ocean. It is after this that Tamiḻnāṭu has shrunk to its present boundaries. Today Tamiḻnāṭu extends from Tirupati mountain in the north to Cape Kumari in the south.[182]

Looking at the map in conjunction with this statement, the young student would have been left with little doubt about the loss of territory that "our Tamiḻnāṭu" had endured over time.

Strikingly, the young student is expected to remember the loss endured by "our Tamiḻnāṭu" through a series of questions that are posed at the end of many lessons, questions such as "What did 'Kumari' refer to in former times? What does it refer to today?" Or "What are the reasons for the shrinking of the boundaries of Tamiḻnāṭu?" "What rivers flowed in the ante-diluvian Tamiḻnāṭu?" "What do you know of Kumarikkaṇṭam?" And my favorite of all, "What are the true researches of Scott, Elliot, [*sic*] and others?"[183] Such questions (which presumably would reappear in school examinations) compel the student to remember—and memorize—the catastrophic loss of territory suffered by "our Tamiḻnāṭu" over time. Further, in contrast to other textbook contexts, in which Lemuria is mentioned in highly speculative terms, the typical Tamil-language reader leaves the young student with little doubt regarding the reality of this antediluvian Tamil homeland and its catastrophic loss. The questions only reaffirm this empirical fact.

So, although in comparison to geology or history textbooks, Lemuria puts in a comparatively late appearance in schoolbooks meant for Tamil language instruction, it is in these that its very existence finds the most sustained support, a support that is invariably expressed in the highly realist and empirical terms of historicism. It is also in these books that Kumarikkaṇṭam is unequivocally presented as a Tamil place with an antediluvian past. And, above all, these textbook labors of loss are quite unabashed in their detailing of the material and political consequences of its drowning for the Tamil present which the young student inhabits.

Tamil labors of loss also find a place in the college curriculum as early as 1908, when Suryanarayana Sastri's pioneering history of the Tamil language, which clearly identified Sclater's lost place-world as "Kumarināṭu" and which detailed the 49 "Tamil" *nāṭu* that were seized by the ocean, was prescribed for use in Madras University for Master's level courses in Tamil for the academic year 1908–9. Over the next few decades, other works that I have mentioned in these pages—such as Purnalingam Pillai's *A Primer of Tamil Literature* (1904) or *Tamil Literature* (1929), Kandiah Pillai's *Tamiḻakam* (1934), and Srinivasa Pillai's *Tamiḻ Varalāṟu* (1927)—were also included in the Tamil curriculum at the University of Madras and Annamalai University. As in schools, it is through instruction in the Tamil language and its literature that Tamil labors of loss circulate in colleges. Every now and then, as the biographies of Somasundara Bharati, Devaneyan Pavanar, and R. Nedunceliyan, among others, demonstrate, students who were exposed to these labors of loss in their classrooms went on to themselves participate in Tamil place-making.

Interestingly, in recent years even textbooks used in college history-teaching and written in Tamil include extensive treatments of Kumarināṭu, in contrast to school history books, which are quite reticent on this score, as I have noted. Here, a 1975 text on the prehistory of Tamilnadu, written by a special committee of historians and litterateurs constituted by the Tamil nationalist DMK government and published by a government textbook agency, stands out for its elaborate discussion of Sclater's lost continent.[184] Thus, the adult student is told that "scientists have not wholly accepted the existence of continents like Lemuria, Atlantis and Mu in the Pacific. Nevertheless, that Lanka was connected with southern India, and that there was a landmass to the south of Cape Kumari are accepted truths."[185] Further, following a detailing of metropolitan labors of loss over Lemuria by the likes of Ernst Haeckel and Scott Elliot, the textbook concludes: "All these opinions show that Dravidians emerged on a landmass to the south of Tamiḻakam; when this region drowned because of a giant flood, they moved north to settle; the region that they thus settled is the southernmost part of today's Tamiḻakam. *These opinions were not just aired by literary scholars or historians. They were voiced by the foremost geologists, ethnologists and anthropologists of their time.*"[186]

Here, the college textbook does not merely present the fact of Kumarināṭu's existence to the student, as does the typical schoolbook. Instead, it invokes "science," presents the pros and cons of arguments about the historicity of Lemuria, details all the evidence in favor of the lost continent, and finally concludes by appealing to the adult student's (devotional) rationality.[187] Having discussed the antediluvian homeland in these terms, the textbook moves seamlessly into a normative historical narrative of the three stone ages and the arrival of the Aryans, without worrying over the inconsistency involved in such a move.[188]

Indeed, this in itself is telling, for in the vast majority of school and college textbooks in which Lemuria puts in a tantalizing appearance, the labors of loss around the antediluvial geography and catastrophic history of the former homeland never disrupts either the normative flow of Indian history or the dominant perceptions of the geo-space of postlapsarian India. In other words, the catastrophic place-making imagination is limited to the prelapsarian past, and rarely, if ever, does it threaten the geopolitical and historical realities of the postdiluvial world to which the rest of the textbook is devoted. So, rather than overturning reality as the student knows it, the occasional and random textbook labor of loss functions as a supplement, offering at best a prelapsarian addendum to the geography and history of the world, whose postlapsarian condition is accepted, more or less, normatively. As such, the subversion of reality attempted by this catastrophic and fabulous imagination and its labors of loss in the realm of textbooks is contained, and thus rendered safe. It remains eccentric. Hence it is allowed to flourish, for it is confined to a long lost past—a world vanished in the shadows of an absolute past, to recall Casey—that not even the most devoted textbook author seeks to recover but only invokes to goad the student into remembering the greatness of the ancestral Tamil people, once upon a time, long long ago, on a homeland that has since disappeared into the ocean.

Further, because these textbook labors of loss are largely confined to courses on language and literature, which occupy the bottom of the hierarchy in the complicated knowledge pyramid of the Indian educational system in both schools and colleges, discussions in the Tamil textbook of Kumarināṭu and its catastrophic drowning are not as threatening as they would be if these had occurred in science classes, or even in history or geography books, with their emphasis on positivist and empiricist information. From these realms of "real" learning that are deemed to matter in the material world of success and achievement, the fabulous Kumarikkaṇṭam is safely kept out. In this sense as well, these textbook labors of loss are eccentric.

Not least, such fabulous place-making in textbooks is allowed to flourish because it is largely limited to the world of Tamil pedagogy, a world that half a century of effort by postcolonial governments notwithstanding, still predominantly caters to the subaltern peoples of the region who are largely dependent on state-supported education. Indeed, the typical school and college student from middle-class and elite families, who almost always has an English-medium education in private schools, would rarely, even never, encounter Kumarināṭu in her classroom since she is less likely to formally study Tamil or read locally produced textbooks. Since Kumarināṭu is most extensively discussed in textbooks written in Tamil, and especially those meant for Tamil language instruction produced by local publishers and the government textbook society, this means that Tamil labors of loss are nurtured, reproduced, and contained within the world of Tamil pedagogy. Regardless of official support from Tamil nationalist governments since the late 1960s, this is still a subaltern world that yearns for respectability and empowerment, not to mention material rewards. Its very existential condition fosters such catastrophic notions of loss, and in turn helps sustain the reigning sensibility of the typical Tamil devotee, who lives a life of loss, forever in exile from an imagined state of plenitude and perfection, of which he can only dream but never ever attain. His labors of loss, too, remain eccentric, oppositional, and off-modern.

Chapter 6

Mapping Loss

The best of all Good Companions to take with you to a strange place is undoubtedly a MAP.[1]

CARTOGRAPHIC LABORS OF LOSS

In 1870, barely six years after it was born in the pages of the *Quarterly Journal of Science,* Sclater's Lemuria found cartographic expression when a map featuring it appeared for the first time in German (Fig. 3).[2] Since then and up until today, when the internet and its world wide web have provided a new opportunity and a new context, maps of this vanished place-world have routinely put in an appearance in Europe, the United States, and India, giving Lemuria a cartographic identity that is as diverse as the discursive identity that I have documented here. Maps of Lemuria vary widely, ranging from those made to scale, with geometrically proportionate representations of the earth's (paleo) surface, to ones that are no more than rough sketches with hastily drawn outlines of present continents and past lands. A triumphalist history of cartography which has typically narrated the story of maps in modernity as a heroic progress toward more scientifically accurate representations of geographical reality would perhaps dismiss many of these as not worthy of scholarly attention. This same history would also probably cast aspersions on these maps, for they are not about the real and present world but cartographies of fantasy that chart fabulous lands beyond the usual range of facts. For in this history, the story of the modern map is written as the eventual victory of empirical good science over irrational myth and fanciful imagination.[3]

Yet, here, I obviously take a different track, following the cue of a revisionist scholarship which defines maps as "graphic representations that facilitate a spatial understanding of things, concepts, conditions, processes or events in the human world."[4] Their significance "derives from the fact that people make them to tell other people about the places or spaces they have experienced."[5] As Brian Harley observes, "Locating human actions in space remains the greatest intellectual achievement of the map as a form of knowledge."[6] Similarly, Denis Cosgrove writes,

> To map is in one way or another to take the measure of a world, and more than merely take it, to figure the measure so taken in such a way that it may be communicated between people, places or times. The measure of mapping is not restricted to the mathematical; it may equally be spiritual, political or moral. . . . Acts of mapping are creative, sometimes anxious, moments in coming to knowledge of the world, and the map is both the spatial embodiment of knowledge and a stimulus to further cognitive engagements.[7]

Accordingly, by the mere fact of their existence, the maps of Lemuria I consider here are instances of what I characterize as cartographic labors of loss undertaken to facilitate a spatial knowledge and cognitive understanding of vanished place-worlds and disappeared lands. Even as they bear testimony to the deployment of modern science in the service of loss, maps of Lemuria are also telltale signs of the modernity of the labors of loss that have produced them. For it is only with modernity that the map emerges as a guarantor of geographical reality, standing in for a given territory even when that territory might not exist, as Jean Baudrillard reminds us.[8] This, then, is one of the primary reasons for the map's popularity in Lemuria's place-making. It enables all those who seek to convey the truth about the vanished place-world to mobilize the certitude guaranteed by modern cartographic practice in the service of their varied labors of loss.

At the same time, the modern science of mapping poses three fundamental challenges for the cartographic labors around Lemuria, the density of which vary across the different discursive formations that I consider here, given their contrary investments in the lost place-world. First, cartography is useful for Lemuria's place-makers because the map helps them bring into visual reality a land that would otherwise have remained invisible and hidden. It allows its reader to "see" Lemuria at one glance,[9] and in relation to other lands, especially the continents, that make up today's world. It encourages the reader to ponder over the size of the lost land, and to compare this to the extent of today's continents. And it persuades the reader to imagine a world that looked very different from the earth s/he is more familiar with from modern atlases and geography textbooks with their visualizations of present-day continents and oceans as we have been trained to recognize them. How, then, could the very same technology, used to make invisible lands visible, also be deployed to document their loss and disappearance? How may the map form be used to persuade the reader to ponder over and wonder about worlds that once existed in deep time, but no more? To simultaneously make a vanished world both appear and disappear using the technology of the map poses a basic conceptual challenge that faces the many cartographies of loss around Lemuria, and one which provokes a range of responses.

Second, the authority of scientific cartography derives from its claim to map land that has been surveyed and empirically measured. As Harley writes, "In 'plain' scientific maps, science itself becomes the metaphor. . . . Accuracy and austerity of design are now the new talismans of authority. . . . The topography as shown in maps, increasingly detailed and planimetrically accurate, has become a metaphor for a utilitarian philosophy and its will to power."[10] In turn, this produces what Anne Godlewska refers to as "the similitude of reality" that is central to the power of the "plain" scientific map.[11] It is this illusion of reality to which maps of Lemuria aspire as well, given that one of their fundamental goals is to demonstrate that the lost place-world did exist, if only once upon a time. And yet, because it is no longer present or available, Lemuria cannot be surveyed, measured, or charted according to the protocols and procedures of scientific cartography. On the one hand, this has been liberating for its place-makers, for they can draw Lemuria in any which way they please in a manner that would be arguably difficult, if not impossible, for real and present lands.[12] On the other hand, how may the similitude of reality be produced, given that the mapping of this lost land is based not on precise surveys and measurement but on intangible imagination and on the powers of the mind rather than of the eye?

But perhaps most challenging of all is the pressure that the cartographic act exerts on the fuzzy place-making imaginations that undergird the discursive production of Lemuria, in which the extent and limits, even the location, of the lost continent are frequently left unspecified. Making a map of Lemuria may bring in a semblance of the much-needed reality, certitude, and visibility to this lost place-world sought by those who labor over its loss. All the same, the very act of drawing maps of the lost land means that Lemuria's place-makers are compelled to translate their fuzzy place-world into the precision of the scale and of the mathematically deduced order of the cartographic grid of latitudes and longitudes, the graticule.[13] They *have* to draw borders to a land that they frequently leave unbounded and undefined in their place-talk. And they *have* to specify its extent and its limits so that their Lemuria, quite literally, can be put on the map. As we will see, the response to this cartographic challenge of conforming to "gridded representations of reality"[14] varies, from those who draw maps to scale and in accordance with the coordinates of meridians and parallels, to those who make no attempt to conform to these demands of scientific mapping, to those who refuse to pin down Lemuria by mapping it at all. Nevertheless, the very act of putting Lemuria on the map, with all its attendant challenges, exposes many of the contradictions and faultlines of the fabulous place-making of this lost world.

PUTTING LEMURIA ON THE MAP

At a very basic level, Lemuria could not have been summoned into existence if it had not been for the map, and especially, the world map.[15] Again and again, Lemuria's place-makers tell us that it was by looking at maps that they were convinced of the existence of land connections where now none exist. They, in turn, invite us to look at maps as they chart out the contours of their lost place-world, and they make use of maps to bolster their arguments and to provide evidence for their place-making contentions. Although he does not tell us this in so many words, it is clear that Philip Sclater had been looking at maps of the distribution of the lemur when he proposed, in 1864, the former existence of a land connecting Africa with India that he so fortuitously named Lemuria.[16] Half a century later, Alfred Wegener was much more forthright, in 1915, when he wrote that "the first notion of the displacement of continents came to me in 1910 when, on studying the map of the world, I was impressed by the congruency of both sides of the Atlantic coasts."[17] A few years before him, in 1909, Edmund Marsden in Macmillan's *History of India for Senior Classes* instructs his young readers to look at a physiographic map of India and then asks them to imagine themselves traveling back in time:

> Countless ages ago, if we could have looked down upon what is now the great tract of land called India, we should have seen nothing but the broad ocean. Looking again after an immense lapse of time we should have seen the huge mass of land now called the Deccan, which had slowly arisen through the water with the Aravalli hills as its north-western boundary. It was, however, only a part of a still larger mass of land which rose up with it, a great southern continent which stretched away far to the south-west—over what is now the Indian Ocean—to South Africa, of which it formed a part.[18]

Similarly, a few decades later, Kandiah Pillai wrote a book entitled *Namatu Nāṭu* (Our Nation), also intended primarily for young children, in which he instructs them:

> Look at India on the map. At its southern end is the cape called Kanyakumari, south of which is a vast sea. Today this is called the Indian Ocean. Our Earth appeared countless years ago. Over these countless years, it has changed many times. Before the Indian Ocean appeared, there was a vast land expanse there. This was called Nāvalantīvu. Europeans called it Lemuria. Gondwana was another name they gave it.[19]

So much so that Owen Rutter's adventure novel *The Monster of Mu*—clearly inspired by Churchward's first book on Mu and its maps—begins with its hero, the traveling ethnologist Colin Dale, receiving an aging scrap of paper, from a dying sailor named Jack Nye, which has a hand-drawn map of the "Island of Mu" with some rough coordinates. When Dale tells the sailor he had never seen Mu on any map, Nye responds, "That's right. Not on the charts. Never likely to be . . . Unless you get there . . ." Dale and his companions do eventually get "there," with the help of charts of the Pacific Ocean which take them to a location "800 miles from East Nor'east by East," about 6,000 miles from Honolulu, as another map printed in the novel informs us.[20]

Colin Dale may have been interested in getting there in search of emeralds and adventure, but most of its place-makers do not draw maps of Lemuria in order to get there, for there is no *there* there. Unlike a large number of modern maps that provide directions to specific places and locations, maps of Lemuria are not primarily utilitarian. Nor are they decorative in the way that wall maps and globes have been for a few centuries now, especially in Euro-America, where they have adorned stately courts, sterile offices, schoolrooms, gentlemen's studies, and even the occasional lady's boudoir.[21] Only in Tamil India, where the density of the cartographic labors of loss is also the most intense, do they appear on the dust jackets of books on Lemuria, adding perhaps to their allure.[22] The primary purpose of most maps of Lemuria, however, is illustrative; they serve to visualize the place-talk that they generally accompany. Although, in many instances the map is not commented upon at all or even acknowledged, its very presence in the place-making publication points to the importance that cartographic visualizations have attained in post-Enlightenment modernity, even while it is a testimony to the fact that the many labors of loss over Lemuria are fundamentally cartographic in their sensibility.

There are numerous signs that Lemuria's place-makers are products of a cartographic common sense that is basically post-Enlightenment in its origins. It is this common sense that leads them to draw maps of Lemuria that are oriented toward the north. When they use projections, it is almost invariably Mercator's, even when their own cartographies frequently undo the world that the famous mapmaker so carefully drew in the early sixteenth century. They also invariably assume that the surface of Earth is naturally constituted by five continents—Asia, Africa, Europe, the Americas, and Australia—even when they invite their readers to imagine a time when these landmasses were not so configured. Yet their maps also show that while present-day continents (which they frequently leave unnamed) have a standard shape that has conferred on them a secure cartographic identity which does not require nominal identification, Lemuria, by virtue of the fact that it is an unmeasured and uncharted paleo-land, can never aspire to this. Assuming numerous forms and configurations, the lost place-world has not attained, even after a century and more of mapmaking, what Benedict Anderson has identified as "logo" status. This happens, he notes, when maps become so familiar, so standardized, and so "instantly recognizable" as to need no explanatory glosses, no place names, no signs for rivers, seas, neighbors, and the like. As "pure sign," "the map-as-logo is infinitely reproducible . . . available for transfer to posters, official seals, letterheads, magazine and textbook covers, table cloths and hotel walls."[23] Because they are about a paleo place-world that cannot be cartographically fixed, maps of Lemuria, even in Tamil India, where the continent has received the support of a modern state and its pedagogical apparatus, never circulate as pure sign.

That a modern cartographic common sense had come to prevail among at least the educated classes of Europe by the time the earliest maps of Lemuria come to be drawn, from the 1870s, is perhaps not surprising. As recent studies have shown, geography (and its accomplice, cartography) benefited enormously from the two most important sociopolitical processes of the latter half of the nineteenth century, nationalism and imperialism, and, in turn, spurred them on. The science of geography "constitutes the taking of possession of the earth, and the intellectual domination of space."[24] This possession and domination proceeded at several levels from the late eighteenth century, but especially from the mid-nineteenth, with the professionalization of geography and with its introduction in metropolitan schools and colleges as a subject of study, often mandatory.[25] As a result, the earth came to be known "as designed for man, with a place for everything and everything in its place."[26] Through direct observation, classification, and comparison, it was measured, mapped, and described in intimate detail. It thus came to be progressively tamed, and hence rendered non-mysterious, un-wondrous, dis-enchanted. Virtually no place was left un-known, no land un-charted. Correspondingly, maps of the world proliferated beginning in the latter half of the nineteenth century, becoming a ubiquitous feature of European public places, in scientific journals and illustrated magazines, and in school textbooks and the classroom.

Thus, by the opening years of the twentieth century Europe had asserted its mastery over the earth, not only literally through the actual occupation and possession of large parts of its inhabited territories, but also (and perhaps more enduringly) epistemologically, through its knowledge practices that provided the legitimate conceptual apparatus with which to talk and write about the physical world and to visualize it with the help of maps and globes. By this time, maps had assumed a centrality in the various physical sciences as well, for formulating hypotheses, for explaining the distribution of plants, animals, and rock formations, and for synoptically presenting vast amounts of data.[27] As these knowledge-forms were progressively disciplined as "science," they also became among the best exemplars of the ocularcentrism of the West, and of the domination of vision as the master-sense of modernity.[28] In the words of the first president of the London-based Geological Society in 1841:

> Words following words in long succession, however ably selected those words may be, can never convey so distinct an idea of the visible forms of the earth as the first glance of a good Map. . . . In the extent and variety of its resources, in rapidity of utterance, in the copiousness and completeness of the information it communicates, in precision, conciseness, perspicuity, in the hold it has

> upon the memory, in vividness of imagery and the power of expression, in convenience of reference, in portability, in the combination of so many and such useful qualities, a Map has no rival.[29]

Not surprisingly, from the latter half of the nineteenth century, as speculations about paleo-spaces and the shifting configurations of continental landmasses began to gather in strength, their proponents, too, resorted to cartographic visualizations to visually bolster their place-making. Maps clothed their place-making with the mantle of scientific authority, even as they enabled the enframing of the paleo-world as picture, "to be viewed, investigated, and experienced."[30] Paleo-cartography is therefore very much an exemplar of the exhibitionary order and the will to display that Timothy Mitchell has identified as constitutive of (European) modernity more generally.

Yet, the densest cartographic labors of loss over Lemuria happen, not in this metropolitan scientific world of high modernity, but on its edges, in Tamil-speaking India, whose maps of Lemuria are also the most contestatory of the numerous circulating images of this lost place-world. The investment in the map form to demonstrate the catastrophic loss of their ancestral homeland by Tamil place-makers is especially surprising because even after about two centuries of the dissemination of a modern geographic consciousness which is essentially European in its origins, it would be fair to say that India is not a cartographic-minded society in which maps are central to its modernity. It is true that the British diligently and assiduously mapped the entirety of subcontinent beginning in the closing decades of the eighteenth century,[31] and that from at least the 1830s geography was taught in schools, and later in colleges, in colonial India.[32] At this point, however, the enduring impact of these attempts is difficult to gauge. We know little or next to nothing about the dissemination of modern mapping practices and technologies, or about the spread of a scientific geographic consciousness among the Raj's subjects. Neither do we have a sense of the spread of cartographic literacy nor of the various uses to which maps have been put by modern Indians in their everyday lives. In other words, the social life of the map in modern India is still uncharted territory.[33]

All one can say is that from at least the middle of the nineteenth century wall maps and globes (in English, and increasingly, in various Indian languages) were introduced into Indian classrooms, although they were never as ubiquitous a feature as in the West, nor are they even today. Also, colonial history and geography textbooks began to include cartographic representations of "India" as the region was progressively surveyed, measured, and mapped by the British. (Indeed, up until today, education remains one of the primary means through which Indians are exposed to cartographic images of their nation and of their world.) From early in the twentieth century, and especially in the decades leading up to Independence and beyond, maps of India were also deployed in Indian nationalist activities, although, as I have elsewhere demonstrated, not necessarily in the disenchanted form favored by the state.[34] In the 1940s and 1950s, Tamil and Dravidian nationalists similarly published and circulated cartographic representations of their imagined Dravidian nation in their bid for separation from India.[35] Maps have also been used by political parties associated with a resurgent Hindu nationalism in recent years.[36] In the postcolonial period, the Indian state, as well as several state governments, have published thematic maps and atlases of various sorts. All this suggests that the modern map, as a technology for representing the spatial realities of the world, even while not ubiquitous, is not entirely an uncommon artifact, especially in urban settings, in various official arenas, and in schools and colleges. It is also clear that the modern map, with its specific cartographic representation of a place called "India" occupying a particular location on the earth's surface, has largely supplanted, at least in political, pedagogical, and public institutional contexts, older precolonial ways of mapping space.[37] To this extent, the modern scientific map in its Indian incarnation has been a highly successful mimic form, in its formal appearance, its north orientation, its adoption of the cartographic grid of latitudes and longitudes, its reliance on geometric projections, its conventions of naming, and the like.

Even given all this, however, the enthusiasm with which Tamil place-makers have embraced the modern map form to convey the spatial truths regarding their lost homeland remains surprising. Why they do this and to what end is therefore one of the principal concerns of this chapter. But before I turn to this, I explore the two other traditions of mapping Lemuria, among Euro-American paleo-scientists and occultists, that precede the rise of Tamil cartographic labors of loss.

MAPPING PALEO LEMURIA

To recall, Lemuria assumes a cartographic identity for the first time in 1870, in a German map that appears in the second revised edition of Ernst Haeckel's *Natürliche Schöpfungsgeschichte.* Reprinted a few years later, in 1876, in English and with some minor variations, the map is revealingly titled "Hypothetical Sketch of the Monophyletic Origin and of the Extension of the 12 Races of Man from Lemuria over Earth" (see Fig. 3).[38] The map in German is of Earth today, with its continents and oceans clearly outlined and roughly drawn to scale. The shadowy contours of "Lemuria" extend from Madagascar through India to the islands of Southeast Asia. Its boundaries are left fuzzy, but it is clear that it is a landmass that lay almost entirely in the southern hemisphere, south of the Equator but north of the Tropic of Capricorn. Fanning out from Lemuria are numerous arrows signifying the movement of the twelve races of man as they moved out of "Paradise" to settle in different parts of the world. Haeckel himself appended a brief explanation to his *Sketch,* which he underscored was "hypothetical" and "provisional":

> The probable primeval home, or "Paradise," is here assumed to be *Lemuria,* a tropical continent at present lying below the level of the Indian Ocean, the former existence of which in the Tertiary period seems very probable from numerous facts in animal and vegetable geography. But it is also very possible that the hypothetical "cradle of the human race" lay further to the east (in Hindostan or Further India), or further to the west (in eastern Africa). Future investigations, especially in comparative anthropology and paleontology, will, it is to be hoped enable us to determine the probable position of the primeval home of man more definitely than it is possible to do at present.[39]

As Haeckel struggles to cartographically translate his tentative labors of loss around Lemuria as the primeval home of mankind, he resorts, first in the 1870 map, to depicting the drowned continent with fuzzy contours. By the time the English version of the map appears in 1876 in color, Lemuria ceases to even be identified as a shadowy landmass, as only its name is printed two times in the vicinity of the modern-day island of Madagascar. Haeckel's cartographic hesitancy echoes his verbal place-making: "The gradual transmutation of catarrhine apes into pithecoid man probably took place in the Tertiary period in the hypothetical Lemuria, and the boundaries and forms of the present continents and oceans must then have been completely different from what they are now. . . . I here, therefore, as in my other hypotheses of development, expressly guard myself against any dogmatic interpretation; they are nothing but *first attempts.*"[40]

The Lemuria of the paleo-scientists, I have noted, was a speculative and tentative place-world to start with, and it remained so for much of its existence in their labors of loss before it disappears entirely in the face of the gathering influence of the theory of continental drift and plate tectonics from the 1960s. Their maps ran the risk, however, of making the paleo land seem much more of a sure thing than they were willing to accord it in their verbal place-making. Haeckel's shadowy Lemuria—now there, and then, not there—bears the marks of this cartographic dilemma that other place-makers deal with through different strategies.

Haeckel's map is unusual among the cartographic labors of loss over paleo-Lemuria. For one thing, it is one of the few paleo-maps principally centered on Lemuria. In contrast, most others are primarily concerned with visualizing a lost Mesozoic world rather than with any one specific former continent, even leaving unnamed the land connection between India and Africa that is graphically represented. In this respect, Haeckel's map is more like the occult and Tamil devotional maps, which tend to be Lemuria-centric, as we will see. Further, although purportedly made by a disenchanted man of science, magical elements quietly enter into Haeckel's map in the guise of identifying Lemuria as "Paradise" (albeit with a query mark signifying doubt and uncertainty), and in the guise of featuring "the Hyberboreans," who are shown occupying the area around the Arctic Circle today but whose parent race, the Polar, dispersed from the drowned Indian Ocean continent.[41]

Finally, Haeckel's is the only map in the entire archive of paleo-cartographic labors of loss around Lemuria which is concerned with the dispersal of the races of mankind out of this vanished place and their subsequent settlement in different parts of the world. They are identified as 1. Papuans, 2. Hottentots, 3. Kaffirs, 4. Negroes, 5. Australians, 6. Malays, 7. Mongols, 8. Polars, 9. Americans, 10. Dravidas, 11. Nubians, and 12. the Mediterranese, and are shown as emerging out of the two parent stocks of U[lotirichi] and L[isotricchi]. In doing this, Haeckel was graphically illustrating his tentative hypothesis that Lemuria had served as "the cradle" in which Man had first appeared and then dispersed (as the continent which had given him birth drowned) to specific localities on Earth's surface, assuming different physiognomic characteristics as he did so. Haeckel's map drew upon an increasingly popular nineteenth-century cartographic practice that showed "ethnographic distributions in terms of territory occupied by a specific group, or groups, of people having common ethnic affinities."[42] As Helen Wallis observes, "Inherent in the production of many of these maps is the presumption that the ethnic groups so identified have a right to a separate or independent cultural and political identity."[43] In the course of the latter half of the nineteenth and the early decades of the twentieth century, as European intellectuals and nationalists became increasingly preoccupied with the question of racial homelands and the dispersal of ur-populations, they also took to drawing maps showing movements of prehistoric peoples, movements that are invariably shown as "arrows that thrust outward . . . using the symbols of a military offensive."[44] Haeckel's map, too, does this, and as a consequence, Lemuria graphically appears as the center of the then-known world, dispatching its inhabitants to unpopulated regions of the earth.

Haeckel, as we have seen, is a key figure in these Lemurian labors of loss, much invoked by both occultists and Tamil place-makers, not to mention colonial administrator-scholars in southern India. His map, however, was not widely reprinted or mimicked, although its concern with showing the dispersal of populations out of Lemuria is echoed in occultist and Tamil cartographies, as we will see. Among paleo-scientists after Haeckel's map appeared, however, the preoccupation increasingly shifts toward graphically representing a lost Mesozoic world made up continental connections and landbridges that once existed but no more. The first of this type of map appeared in a German publication of 1887 called *Erdegeschichte* (History of Earth), by the Austrian Melchior Neumayr (see Fig. 1).[45] Aiming to illustrate the world as it once existed in the Jurassic period, Neumayr's map showed the outlines of today's continents, drawn to scale, whose carefully drawn boundaries are however undone by three large landmasses that transgress them. These are named the Brasilian-Ethiopian, Sino-Australian, and Neo-Arctic Continents. Snaking out from the southern end of the Brasilian-Ethiopian continent (which extends across present-day Africa and southern America) is the "Indo-Madagascar Peninsula," linking Madagascar to southern India across the Indian Ocean. Although relegated to the status of a "peninsula," instead of a whole "continent" that was the very center of the inhabited world, Neumayr's Indo-Madagascar Peninsula, with its well-articulated outline, has a much more assured cartographic presence than Haeckel's shadowy Lemuria.

Historians of paleogeography have given Neumayr's map a lot of importance in the development of the cartographic tradition of representing the vanished pasts of the earth, even characterizing it as the "first paleogeographic map" of the world.[46] Ursula Marvin observes that "Neumayr's map inspired countless other scientists to reconstruct the lands and seas of past geologic periods, an occupation that is continuing apace," and notes also that its giant Brasilio-Ethiopian continent was the direct source of inspiration for Eduard Suess's influential concept of Gondwanaland.[47] And, indeed, soon after Neumayr's map was published, other maps of now-vanished paleo-continents began to be routinely published in German, French, and English, several of which featured a land-connection between Africa and India, sometimes explicitly named Lemuria. So, in 1894, the prestigious *Geographical Journal* published "Sketch Map showing Approximate Distribution of Land and Sea at the Close of the Jurassic Period as Compared with that of the Present Day."[48] Centered on Madagascar, the map also outlined present-day India, a part of the Arabian Peninsula, and the east coast of Africa. A shadowy "Indo-African Continent," its borders left fuzzy, snakes its way south from the Arabian Peninsula across the east coast of Africa to Madagascar and beyond to peninsular India. This is the first map in English in which Lemuria finds a cartographic presence. Its author, Richard D. Oldham (1858–1936), worked for the Geological Survey of India between 1879 and 1903, and the essay in which the map appeared argued that the subcontinent assumed its present configuration after a series of earth movements and volcanic eruptions caused the Himalayas to erupt, which then led to the gradual submergence of the Indo-African continent by the end of the Cretaceous and the beginning of the Eocene. These "great earth movements," Oldham concluded, "led to a most extensive reshaping of the surface, and the radical change of the geography of India, to that represented in the more familiar form found in modern Atlases."[49] Oldham's map (as, indeed, Neumayr's before his), with its outlines of present-day continents and past lands, persuades the reader to compare the world as it appeared in modern atlases with a world that had once existed, long ago in deep time.

There are several other examples from early in the new century of this *syn-*

optic strategy of simultaneously representing the present-day distribution of the continents and oceans along with their (imagined) past configuration.[50] The synoptic strategy furthers these paleo-cartographic labors of loss in two ways. First, by locating within a single frame the configuration of today's continents with the outlines (shadowy or otherwise) of the continents of the past, these maps implicitly invite their readers to compare the present and the past, and to consider if not the loss of land, then at least its shifting configurations. Secondly, the juxtaposition of present and past continents is also matched by the enormous temporal conflation that these maps accomplish by simultaneously presenting today's continents with those of the Mesozoic, *as if* these coexisted simultaneously. Events occurring over vast geological time are accelerated here into the single frame of the map. Richard A. Fortey observes that such paleo-maps are "essentially a modern geography taken back in time" and are symptomatic of the deep reluctance of paleo-scientists to give up the idea of the fixity of present-day continents at a time when the very idea of drifting landmasses was dismissed as "fantastic."[51] I suggest that the very inclusion of the outlines of the continents of today is an attempt to persuade the reader to speculate about disappearances of lands which no longer exist but which the accompanying place-talk sketches out in detail.

From the early years of the twentieth century, this synoptic operation is supplemented by another that I call *sequential,* which features a series of maps showing the transformation of the distribution of land and oceans over time, leading up to the world attaining its present configuration.[52] This sequential strategy is used to great effect in a map that has attained near-iconic status in these labors of loss, and in paleogeography more generally. Published originally by Alfred Wegener in 1922 in German to illustrate the breakup of his primeval continent of Pangaea, this map in English, from 1924, cartographically narrates the fate of paleo-continents in the Upper Carboniferous, Eocene, and Older Quaternary periods (see Fig. 2). Wegener's map is a "three-part graphic narrative" which adopts "the temporal sequence of three maps" and "accelerates hundreds of millions of geologic history to portray the breakup of the supercontinent and the horizontal displacement of the individual continental blocks to their present positions."[53] In the legend accompanying the map, Wegener notes that he "has included present-day outlines and rivers only for the purpose of identification."[54] These present-day outlines also draw the reader's attention to the shifting configuration of the earth's surface which Wegener's theory of continental displacement narrativizes at great length. Wegener's monograph also included a cartographic representation of what he called the "Lemurian Compression," as the "long connecting portion [which linked India to Asia] was more and more folded together through the continuous gradual approach of India [from the south] to Asia" to produce the Himalayas.[55] As we saw, Wegener's theory of continental displacement had no room for a submerged Lemuria. Nevertheless, his map continues to be reprinted over the years by Lemuria's place-makers, especially on the margins of metropolitan science such as Tamil India, probably because it draws attention to the shifting configurations of the earth's landmasses, their mobility, and the vast differences in their extent over geological time.

Meanwhile, even as Wegener battled it out with his colleagues to find acceptance for his displacement theory, place-making of submerged continents and drowned landbridges continued to flourish as we saw, and found cartographic expression in sequential maps and atlases that were published from the 1920s.[56] These maps chart the appearance of Lemuria toward the end of the Paleozoic, in the Permian period; its continued expansion through the Mesozoic (reaching its largest extent during the Jurassic); and its eventual disappearance sometime during the early Tertiary. Occasionally, some maps also show the dispersal of prehistoric animals of various sorts, graphically reiterating Lemuria's primary identity in these paleogeographic labors of loss as a faunal highway.[57]

From the late 1960s, as paleogeographic labors of loss over a submerged Lemuria die a quiet death, as theories of continental drift and plate tectonics gain ground, fewer maps featuring Sclater's lost place-world appear in the metropole.[58] In the cartographies of drift published in the last few decades of the twentieth century, there is no place at all for a drowned Lemuria amid vast paleo-continents rifting apart and subsequently fusing with other landmasses in deep time.[59] Nonetheless, these maps perform their own cartographies of loss, as Peter Whitfield observes, for with their graphic representation of lands which vanished millions of year ago they "have an uncanny and disturbing power, undermining the apparently fixed and permanent structures of the external world."[60]

What is the significance of these paleo-scientific maps for cartographic labors of loss over Lemuria? To recall, a majority of these maps are not primarily or centrally concerned with Lemuria, but with illustrating a lost Mesozoic time of former landbridges and continental connections. Accordingly, Lemuria has frequently an incidental presence in a much grander cartographic narrative about a disappeared world and a vanished time. To paraphrase the American geologist Charles Schuchert, "dressing and redressing the features of our Mother Earth so as to see her correctly in her evolving form"—this is the primary intent of the paleogeographic map form.[61] Nonetheless, by graphically capturing on a map the lost place-worlds of deep time, these paleo-scientific maps clear the ground for other cartographic labors of loss that are much more explicitly concerned with Lemuria. They also popularize the cartographic operations I have characterized as synoptic and sequential, which are put to dramatic new uses in occult and Tamil labors of loss in the century that follows.

I have insisted that paleogeographic labors of loss are the most tentative in their mapping of a drowned Lemuria, as the heavy hand of scientific caution tempers any extravagant flights of cartographic fantasy beyond the usual range of facts. Yet, paradoxically, in their graphic representations of the former place-world, these maps are the most precise in their delineation of Lemuria, fuzzy and shadowy though its form and shape might be. Thus, the maps of paleo-Lemuria are invariably drawn to scale, frequently show the grid of latitudes and longitudes on which its rough location is charted, and outline its extent on the basis of positivist data provided by fossil finds, the geographic distribution of animals, the terrestrial formation of the ocean floors, and so on. A striking example of this is offered in this detailed explanation provided by the American geologist Bailey Willis to explain why his "Africa-India Isthmus" assumes the shape that it does in a color map that he published in 1932—and which very nicely demonstrates the density of work involved in these cartographic labors of loss:

> There is biologic evidence of migrations between Africa and India which requires the existence of a land connection at least in Permian time. To define the probable path we may examine the bathymetric map to discover the positions of the deeps and to trace the submerged ridges between them. . . . The submerged ridge which may be regarded as the trace of the former isthmus between Africa and India is tortuous, but well defined. It runs from Africa east to Madagascar, thence northeast towards minor islands to the great arcuate ridge of the Seychelles, which it follows southeastward for some 500 miles to the broad, barely submerged bank, Saya de Melha. From that plateau it crosses a channel that is 12,000 feet deep to the long swell which supports the Chagos, Maldive and Lacquedive islands and which extends northward to the western side of the Indian continental peninsula. The general form of this isthmus is that of the capital letter N. . . . This isthmian link may best described as the Africa-India ridge or isthmus. In tracing its outlines we are governed by the location of the deeps, which isostasy forbids us to cross, and guided to some extent by the actual relief. The bathymetric contours are not entirely reliable [however]. [62]

Constrained as paleo-scientists like Willis are by the material evidence that their sciences offer them for the distribution of lands and oceans in times past, their Lemuria is inevitably not an ambitious continent, invariably limited as it generally is to that section of the Indian Ocean that extends between present-day Africa and India.

What happens, however, when a fabulous place-making imagination is given full rein, unconstrained by material evidence, and when there are virtually no holds barred in visualizing Lemuria? I turn to occult maps of Sclater's lost place-world published in Euro-America from the closing years of the nineteenth century to answer this question.

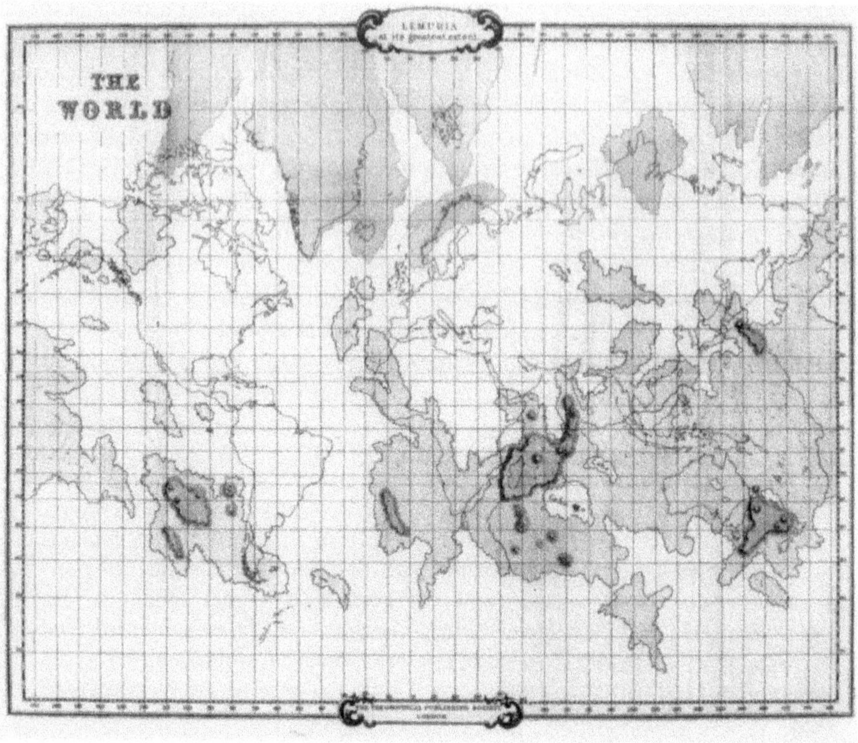

Figure 5. W. Scott-Elliot, Lemuria at its greatest extent. The lightly shaded lands north of the 60th parallel represent the remnants of Hyperborea; all other shaded territories represent Lemuria. From Scott-Elliot, *The Lost Lemuria with Two Maps Showing Distribution of Land Areas at Different Periods.* By permission of the Theosophical Publishing House, London.

MAPPING OCCULT LEMURIA

The most extraordinary aspect to occult cartographies of loss of Lemuria is the insistence that they are based on maps which had been *actually* produced on the drowned continent itself, and which had, since its submergence, been preserved in secret archives that the occultist painstakingly recovers through acts of astral clairvoyance and telepathy. This claim is first advanced in 1896, when the English Theosophist Scott-Elliot published his *The Story of Atlantis,* which included four maps of the world for a period ranging from 1,000,000 years ago to 9,654 B.C. The foreword to the monograph by the Anglo-Indian Theosophist Alfred P. Sinnett noted proudly that Scott-Elliot had had access to "some maps and other records physically preserved from the remote periods concerned," and hence "historical research by means of astral clairvoy-

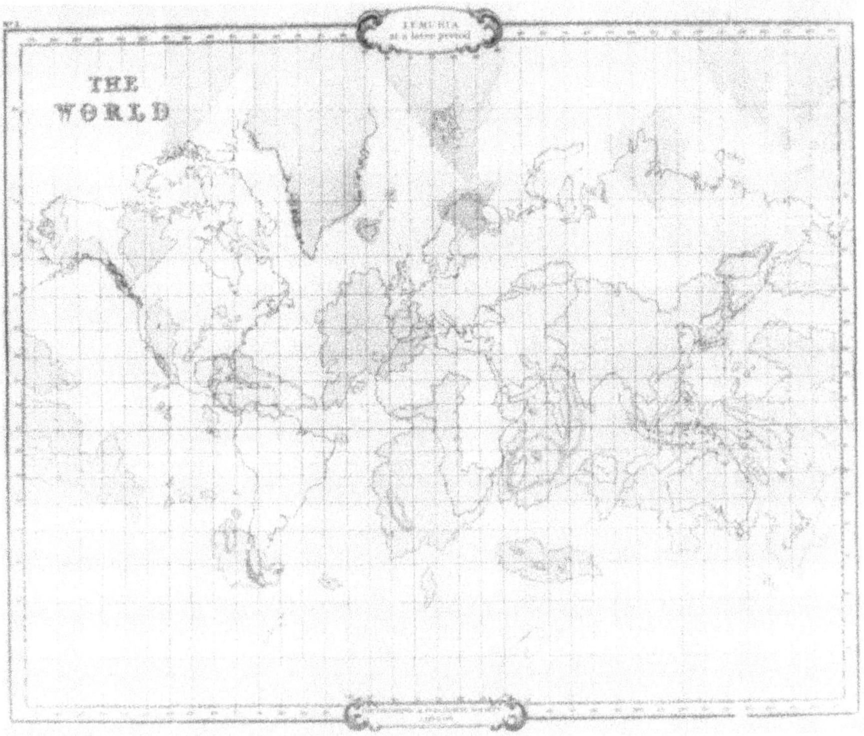

Figure 6. W. Scott-Elliot, Lemuria at a later period. The lightly shaded lands north of the 60th parallel represent the remnants of Hyperborea; all other shaded territories represent Lemuria. From Scott-Elliot, *The Lost Lemuria with Two Maps Showing Distribution of Land Areas at Different Periods*. By permission of the Theosophical Publishing House, London.

ance is not impeded by having to deal with periods removed from our own by hundreds of thousands of years.[63] Scott-Elliot himself, observing that his book had been based on occult records to which he had been privy, insisted that "it has been the great privilege of the writer to be allowed to obtain copies—more or less complete—of four of these [maps]. All four represent Atlantis and the surrounding lands at different epochs of their history."[64]

A few years later, when he published his *The Lost Lemuria with Two Maps Showing Distribution of Land Areas at Different Periods* in 1904, Scott-Elliot again reiterated, as in his earlier monograph on Atlantis, that he had obtained the two maps that were included in this work (Figs. 5 and 6) from occult records that he had been "privileged to obtain."[65] At the same time, he acknowledged that while "it was by mighty Adepts in the days of Atlantis that the Atlantean maps were produced, . . . we are not aware whether the Lemurian maps were fashioned by some of the divine instructors in the days when Lemuria still existed, or in still later days of the Atlantean epoch."[66] Because of this, he charmingly conceded, it was possible that the maps of Atlantis were more accurate than the maps of Lemuria, although none of them "were correct *to a single degree* of latitude, or longitude":

> In the former case [of maps of Atlantis] there was a globe, a good bas-relief in terra-cotta, and a well-preserved map on parchment, or skin of some sort, to copy from. In the present case [of maps of Lemuria] there was only a broken terra-cotta model and a very badly preserved and crumpled map, so that the difficulty of carrying back the remembrance of all the details, and consequently of reproducing the exact copies, has been far greater.[67]

Nonetheless, based as these maps are on "archaic originals," their transcriber "believes that they may in all important particulars, be taken as approximately correct."[68]

A few decades later, James Churchward, the Anglo-American occultist who moved Mu from the Atlantic to the Pacific and who claimed for it the status of the Garden of Eden, published a map in 1931 titled "South America showing the ancient Amazon sea and canals connecting it with the Pacific Ocean from a tablet 25,000 years old in one of the western monasteries of Tibet." The map, which charts the dispersal of the Negroid race out of Mu toward what is now South America and beyond, is based, Churchward insisted, on an "ancient Oriental Map [from] 15,000 to 20,000 years ago."[69] Similarly, when the New Age psychic Christine Hayes published her book *Red Tree: Insight into Lost Continents, Mu and Atlantis, as Revealed to Christine Hayes* in 1972, she included three roughly drawn maps (two of which showed Mu and Atlantis as they were 8,000,000 years ago and 2,000,000 years ago) which she insisted "are not artist-drawn illustrations. They are free-hand originals drawn exactly as they were revealed to the author. No effort was made to have these redrawn or touched up as we feared the author's spontaneity would be lost. Their value lies, not in being executed in a professional manner, but rather in the fact that these illustrations are totally original in concept."[70]

Hayes writes that since the time she had turned twenty she realized that she "could delve into the intricacies of my mind. . . . I am able to reach back through the hovering energy waves created by the past and witness remembrances, loves, sorrows of men and women in an age on this earth that has long passed from us. I am not only an observer but a rather shadowy participant in the customs and rituals of nations forfeited in cataclysms and ignorance long ago."[71] Her maps as well are products of these occult technologies of listening and channeling.

So, in contrast to the paleo-scientists' maps of Lemuria, which are painstakingly reconstructed from the disenchanted empirical evidence provided by rock formations and fossil finds, limited though these might be, occult cartography is based on enchanted acts of recovery of racial memories and astral sedimentation. The maps of Lemuria that occultists published had been produced in situ, so to speak, but they had either been stored in secret archives that had been preserved over the centuries after the continent drowned, or they were frozen in racial memories. Through occult strategies of listening and reading, these had been recovered and now presented to the modern public. Archaic though these maps might be in origin, they are revealingly modern in their north orientation, their conventions of naming, the shapes of today's continents that they depict, and so on. A large majority of these occult cartographies are not drawn to scale, however, and Lemuria frequently looms large in them, dominating the space of the entire map. A mere landbridge in the paleogeographer's maps, it is transformed into a territory that is ambitiously continental in occult cartography.

The first time Lemuria put in a cartographic appearance among occultists was in 1896, when Scott-Elliot published a series of colored maps on Atlantis which showed the progressive submergence of Plato's hapless land, from its greatest extent 1,000,000 years ago to its catastrophic final disappearance in 9,654 B.C. The first of these maps, entitled "Atlantis at its Prime," illustrated "the world about 1,000,000 years ago, during many previous ages, and up to the catastrophe of about 800,000 years."[72] The map illustrates Earth a million years ago. "Scattered fragments of what became the continents of Europe, Africa and America as well as remains of the still older, and once wide-spread continent of Lemuria are shown on this map. The remains of the still older Hyperborean continent which was inhabited by the Second Root Race, are also given."[73] Because Atlantis emerged out of Sclater's vanished place-world in Theosophical labors of loss, Lemuria merits a presence in the earliest maps of Atlantis, where it is shown occupying a spot "at latitude 7 [degrees] north and longitude 5 [degrees] west, which a reference to any modern atlas will show to lie on the Ashanti coast of today."[74] It was on this spur of Lemurian land that the Rmoahals, the first of the seven Atlantean sub-races, emerged in Theosophical place-making.

In 1904 Lemuria invited the Theosophists' primary attention, when Scott-Elliot published two more maps, entitled "No. 1: Lemuria at its greatest extent," and "No. 2: Lemuria at a later period" (Figs. 5 and 6). He observed that the first map illustrates the earth's configuration from the Permian through the Triassic into the Jurassic periods, while the second map represented the earth's configuration through the Cretaceous into the Eocene.[75] In contrast to many a paleo-scientist who rarely comments on the map he appends to his verbal labors of loss, Scott-Elliot eloquently describes his:

> From the older of the two maps it may be seen that the equatorial continent of Lemuria at the time of its greatest expansion nearly girdled the globe,

> extending as it did from the site of the present Cape Verd Islands a few miles from the coast of Sierra Leone, in a south-easterly direction through Africa, Australia, the Society Islands and all the intervening seas, to a point but a few miles distant from a great island continent (about the size of the present South America) which spread over the remainder of the Pacific Ocean, and included Cape Horn and parts of Patagonia.[76]

Remarkably, here, as in the description that follows of the second map, the occultist takes recourse—claims of his maps being based on "archaic originals" notwithstanding—to modern maps and atlases to substantiate the location and extent of his lost continent:

> A remarkable feature in the second map of Lemuria is the great length, and at parts the great narrowness, of the straits which separated the two great blocks of land into which the continent had by this time been split, and it will be observed that the straits at present existing between the islands of Bali and Lomboc coincide with a portion of the straits which then divided these two continents.[77]

While his first map of Atlantis in 1896 shows the remains of Lemuria, Scott-Elliot's second map of Lemuria in 1904 correspondingly shows the nucleus of what would become the large continent of Atlantis once Lemuria was drowned out of existence.

Scott-Elliot's maps have had an influential afterlife. Soon after they were published in English, they were translated into German and published by the Ariosophist Jorg Lanz van Liebenfels in his *Die Theosophie und die assyrischen 'Menschentiere' (Theosophy and the Assyrian Man-beasts)* (1907), and by Guido von List in his *Die Ursprache der Ario-Germanen (The Proto-Language of the Ario-Germans)* (1914).[78] From 1941, as we will see, they have been repeatedly reprinted—with some important transformations—in India. And in recent years they have also been republished in many freelance scholarly publications on lost continents and lost civilizations. Their enduring value is perhaps as much based on their aesthetic appeal, their concession to scientific precision in the use of the grid of latitudes and longitudes, and their conformity to scale, as in their evocative illustration of Lemuria as a vast continent, filling up the known world, once upon a time.

The only other Theosophist to draw maps was the obscure K. Browning, who in a short pamphlet entitled *Lemuria and Atlantis: Two Lost Continents* included a series of roughly drawn maps on the inside cover. Using the same sequential strategy that Scott-Elliot had deployed in his maps of the two lost continents, Browning similarly also included five maps, entitled "The World Today," "Lemuria over 1,000,000 years ago," "Atlantis 600,000 years ago," "Atlantis 100,000 years ago," and "Atlantis before the 'Flood', 9,654 B.C.," which charted the progressive loss of land over time and the reconfiguration of the earth's continents to assume their current forms and shapes.[79]

Post-Theosophical occultists like Wishar S. Cervé also make effective use of the sequential strategy in their cartographic labors of loss. His *Lost Lemuria,* published in 1931, included three maps, arranged sequentially, for a period ranging from 200,000 years ago, when the continent was at its greatest extent, to about 50,000 years ago, when it began to finally submerge (see Fig. 7). Using a partial grid (of only meridians but not latitudes), Cervé's maps first show Lemuria as a worldwide continent, extending across the Pacific and Indian Oceans about 200,000 years ago (see Fig. 7, map no. 1). It then progressively becomes a Pacific continent about 82,000 years ago as its connections with what is today Asia and Africa break (see Fig. 7, map no. 2). In map no. 3, which illustrates the world around 50,000 years ago, the continent is illustrated as "submerged Lemuria." Importantly, where Lemuria had formerly stood in the middle of the Pacific without any contact with the Americas in map no. 2, now parts of North America, especially California, are incorporated into the eastern edge of the submerging continent. As the continent finally disappears 15,000–12,000 years ago, only California remains above water, leading the author to insist: "By studying the maps in this book . . . one will see that the western portion of the United States is a remnant of the submerged continent of Lemuria. Here we have the oldest of living things, the oldest of cultivated soil, and the more numerous relics of the human race which had reached a higher state of cultural development and civilization than any other races of man."[80] A few years later, when the American occultists of the Lemurian Fellowship began publishing their maps of the former continent, once again, California and other parts of the west coast of North America are shown as part of Lemuria's eastern seaboard.[81]

Post-Theosophical occultists are also concerned with graphically illustrating their place-making contention that Lemuria was the original Paradise, the Garden of Eden where Man had first appeared and then dispersed to populate the rest of the world. James Churchward, in his multivolume documentation of the trials and tribulations of Mu, included a number of maps in which "the motherland" is shown at its greatest extent as a continent that dominates the Pacific, incorporating within its borders present-day Easter Island, the Marquesas, Tahiti, Samoa, Fiji, Hawaii, and the Ladrones. Other maps show the dispersal of various races and populations out of Mu, with arrows pointing to the directions in which they were headed to settle the entire world.[82] The occultist claim that Lemuria was inhabited by the first humans also finds expression in the maps that accompanied the many publications of the Lemurian Fellowship. These are almost the only occult maps that graphically illustrate a familiarity with the continent's topography, as mountains, rivers, and the twelve river valleys where the principal clans that constituted the Mukulian Civilization, "mankind's first and greatest," were settled.[83] The maps of the Lemurian Fellowship are unusual

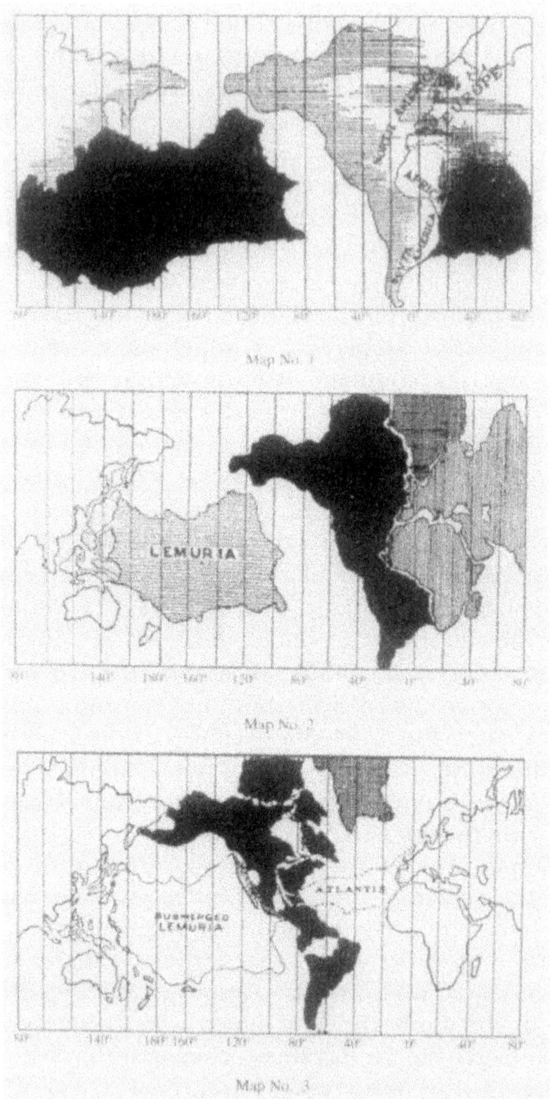

Figure 7. Wishar S. Cervé, Maps showing the shifting configurations of Lemuria. Map No. 1 is the world about 150,000 years ago; the area in black represents Lemuria. Map No. 2 shows the world around 82,000 years ago, when Lemuria was a much smaller continent located entirely in the Pacific Ocean. Map No. 3 is the world as it appeared around 50,000 years ago, when Lemuria began to submerge. Reprinted from Cervé, *Lemuria: The Lost Continent of the Pacific (with a Special Chapter by Dr. James D. Ward).* Reproduced with permission of the Rosicrucian Order, AMORC.

in this regard, for in the vast majority of occult maps Lemuria appears as terra incognita, unknown and uninhabited, even though the published place-talk in which they come embedded suggests otherwise.

Occult cartographies of loss, especially in their post-Theosophical phase, are primarily concerned with illustrating the vanished worlds of the Pacific Ocean rather than those of the entire world. Further, in contrast to the paleo-scientists' maps, these occult maps are generally Lemuria-centric. Lemuria is frequently the only paleo-continent of any consequence (although occasionally Atlantis shows up as well, vying for the reader's attention). This helps its occult place-makers to graphically illustrate one of their principal contentions, that Lemuria was the home of the world's first civilization, the motherland of all motherlands.

That occultists are victims of the modern imperative to include maps in their publications is perhaps not surprising if we recall the conflicted intimacy between their labors of loss and the physical sciences. This compels them to adopt many of the protocols of scientific cartography in creating their own maps, as well as leads them to use modern atlases and maps to bolster their fabulous place-making that putatively derives its inspiration from astral clairvoyance and telepathic readings of racial memories. Nevertheless, their implication in science and in scientific cartography has meant that they ultimately fail to represent Lemuria as Paradise, their enchanted intentions in this regard notwithstanding. There is very little in the appearance of these maps, in the iconography that they adopt, or in their symbolic makeup that would suggest that Lemuria is the lost Pacific Eden of the occultists' place-talk. As Alessandro Scafi, among others, has observed, there was a rich tradition of mapping the Earthly Paradise or the Garden of Eden in the premodern cartographic practices of Europe, until about the sixteenth century. This tradition, however, fades into oblivion with the rise of secular and scientific cartography, with its realist preoccupation with mapping the world impersonally, homogeneously, disenchantedly. Scafi thus observes that the disappearance of the Earthly Paradise from world maps "points to the shift from medieval to modern thinking, from a holistic to a fragmented view of reality, from a mapping which sought to penetrate the mystery of the whole universe beyond human boundaries to a mapping which is constrained strictly within the frameworks of analytical thought and Euclidean geometry, and from cosmography to geography."[84]

The Earthly Paradise was not the only victim of this shift. In medieval Europe's *mappae mundi* the center was unambiguously located at Jerusalem, which provided the key orientation to these maps. In the new world maps that came to be drawn, from the mid-sixteenth century, based on Ptolemaic geometry and Mercator's projection, the central focus is on the equator at zero. This "carries no suggestion of 'home,'" observes John Gillies, who goes on to note that "a modern refusal to privilege a sacred omphalos defined the very possibility of [the] new geography."[85] Following Foucault, Gillies usefully suggests that the difference between the medieval and modern map lies in the fact that the former "is a space of 'emplacement' characterized by 'localisation,' the other 'an open space' characterized by 'amorphic extension.' "[86]

However much they might be interested in recuperating archaic traditions and primeval pasts, modern occultists are primarily inhabitants of a life-world in which this paradigmatic shift has taken root. Even as they resist the progressive de-theizing of Earth, they are at the same time dependent on modern scientific cartography with its impersonal and homogeneous Euclidean grid that enables them to graphically represent Lemuria as real and true. Caught in this bind, occult cartography misses the opportunity to present the lost continent as the sacred Paradise of their place-talk. A big gap opens up between verbal and cartographic labors of loss, a gap that is ultimately incapable of being bridged. Occult cartographies are ultimately not successful because nowhere in their maps does Lemuria appear as modernity's new Paradise, its promised Eden.

MAPPING KUMARIKKAṆṬAM

Maps of Lemuria begin to be drawn last of all in the Tamil country, but the cartographic labors here are also the most spectacular in their diversity, in the intensity of their detail, and in their investment in visualizing loss. To recap, the maps of the paleo-scientist are principally concerned with visualizing a vanished Mesozoic time when the earth's surface was configured differently from what it is today. The occultists, especially in the post-Theosophical phase, are primarily focused on graphically relocating Lemuria to the Pacific and in illustrating it as the submerged motherland, the Paradise from where the rest of the world was settled. Tamil cartographic labors are invested, however, in visualizing Lemuria as a lost *Tamil* place that was also the ancestral homeland of the Tamils, from where they fanned out to populate Earth. The graphic illustration of Lemuria as an intimate Tamil home-place that had once existed but, alas, no more—this is the challenge that faces Tamil cartographies of loss.

The first attempt to cartographically visualize the loss of Tamil territory can be dated to late 1916, when a map was published by a Tamil scholar named S. Subramania Sastri in *Centamiḻ*, a leading Tamil literary journal and a forum for the expression of many a Tamil devotional idea from 1902, when it first appeared.[87] Interestingly, this map was drawn not to lend support to the idea of an ancestral Tamil homeland that had extended far south of present-day Kanyakumari, but to disenchantedly attack the extravagance of such claims. So, Tamil maps of Lemuria begin on a note of historicist critique rather than one of devotional celebration. The ostensible aim of

Subramania Sastri's map is to cartographically recast the literary information provided by the medieval commentator Adiyarkunallar regarding the extent of the antediluvian territory that was lost to the ocean's fury, as well as its topography. In true historicist fashion, Subramania Sastri's intent, however, appears to have been to use the map form to question the labors of loss undertaken by the venerable doyen of Tamil literary studies, Arasan Shanmugham Pillai (1868–1915), who a few years earlier, in 1905, had insisted that the land that drowned (which he, like many other Tamil placemakers, named "49 Tamilnāṭu") amounted to thousands of miles.[88] Subramania Sastri, on the other hand, contended that barely the equivalent of a modern *tāluk* (a subdivision of a district, generally not larger than a few hundred square miles) disappeared into the ocean.[89] The "map" *(paṭam)* that Subramania Sastri appended to illustrate the validity of his measurements over those of Shanmugham Pillai's is really a schematic diagram, although it follows the protocols of modern cartography in its north orientation, its representation of southern India as a peninsula, and in its use of standard cartographic symbols to represent rivers and mountain ranges. The diagram shows two nested territories, both curiously peninsular in shape. The smaller of the two territories, left unnamed, represents the antediluvian extent of the Tamil country as Adiyarkunallar intended, according to Subramania Sastri. The river Kumari flows just a little south of present-day land's end, constituting the border of the antediluvial territory that was lost to the ocean. In contrast, the larger of the two territories—identified as "49 Tamilnāṭu"—extends far to the south of Kanyakumari to a distance identified as "700 *kāvatam*." This territory represents Shanmugham Pillai's extravagant (mis)interpretation of Adiyarkunallar's more modest claims, in Subramania Sastri's reading.

There are a number of problematic elements in Subramania Sastri's map, including the absence of scale that would have enabled the reader to graphically compare the extent of the antediluvial land as specified in these rival cartographic labors of loss. Also, tellingly, the shape of the antediluvial territory (as imagined by Shanmugham Pillai in Subramania Sastri's reading) is a literal re-creation of the peninsular form of the present-day Indian subcontinent.[90] These notwithstanding, it is important to note Subramania Sastri's early effort to translate Adiyarkunallar's medieval literary imagination onto the modern map form, and to cartographically visualize the loss of Tamil land. This preoccupation with cartographically demonstrating the loss of the ancestral homeland only picks up momentum in the subsequent decades.

At a very basic level, this preoccupation is apparent in the very titles chosen for the Tamil maps by their makers. As in Euro-America, several maps of Lemuria remain untitled, but many others are suggestively named to draw attention to the loss of Tamil land to the ocean in a manner that does not

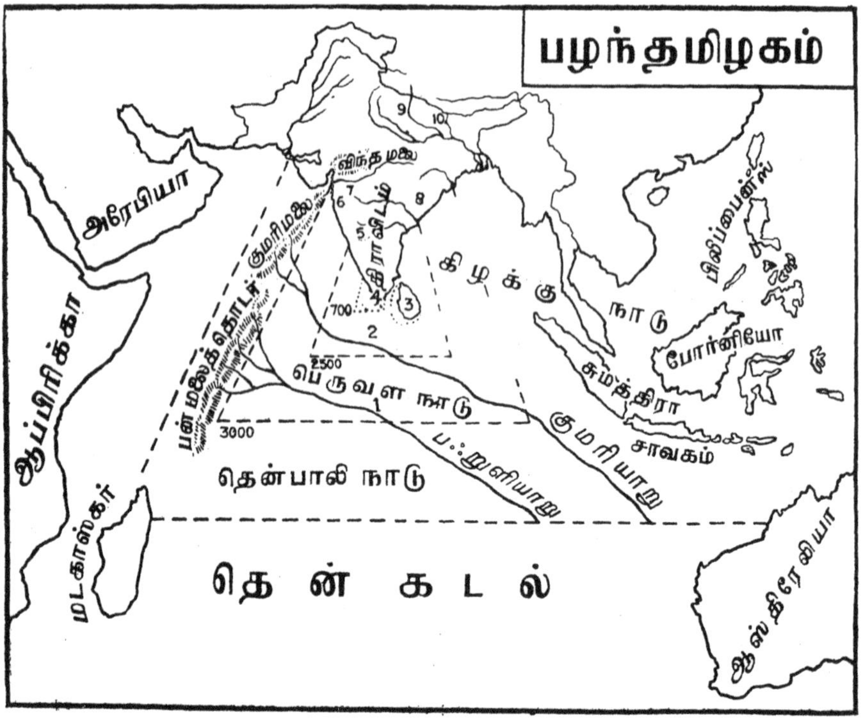

Figure 8. Pulavar Kulanthai, Pa<u>l</u>aiya Tami<u>l</u>akam [Ancient Tamil Home-place]. This map, in Tamil, shows the modern Indian coastline superimposed on the ancient submerged landmass of Lemuria (designated here by the outermost dotted lines), reaching at its greatest extent to Madagascar in the west and Australia in the east. The dotted lines marked 3000, 2500, and 700 represent the progressive loss of land to the ocean. Callouts 1 and 2 represent the lost cities of To<u>n</u>maturai and Te<u>n</u>maturai on the banks of the rivers Pah<u>r</u>uḷi and Kumari, whose sources lay in the mountain ranges along the western edge of Lemuria. From Pulavar Kulanthai, *Irāvaṇa Kāviyam.* Reproduced with permission of Vela Patippakam, Erode.

readily happen in the paleo-scientists' or the occultists' cartographic labors. So, one of the earliest maps of the antediluvial homeland, published in 1927, is captioned "Puranic India before the Deluges" and "Puranic India after the Deluges."[91] A few years later, Pulavar Kulanthai's map of 1946 was entitled "Pa<u>l</u>aiya Tami<u>l</u>akam" (Ancient Tamil Home-place) (see Fig. 8).[92] Similarly, N. Mahalingam's English map of 1981 is labeled "India in 30,000 B.C.: Tamilnadu" (see Fig. 9).[93] More recently, the title of a Tamil map of 1995, "Kumarikkaṇṭam Seized by the Ocean,"[94] echoes that of a 1977 map, in Tamil, published by R. Mathivanan, which is called "Kumarināṭu Seized by the Ocean,"[95] and resonates with the name of another map, more recently

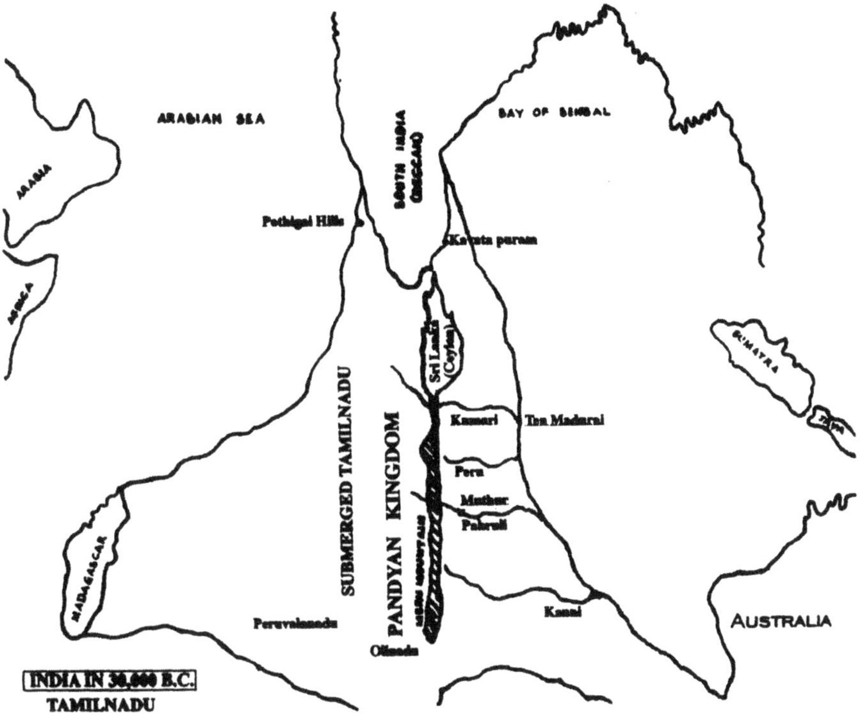

Figure 9. N. Mahalingam, India in 30,000 B.C.: Tamilnadu. From Mahalingam, *Gems from the Prehistoric Past.* Reproduced with permission of N. Mahalingam.

authored by Mathivanan in 1991, called "Vanished Lemuria Continent" (see Fig. 10).[96] A map of 1992, published in Tamil, is named "The Lemuria Continent which was destroyed by the ocean,"[97] while a map, in English, of 1993 is called "Lemuria or The Lost Kumari-Continent or Navalam Island," its all-encompassing title revealing the three principal sources that feed Tamil labors of loss—European science, Tamil literature, and the Sanskritic *Purāṇas*.[98] Thus, Tamil cartographies of loss are forthright in their announcement of the principal place-making claim of Tamil devotion: that there was an ancient Tamil land that was lost in floods caused by the ocean's fury.

They are also explicit in drawing attention to the fact that Lemuria was no impersonal or homogenous paleo or occult place-world, but a *Tamil* homeland, by the naming operations they perform on the land illustrated in their maps. So, while some Tamil maps do label the lost land as "Lemuria" (or its Tamil incarnation, "Ilemūria,"), or even "Gondwanaland," a majority use the putative Tamil toponyms of Kumarināṭu or Kumarikkaṇṭam. Sivasailam Pillai's textbook map of 1951 (see Fig. 4) even identifies

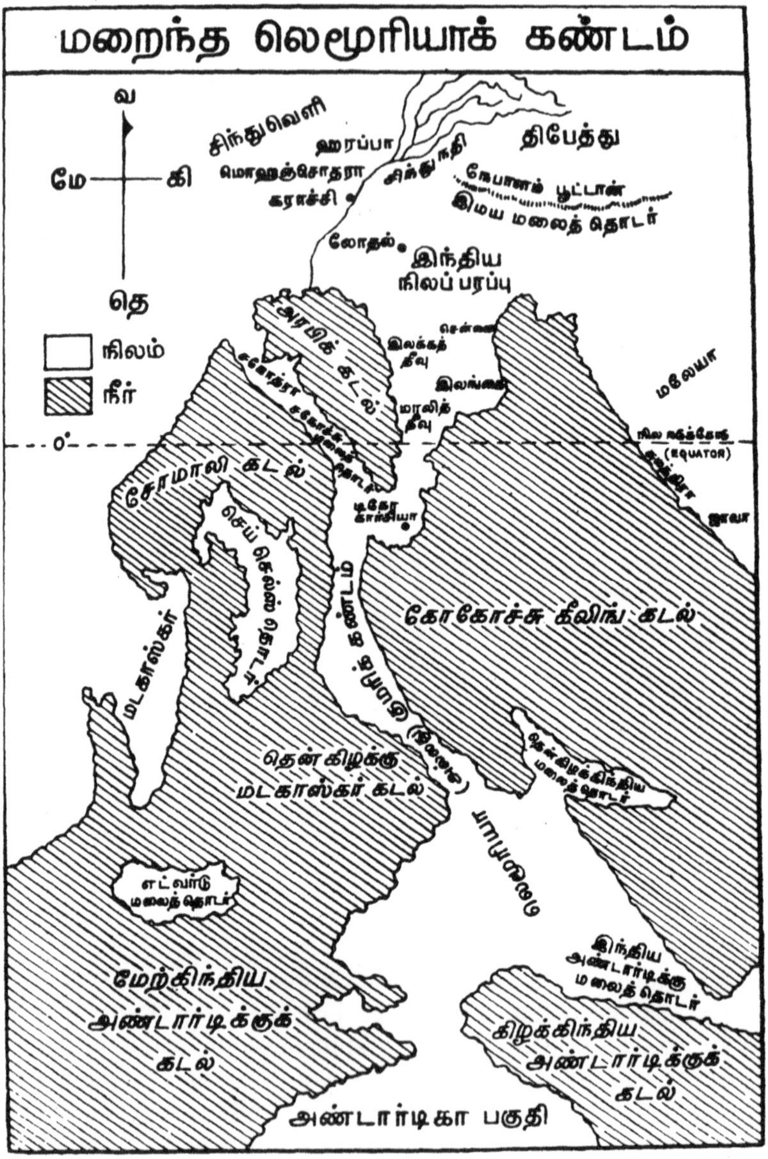

Figure 10. R. Mathivanan, Maṟainta Lemūriak Kaṇṭam [Vanished Lemuria Continent]. This map, in Tamil, shows the reconstructed Indian peninsula to the north, with a landbridge connecting to the large landmass of Lemuria in the south, terminating in Antarctica. The shaded portions represent the oceans surrounding Lemuria. From Ramachandran and Mathivanan, *The Spring of the Indus Civilization.*

the hapless land as "Kumarikkaṇṭam seized by the ocean."[99] In maps published in the 1940s and 1950s—when the extent and limits of the lived homeland of Tamilnadu were being actively debated in the context of the creation of new states within the new Indian union—the lost homeland is strategically identified as "Tamiḻnāṭu," "Tamiḻakam," and even, in one case, as "namatu nāṭu" (our nation).[100] Maps published in more recent years have also named the mapped territory as "Pandyan Kingdom" (see Fig. 9), and as "Pandya realm seized by the ocean," cartographically connecting the lost homeland to the name and memories of one of the Tamil country's most ancient dynasts.[101] Despite the anti-Sanskritic Tamil nationalist milieu in which some maps are drawn, even the Puranic name for the subcontinent, Jambudvīpa, puts in an appearance in its Tamil manifestation, "*nāvalantīvu*" (island of the rose apple).[102]

Many scholars have demonstrated that the cartographic naming of territory is not a neutral, innocent process but has enormous political and ideological consequences.[103] In European colonial maps of the non-Western world, not only are native place-names frequently misspelled, but often they are totally replaced by new Western toponyms. As Harley documents, through the imposition of European place-names over other territories land is claimed from indigenous populations and reordered in accordance with European spatial and cartographic logic. The complicity of the map in the appropriation of other lands is consistent across the colonial world.[104] Through such acts of cartographic appropriation, native place-names are "written out of history," and, in turn, natives "became aliens in their own land."[105] In response, one of nationalism's most important gestures is to cartographically replace such colonial toponyms with native names, many of which are frequently neologisms. The renaming of Lemuria as Kumarikkaṇṭam or Kumarināṭu in Tamil maps is very much an illustration of this cartographic version of the invention of tradition. Each such act of (re)naming constitutes a form of "toponymic possession" and "territorial consecration" through which a new place called Kumarikkaṇṭam is cartographically incorporated into a Tamil horizon of meaning and memory.[106]

That this act of cartographic (re)nomination is clearly intended to (re)claim Lemuria for a Tamil project of loss is clear from three striking examples. In 1941 K. Appadurai (1907–89) published the first monograph in Tamil on Lemuria, revealingly titled *Kumarikkaṇṭam allatu kaṭal koṇṭa teṉ-ṉāṭu (Kumarikkaṇṭam, Or the Southern Land Seized by the Ocean)*, a book that continues to be in print still today.[107] Although the bulk of the book is based on the life and times of the Tamils of Kumarikkaṇṭam, which is, in turn (bizarrely enough), modeled on the life and times of Cervé's Pacific Lemurians (a fact left unacknowledged by the author), the two maps included in the monograph are derived from the Theosophical cartographies of Scott-Elliot. Entitled "Ilemūria: Paṇṭai Nilai" (Lemuria: Ancient State) and

"Ilemūria: Kuṟukiya Nilai" (Lemuria: Shrunken State), they are near-translations into Tamil of Scott-Elliot's 1904 world maps that showed the giant continent of Lemuria sprawled virtually across the entire earth. But there is one important difference. The continent—which is left unlabeled by Scott-Elliot—is now explicitly identified as "Kumarikkaṇṭam," the Tamil orthography boldly etched across the part of the map that lies directly to the south of present-day India. And it is in this form that Scott-Elliot's original occult maps have circulated in Tamil India since the 1940s, as Appadurai's cartographic (re)creations have been reprinted several times, including in the 1975 college textbook published by the state government that I discussed at the end of the previous chapter.[108]

Another example of cartographic appropriation for the cause of Tamil labors of loss is a recent map that appears in an obscure introductory book on Tamil culture and history called *The Land and People of Tamilnadu: An Ethnographic Study*, by B. S. Chandrababu (1996). The by-now mandatory, albeit routinized, nod toward Lemuria is accompanied by a map (in English) showing the familiar outline of the Indian peninsula bounded on the north by an unnamed sea, further north of which are the edges of two land formations named "Europe-Syria" and "Siberian Asia." India, parts of the Arabian peninsula, and the east coast of Africa however, are joined in a giant territorial formation that is boldly labeled "Kumari Continent." The map is not drawn to scale and does not use the grid of latitudes and longitudes, so it is difficult to measure the extent of the continent or to pin down its exact location, but Kumari occupies the entirety of the map.[109] What makes this map interesting for the purposes of my argument is that it is a rough reproduction of an original that was first published to illustrate the exciting underwater revelations of the Anglo-Egyptian John Murray Expedition to the Indian Ocean, which mapped some 22,000 miles of the ocean floor (1933–34).[110] The original map, however, identified the land that Chandrababu's map names "Kumari Continent" as "Gondwanaland," and it is clear that its author (Professor J. Stanley Gardiner) saw this as a Jurassic landmass teeming with dinosaurs and other prehistoric fauna, a far cry from the heroic Tamil ancestors of Chandrababu's fabulous imagination. The London-based *Daily Mail* carried the following description, based on an interview with Gardiner: "Gondwanaland belonged to the reptilian period, and was the home, no doubt, of monstrous scaly reptiles. . . . The 10,000-ft high ridge, which runs south-west towards Socotra, was obviously a continuation of the Aravalli range (in Ajmere, India) and other mountains. There is a deep gully to the south-east, and it seems certain that it formed, in ages long ago, an extension of the bed of the River Indus. One can only deduce . . . that the whole vast tract of land, and part of the Indus, went down head-first, so to speak, in a tremendous volcanic upheaval."[111] This attractive map—accompanied by illustrations of typical Mesozoic flora and fauna—was subsequently reprinted in the *Illustrated Weekly of India* on July 29, 1934. It was reproduced, first in 1940 by Devaneyan Pavanar, and then in Tamil in 1975 in the government-published college textbook on Tamilnadu prehistory, in both instances *without* the identification of the submerged continent as Gondwanaland. Chandrababu's republication in 1996 of the map, with its unambiguous identification of the now-vanished landmass as Kumari Continent, completes the process by which a paleo place-world of European science—Suess's Gondwanaland—has been appropriated into Tamil regimes of remembering loss.

My final example of claiming through (re)naming is the famous map of the Indian Ocean floor that was originally published in October 1967 in *National Geographic,* a magazine that has considerable circulation in India, especially among its English-educated reading public. Simply titled "The Indian Ocean Floor," the map was conceived by Nuerman Bunstead, based on bathymetric studies of the Indian Ocean floor by Bruce C. Heezen and Marie Tharp of the Lamont Geological Observatory in New York City. The published map, which was based on a painting of the Indian Ocean floor by the Austrian landscape artist Heinrich Berann, was commissioned by the magazine to illustrate an article on the multinational, multiship International Indian Ocean Expedition, which mapped the floor of the Indian Ocean in 1959–65.[112] The map—as indeed the findings of the expedition, which were later published[113]—revealed a spectacular series of underwater ridges and mountains running down the western half of the Indian Ocean in an inverted-T formation. An essay subsequently published in *National Geographic* and evocatively titled "Indian Ocean Unveiled on a Dramatic New Map: Science Explores the Monsoon Sea" noted of the map:

> From [Berann's] painting, in high relief, jump features of earth's wrinkled face that were unknown a few years ago: submarine mountain ranges, mid-ocean "micro-continents," undersea river valleys and plains, and jagged ridges and clefts in the sea bottom that support startling new concepts about earth's geologic past. The over-all view is much as if an astronaut, orbiting far out in space, were to look down upon a world drained of all its water.[114]

The map also added to the emerging consensus around plate movements and the drift of continents, and put to rest any residual notion of vast drowned continents like Lemuria among metropolitan scientists.[115] In the Tamil country, however, among some of its place-makers, the map only rekindled and reconfirmed their belief in a drowned Kumarikkaṇṭam now lying at the bottom of the Indian Ocean.[116] Not surprisingly, when the state government commissioned its documentary film on Lemuria in 1981, the map's underwater mountains and ridges were featured, and identified as the former "Lemuria Continent" and as "Kumarikkaṇṭam." A recent school atlas in Tamil has even reprinted this iconic map with a significant difference: the underwater ridge immediately south of Kanyakumari is unambiguously identified as "Kumarikkaṇṭattoṭar" (Kumarikkaṇṭam range).[117]

In addition to claiming through renaming, another cartographic operation through which an impersonal and homogeneous Euro-American place-world called Lemuria is converted into the intimate Tamil homeland of Kumarikkaṇṭam is the use of Tamil orthography. It is true that the occasional paleogeographical map of Gondwanaland circulates as such—without any modifications—in Tamil publications.[118] And in recent years even metropolitan maps illustrating the drifting apart of continents, following the theory of plate movements, have been reprinted in Tamil books on Lemuria, although the theory disavows earlier, fixist conceptions regarding drowned landbridges and submerged continental formations.[119] Maps of Lemuria in English, as well as bilingual maps (in English and Tamil) have also been published in south India.[120] Nonetheless, the majority of Tamil cartographic labors of loss over the drowned Lemuria are, appropriately enough, in Tamil. The use of Tamil orthography has meant, first of all, that the maps of the Euro-American paleo-scientist and occultist, when reproduced in the Tamil country, are almost the same but not quite.[121] The use of Tamil orthography—even if only to respell Lemuria as "Ilemūria"—has meant that Sclater's vanished place-world is now accommodated within a Tamil horizon of meaning and memory. Further, the very translation into Tamil begins to undermine the representational authority of maps of Lemuria produced in Euro-America that had hitherto been encoded in English or other European languages. But above all, the use of Tamil orthography has meant that only readers who are conversant in Tamil can understand maps of the disappeared Tamil ancestral home-place. While this may set limits on the consumption of Tamil cartographies of loss by non-Tamils, it also means that a closed and intimate Tamil community united around a shared labor of grief concerning the submerged homeland is produced through modern mapping practices.

The intimacy of Kumarikkaṇṭam is also intimated by Tamil place-makers through the cartographic operation of filling up the empty spaces of Euro-American maps with rivers, mountains, regions, and cities, the names of which are derived, to recall, from both ancient Tamil literature and the Sanskrit *Purāṇa*s. Neither fixity nor precision, however, constrains the graphic representation of this fabulous topography of loss. The antediluvian rivers Kumari and Pahṟuḷi appear as early as 1916 in Subramania Sastri's map, flowing outside the boundaries of present-day India but helping to define the limits of the lost homeland. These rivers become a standard feature of almost all maps published by Tamil place-makers, and by mid-century are supplemented by others, such as the Kaṉṉi and the Pēraṟu (see Figs. 4, 8, and 9).[122] Similarly, the antediluvian mountain called Kumarikkōṭu also appeared as early as 1916, and in maps published especially after the 1940s, elaborate mountain ranges (such as Paṉmalaitoṭar or Kumari Malai) are shown running along the western edge of the lost homeland (see Figs. 4 and 8).[123] Even the axial Puranic mountain Meru puts in an appearance in the occasional map, its Sanskritic associations notwithstanding (see Fig. 9).[124] The ocean, the principal villain of Tamil labors of loss, comes to also be identified and named, beginning with Purnalingam Pillai's 1927 maps in English which show "The Eastern Sea," and "The Southern Sea" as bounding the antediluvial land. Pulavar Kulanthai's Tamil map of 1946, too, identifies Teṉ Kaṭal, "the southern sea" (see Fig. 8). Later, other seas find cartographic accommodation, as in Chidamabaranar's 1948 maps which identify the "Toṉṟu Mutir Kaṭal" (Primeval Ocean) to the northwest and "Toṭukaṭal" (Adjacent[?] Sea) to the northeast of the former homeland. Mathivanan's 1991 map offers the most elaborate geography of the fabulous oceans surrounding the lost Kumarikkaṇṭam in its naming of the Somali Sea, the Southeastern Madagascar Sea, the West Indian Antarctic Sea, the Eastern Indian Madagascar Sea, and the Keeling Sea (see Fig. 10).[125]

Tamil cartographies of loss, however, are not merely content with identifying the topography of the lost homeland; they also underscore that Kumarikkaṇṭam had been an inhabited place, once upon a time. They do so by naming the various regions where the ancestral Tamils had formerly lived, beginning as early as 1916 when Subramania Sastri isolated the antediluvial region of Teṉpālimukam in his map. Purnalingam Pillai's 1927 map, "Puranic India before the Deluges," showed an intricate network of regions with names such as Viramahendrapuram, Mathurai Nadu, and Yema Puram, drawn from a mélange of premodern Tamil and Sanskrit literary works.[126] Pulavar Kulanthai's 1946 map shows the entire antediluvian land as divided into three zones, the southernmost of which is Teṉpālimukam, north of which lies the fertile Peruvaḷa Nāṭu, extending between the Pahṟuḷi and the Kumari Rivers. To the northeast lies Kiḻakku Nāṭu ("the eastern territory") (see Fig. 8). A few years later, in one of Chidambaranar's 1948 maps, Peruvaḷa Nāṭu reappears, although flanked by two other regions named Oḷināṭu and Kumarināṭu. The most elaborate geography of lost regions is provided in R. Mathivanan's 1977 map—which also reappears in the government's 1981 documentary film—in which the "49 Nāṭu" of the medieval commentator Adiyarkunallar find elaborate spatialization and cartographic visualization. Finally, the principal antediluvial cities of Kumarikkaṇṭam, where all the action was—Teṉmaturai and Kapāṭapuram—are also variously marked on the maps, beginning with Pulavar Kulanthai's in 1946 (see Fig. 8).

In a provocative analysis of the European mapping of Australia as a vast empty space, Simon Ryan notes that "the cartographic practice of representing the unknown as a blank does not simply or innocently reflect gaps in European knowledge but actively erases (and legitimizes the erasure of) existing social and geo-cultural formations in preparation for the projection and subsequent emplacement of a new order."[127] Politically, he writes, world maps of the so-called European age of exploration and since, with their vast blank spaces, acted as "incitement to the alteration of ownership." Blank space intimates that "there has been no previous history, but also teleologically constructs the future as a place/time for writing."[128] If Euro-American paleogeographic maps of Lemuria, with their blank spaces, suggest that the continent was uninhabited by humans or without any lived history, the Tamil maps of the vanished place-world intimate the opposite. Indeed, K. Appadurai's 1941 maps of the continent, which, to recall, were almost identical to Scott-Elliot's Theosophical cartographies, clearly note that the shaded part representing Lemuria was "inhabited by people." Even more dramatically, in the maps used in the 1981 documentary film, the camera passes in detail over the various named regions of Kumarikkaṇṭam, and finally zooms in on its capital city of Teṉmaturai, cinematically depicted as thronging with people and resounding with the song of Tamil. So, for Tamil cartographic labors of loss, Lemuria is no terra incognita. Rather, through the well-established cartographic practice of naming and claiming, the uninhabited Euro-American place-world of paleogeography is transformed into Tamil territory, the foundational homeland of Tamil literature, namely, Kumarikkaṇṭam. Following Ryan, I, too, therefore suggest that the Euro-American blank maps "incite" and license Tamil place-makers to fill up their empty spaces with Tamil regions, rivers, mountains, and cities.

Further, by inserting names drawn from ancient literature—primarily Tamil, but also Sanskritic—into the formal frame of a modern scientific map, Tamil cartographies of loss also suggest that "native" geographical knowledges and conceptions of land, as embodied in textual traditions, have the right to share cartographic space with hallowed Western knowledges based on modern empirical science. In their very attempts to bring together radically different, even conflicting, conceptions of space and views of the earth within a single frame, these Tamil maps are resistant cartographies, refusing to participate, as normative maps (both modern Western and Indian) routinely do, in the delegitimation of ancient and archaic geographical conceptions disavowed as fabulous in scientific modernity. In this regard, these maps offer a good example of what Walter Mignolo has characterized in his comparable discussion of Amerindian maps as a confluence of "co-existing territorialities."[129]

As important as the goal of demonstrating that Kumarikkaṇṭam was no terra incognita, but rather an inhabited and intimate Tamil place, is the imperative to graphically illustrate the enormity of loss suffered by the Tamil people when the ancestral homeland finally disappeared into the ocean. This imperative is met by drawing maps in which Kumarināṭu is invariably shown as a large landmass that extends far south of (British) India, expanding it enormously. In contrast to the maps of the paleoscientists and some maps of the occultists, Tamil cartographies of loss are rarely drawn to scale, nor do they conform to the geometrical grid of latitudes and longitudes.[130] They thus have a lot of freedom in giving full rein to the size of the lost homeland. This was certainly the case with the very first map of the antediluvian land, which was published in 1916 by Subramania Sastri, and this is especially true of most subsequent maps, in which the lost homeland invariably looms far larger than the lived homeland of India/Tamilnadu, covering almost the entirety of the southern hemisphere in some cases, and extending, in one striking instance, all the way to Antarctica (see Fig. 10).

Even as the lived homeland is shown much smaller than the lost homeland, these maps—as indeed the place-talk around Lemuria in the Tamil country—radically decenter "India," which is frequently placed at the vanishing point of the map—there, but not quite. The relative insignificance of India, as compared to the real importance of Kumarikkaṇṭam, is graphically reiterated in maps that show the present-day subcontinent as just a region of the lost homeland extending far to its south,[131] or that show the nation-state as a northern supplement to the vast elsewhere that occupies the center of the map (see Fig. 9).[132] As Walter Mignolo, following Rudolf Arnheim, has noted in another context, a commonplace strategy adopted by contestatory cartographies—especially subaltern mapping practices that take on the power of established states—is to rewrite the center, displacing it from its familiar place. The very act of relocating the center mobilizes alternate readings of the space occupied by contending and competing territorialities.[133] In Tamil cartographies of loss, the center is shifted to the elsewhere of Lemuria, whose space is invariably filled in with the names of antediluvial rivers, mountains, regions, and cities. The space outside and beyond the lost homeland, however, is left blank, its otherwise familiar geography (known to us from modern maps and atlases) erased and displaced by the geography of elsewhere (see, e.g., Fig. 9). To recall, in modern cartographic practices, blankness suggests unknown and even uninhabited spaces—terra incognita. Thus, frequently, in these maps it is Lemuria that is the known and inhabited territory—and the center of the known world. India, on the other hand, not to mention the rest of the world, is reduced to blank nothingness. All the same, because India and one of its regions, the Tamil country, is still the lived homeland of Tamil speakers, it cannot be entirely done away with. The compromise finds visual representation in India occupying the vanishing point of the map—there, but barely so. The (Tamil) reader of the map is always thus reminded that this is what postdiluvial Tamil speakers had to live with, after the loss forever of their beloved ancestral homeland.

The radical decentering of India (which from the Indian nationalist perspective ought to be perceived as the true homeland of all loyal citizens) is also accompanied by the dissolution of its known boundaries. So much so that in Purnalingam Pillai's map from 1927, as in Mathivanan's 1991 map (see Fig. 10), India is barely recognizable, its familiar cartographic outline (familiar, that is, to all those who have looked long enough at modern maps, globes, and atlases) totally undone. The clear delineation of boundaries is one of the primary agendas of modern state cartography, if not its sine qua non. As Thongchai Winichakul writes perceptively, "Boundary lines are indispensable for a map of a nation to exist—or, to put it in another way, a map of a nation presupposes the existence of boundary lines."[134] In the subcontinent, both the British colonial and the Indian national state have been dedicated to the boundary-making enterprise over the past two centuries, producing maps that have transformed India from inchoate "Hindu" space into a known and identifiable geo-body conforming to the precisely surveyed and measured Euclidean place of modern scientific cartography as administered by the Survey of India.[135] Yet, Tamil cartographic labors—even when performed by those whose primary political sympathies lie with India—question the very territorial integrity of the nation-state, and point instead to a time when its present, and even historical, boundaries had not mattered at all, or even existed. Even more subversively, they suggest that India's present boundary circumscribes a territory that is far more limited in area and reach, in contrast to the ancestral Tamil homeland, which reaches, in several instances, all the way to Australia and to (but not including) Africa, and in one instance, even to Antarctica (see Figs. 9 and 10). Once again, the modern Tamil reader of these maps is reminded that his lived homeland is a mere remnant—and a paltry one at that—of a much larger lost homeland, a global continent no less.

Finally, the enormity of loss is also cartographically produced by leaving the borders of the lost homeland undefined and fuzzy, thus graphically reiterating the vastness of its extent—as if it stretched indefinitely over the entirety of earth's surface. A striking example of this is a map that appeared in 1945 in Kandiah Pillai's book, *Namatu Nāṭu (Our Nation),* intended for young children. The shaded parts of the map, named "Ancient Nāvalantīvu or Lemuria," represent the antediluvian homeland of Kumarināṭu, which virtually covers the entire southern hemisphere. While its northern boundaries are delineated, the southern limit is left unspecified. Similarly, in Sivasailam Pillai's 1951 school map, the limits of the ancestral homeland remain unspecified, in contrast to the clearly delineated postdiluvian Tamil country (see Fig. 4).

By the time maps of Lemuria began to be published in Tamil-speaking India from 1916, Tamil place-makers who were cartographically literate would have been familiar with the modern mapped image of the subcontinent, which they would have encountered in various geography and history textbooks from the later decades of the nineteenth century, not to mention various Tamil devotional tracts and publications from the early years of the twentieth century. They would also have been familiar with the relatively small space occupied by the Tamil-speaking region within the entity called "India" on such maps. Such a cartographic image, based though it may be on scientific surveys, starkly contrasted with the useful truths of ancient Tamil literature (that modernity had helped salvage and popularize), which intimated that in the distant past a vast landmass had stretched far to the south, where learned Tamil academies had flourished before being swallowed up by the ocean. So the normative cartographic image of India, scientific though it may be, had to be contested if the wisdom of ancient Tamil literature—so necessary for Tamil devotion, Tamil nationalism, and Tamil modernity—was to survive instead of being dismissed as naive and false, or fabulous. From this point of view as well, the Tamil maps of Lemuria are resistant cartographies, hybrid formations which use the cover of the prestigious scientific paleographical knowledges of the modern West to update the geographical truths salvaged from ancient Tamil literature. Pace Brian Harley, who contends that "maps are preeminently a language of power, not of protest," and that "the social history of maps . . . appears to have few genuinely popular, alternative, or subversive modes of expression,"[136] I suggest that the Tamil maps of Lemuria, produced on the margins and in the interstices of global science, are very much expressions of protest over and resistance to the dominant ways of representing the Tamil-speaking region in the normative cartographies of both the British empire *and* the Indian nation-state.

In addition to being subversive and resistant, Tamil maps of Lemuria are, above all, examples of *catastrophic* cartographies, for their ultimate goal is to demonstrate the catastrophic loss of land.[137] And not just any land, not just the impersonal, homogeneous place-world of paleo and occult cartographies—but the ancestral birthplace of the Tamil community. This is where the cartographic labors of loss of Tamil devotion are at their fabulous best, as they masterfully redeploy both the synoptic and sequential strategies of Euro-American maps for the Tamil cause.

The synoptic strategy is well illustrated by Pulavar Kulanthai's 1946 map, which juxtaposes the modern cartographic contours of "India" with a vast, unbounded space to the south, filled with the Tamil names of regions, rivers, and mountains. The progressive loss of Tamil territory over time—figured around 3031, 2531 and 731 B.C.E.—is also indicated on the map with the help of dotted lines (see Fig 8).[138] Sivasailam Pillai's schoolbook map of 1951 synoptically juxtaposes a vast "Kaṭalkoṇṭa Kumarikkaṇṭam" (ocean-seized Kumarikkaṇṭam) with a much-smaller, postdiluvial Tamil realm, itself a minor part of the Indian subcontinent (see Fig. 4). In

Mahalingam's 1981 map entitled "India in 30,000 B.C.: Tamilnadu," post-diluvial India is explicitly juxtaposed against a much larger, "submerged Tamilnadu" that extends all the way to the island of Madagascar in the west and Australia in the east, with its southern borders left unspecified (see Fig. 9).[139] In all these cases, the lost homeland is placed side by side with the lived homeland, inviting the reader to contemplate and ponder—and grieve over—the disappearance of territory.

Sequential maps showing the progressive loss of land over time began to be published from 1927, when M. S. Purnalingam Pillai drew a series of three maps entitled: "Puranic India before the Deluges" (which shows a vast peninsular entity stretching far south into the Indian Ocean); "Puranic India after the Deluges" (which shows a much more shrunken India, with fragments of the former landmass scattered in the ocean); and finally, "The Tamil country in 1st century A.D." (which shows a limited and fixed place confined to a small region of peninsular India). Similarly, Chidambaranar published a series of maps in 1948 which illustrated the loss of ancestral Tamil land over a period ranging from 350 million years ago to 3105 B.C.[140] A variation of this strategy is the publication of twin maps, the former showing Kumarikkaṇṭam at its vastest, the latter showing the lived homeland as a shrunken continent. K. Appadurai deployed this strategy to great effect in 1941, as did one Rajasimman, whose brief essay on Kumarikkaṇṭam in 1944 included two maps, printed side by side and suggestively titled "After seizure by the ocean" and "Before seizure by the ocean."[141]

Many maps of Lemuria published in Euro-America also invite the reader to consider the loss of territory as the continent shrinks over time to vanish forever. What makes the loss of territory in the Tamil maps seem so immediate and intimate is the fact that the cartographic labors of loss of Tamil devotion had, from the time they first started, converted the impersonal place-world of Lemuria in Euro-America into a familiar Tamil home-place. The poignancy of the loss is further underscored by maps that showed the dispersal of the original inhabitants of the ancestral homeland. The first of these appeared in Kandiah Pillai's books from the 1940s, which use the well-worn cartographic operation of drawing lines and arrows to follow the migration of the inhabitants of the doomed Kumarināṭu to Australia, Africa, China, America, Europe, and so on; in other words, to the entire world.[142] Other maps published subsequently similarly present the antediluvian Kumarikkaṇṭam as the homeland of all the world's peoples before their dispersal to the far corners of the earth.[143] These are cartographies of exile in that they graphically illustrate ancestral Tamils as an *urvolk,* dispersing from an *urheimat,* speaking an *ursprache.*[144] In turn, Kumarikkaṇṭam is cartographically recast as the original motherland, rivaling the Mu of the occultists of Euro-America, or the ur-Aryan homeland of the Indo-Europeanists. In these maps, the geo-body of the modern Tamil nation not only extends far out into the Indian Ocean, but also way back into deep time.[145] The possibility that the modern Tamil nation is a recent invention is artfully denied, as these maps fashion for it a biography with vast historical depth.[146] Tamil readers of these catastrophic cartographies are, however, inevitably brought back to earth, so to speak, because the very same maps also starkly point to their confinement, today, in the much smaller lived homeland of India/Tamilnadu, a far cry from the fabulous lost homeland fatally destroyed by the ocean. As in the maps of modern Siam that Thongchai Winichakul so deftly analyzes, "a geo-body which had never existed in the past was realized by historical projection; . . . the agony [of loss of territories] is visually codified by a map. Now the anguish is concrete, measurable, and easily transmittable."[147]

MAPPING LOSS IN THE TAMIL MODERN

How may we account for the popularity of the map form in Tamil devotion's place-making? This popularity is especially surprising because in the Tamil everyday, modern Tamil speakers do not customarily resort to maps. Why, then, would Tamil's devotees find the map useful, or even necessary, to perform their labors of loss around a doomed Lemuria? To answer these questions, I explore three aspects of the modern map: as a sign of modern science; as a spatial surrogate of possession; and as a tool that enables the exercise of mastery over the world.

David Harvey has noted that European modernity is predicated on "the domination of nature as a necessary condition of human emancipation. Since space is a 'fact' of nature, this meant that the conquest and rational ordering of space became an integral part of the modernizing project."[148] Produced systematically and scientifically through empirical observation, survey, and measurement, the modern map emerged as the paradigmatic technology for the rational organization of geographical space in Europe. In turn, it became an emblem of confirmation of the rational and scientific spirit of the modern West. It was also one of the principal technologies through which the West conquered the rest, its cartographic conquest of other conceptions of space both anticipating and consolidating the worldwide establishment of European hegemony. Almost everywhere in the modern world—in the Americas, China, Thailand, India—the triumph of the European over the non-European inevitably found cartographic expression. The map, Benedict Anderson observes, "profoundly shaped the way in which the colonial state imagined its dominion—the nature of the human beings it ruled, the geography of its domain, the legitimacy of its ancestry."[149] Along with other emblems of modern European science, such as the compass and the chronometer, the map was a key feature of the iconography of empire. The scientific map became a metonym for colonial modernity as "triangulation by triangulation, war by war, treaty by treaty, the alignment of map and power proceeded."[150]

The scientific map also became one of the "gifts" of Europe to the rest of the world, whose very lack of possession of such an entity was deemed to reflect their inability to generate abstract, rational representations of space. This is beautifully illustrated by the elaborate cartouche affixed to James Rennell's *Map of Hindoostan,* the first map of India based on scientific survey and measurement, published in 1782, soon after the territorial conquest of the subcontinent by the British began. The cartouche, according to Rennell himself, shows Britannia "receiving into her Protection, the sacred Books of the Hindoos, presented by the Pundits, or Learned Bramins." At her feet lie what look like the tools of the surveyor, and behind her on a pedestal stands a lion with its foot resting on a sphere. The cartouche confirms the circulation of power/knowledge that we have learned to associate with modern empires. In return for the sacred textual knowledges of the East, symbolized by the "Shaster" (learned books) that the Indians give to Britannia, she gives them her latest achievement, Rennell's map of their land, the cartographic embodiment of the new rational, scientific ordering of ("Indian") space.[151] The publication of Rennell's map, and of others like it by the colonial state over the next two centuries, also meant that any discourse on space and place, if it had to secure credibility, had to be necessarily cartographic. Labors of loss around Lemuria, being primarily about space, albeit about one that no longer exists, had to necessarily everywhere rely on the modern map if they had to convince their audience(s) of the reality of the lost place-world. To recall Baudrillard, in modernity, "it is the map that engenders the territory."[152]

For Tamil place-makers, the map is especially essential because the truth about their lost homeland was at least partly based on the non-empirical and non-scientific knowledges embodied in their ancient poems and their medieval commentaries. They needed the map to establish the scientific credentials of their own place-making, so necessary for legitimacy in a colonial and postcolonial world. At the same time, European science and its cartographic labors of loss around Lemuria only served to confirm in this case the ancient truths of Tamil literature, as devotee after devotee insisted, as we have seen. The very technology through which the Tamil homeland had been cartographically limited to its present place in the postdiluvial world could also be used to map a time when it had not been so confined. As Thongchai has noted in his study of Thai cartography, the modern map "anticipate[s] spatial reality, not vice versa. . . . A map [is] a model for, rather than a model of, what it purport[s] to represent."[153] Brian Harley similarly notes for colonial America that "the map is *not* the territory; yet it *is* the territory. In America, cartography is part of the process by which territory becomes."[154] The map form thus allows Tamil place-makers to summon into existence a drowned homeland that is no longer visible to the eye, but which they wish to convince everyone did exist, once upon a time. Maps are crucial to the process by which Kumarikkaṇṭam becomes.

And this leads me to the second reason why maps become critical to Tamil devotion's labors of loss. They serve as "surrogates of space,"[155] particularly necessary when the space in question is no longer not just visible to the eye but unreachable except through fabulous flights of imagination. In lieu of the actual possession of this place-world, Tamil speakers are offered instead a cartographic vision, which is enabled through the conversion of the impersonal homogeneous paleo and occult landmass of Euro-America into an intimate Tamil home-place. The rise of the modern map, with its geometrically delineated grid for locating specific places, coincides with, and indeed helped cement, new attitudes toward land and property ushered in by capitalism in Europe. The map not only helped define individual property rights in and ownership of land, but also confirmed territorial boundaries of emergent nation-states. It both anticipated and consolidated European possession of distant lands on globes and atlases throughout the age of empire. Given the association between maps and territorial possession and proprietary claims on land, an association that their native subjects no doubt learned the hard way from the British in colonial India, it is perhaps not surprising that Tamil place-makers, too, resorted to the map to establish their territorial claim to Kumarikkaṇṭam, even though it was a land that no longer existed except in their imagination and on paper. In lieu of physically occupying the continent, the map enabled them to nominally claim it, for the map has come to serve in modernity as a prelude to actual possession.

Finally, if I may paraphrase Heidegger, the fundamental event of the modern age may be seen as the conquest of the world as map.[156] The staging of the world on a map before one's eye as an enframed whole that can be ordered, secured, rendered knowable and, ultimately, masterable, underscores the European domination of the globe that is such a diagnostic feature of its modernity.[157] A map of the entire world, in its totalizing presentation of the globe on a "coldly geometric" and "systematic" grid, exemplifies what Donna Haraway calls "the god-trick of seeing everything from nowhere."[158] If the imperative to draw scientific maps, with their Ptolemaic grid and their celebration of perspectivism, is so closely implicated in the exercise of Western mastery, what can we say about the mapmaking efforts of colonial and postcolonial subjects? Are they, too, through the modern map, attempting to exercise mastery over the world? If so, what kind of mastery could this be, given the political and economic realities within which they live and produce such maps?

I have suggested that the Tamil maps of Lemuria are mimic productions, imitating in form and sometimes in content, the cartographic productions of modern Europe. Yet, mimicry, as Homi Bhabha has recently reminded us, "is constructed around an ambivalence; in order to be effective, mimicry must continually produce its slippage, its excess, its difference." Mimicry, especially in the colonial world, operates to produce "a subject of difference that is almost the same, but not quite." Like the mimic men Bhabha comments on, who are the product of "the ambivalent world of the 'not quite/not white,'" the Tamil mimic maps of Lemuria, produced "on the margins of metropolitan desire," are also, "almost the same, but not quite."[159] They harbor in themselves, through the production of difference, the possibilities of cartographic resistance and subversion. This resistance finds expression in the transformation of an impersonal, homogenous, Euro-American place-world into an intimate, lived, home-place underwritten by the ancient names and truths of Tamil literature. As such, they disarticulate and dismember one of the founding objects of European modernity, the scientific map with its acclaimed rational, accurate, and objective representation of geographical reality.

These maps (and the verbal labors of loss that sustain them in the Tamil country) are produced in the aftermath of colonial conquest, when the entire world had come under European mastery, when there was nothing much left to conquer and possess, as the victorious cartographies of empire only underscored. In such a context, the mapping of vanished continents, and the claim to possess these as one's own, allow the colonial and postcolonial subject to aspire to mastery as well, but over disappeared place-worlds that European science and modernity had helped reconstruct through the modern technology of cartography. In Tamil India, therefore, as in the modern West, the map enables the exercise of mastery over the world, even if it is only a world that is now vanished forever in the mists of deep time.

Chapter 7

Laboring against Loss

Yes! Yes! That's very well put. . . . I rediscover precious things through [joyful] nostalgia. And in that way I feel that I never lose anything, that nothing is ever lost.[1]

DISENCHANTED DISMISSALS, DISCIPLINARY DISAVOWALS

All this is to not say that these labors of loss over a vanished Lemuria are celebrated by one and all, even in Tamil-speaking India, where so many have written so passionately about Sclater's disappeared world in the many variegated ways that I have documented here, where it is taught in schools and colleges and invoked in public speeches in as lofty a place as the state's legislature, where some folks have even named themselves after it, and where a documentary film financed by the government has been produced. As would be evident from scattered comments in the preceding pages, and especially from my discussion in chapter 5 of historicism's tendentious encounter with the antediluvian memories of ancient Tamil poetry and its commentaries, there have been those who have outright refused to participate in the pleasures of fantasizing about a prelapsarian world, preferring instead to take refuge in the rationalist disciplines of their day. The pressure to do so is acutely felt, particularly in colonial and postcolonial India, where one's credentials as modern—and hence, as rational and as civilized—are established by adhering to the parameters and protocols of the new sciences and disenchanted knowledge-formations that have been ushered in from the metropole.

The disavowal of Tamil labors of loss around Lemuria began very early indeed, in the opening years of the century that followed the Victorian birth of Sclater's drowned place-world. In a scathing review of Suryanarayana Sastri's *Tamiḻmoḻiyiṉ Varalāṟu* (which, to recall, was the first sustained effort to link the metropolitan paleo-scientist's Lemuria with a submerged Tamil home-place called Kumarināṭu), published shortly after the book appeared in 1903, the Reverend C. H. Monahan took the influential Tamil scholar-devotee to task for not abiding "by the principles of scientific philology." Instead,

> he allows an amount of weight to mythological elements which renders his judgement on the antiquity of Tamil to my mind almost worthless. He practically accepts the story dear to some Tamilians that in ancient times land extended south of Cape Comorin for some 7,000 miles (!), which was divided into 49 Tamil countries. In this land were South Madura and other places where Tamil flourished. This country now submerged by the Indian Ocean was the cradle of the human race, and its language was Tamil (!). Haeckel is quoted as authority for this opinion. All this is a sore tax on one's power of belief. But one fairly gasps when one reads the following . . . : "The Indian Ocean contains 25,000,000 square miles. From this it follows that it is somewhat less than 1,600,000 miles long and 1,600, 000 miles broad (!). Accordingly of this 1,600,000 miles length, 7,000 miles of land must have been swept away by the sea (!!!)." One would have supposed the numbers to be due to a mere slip of the pen but for the serious argument based upon them. Moreover, we are informed that the Early Tamil Academy lasted 4,440 years and the Middle Academy 3,700. These geographical and historical (?) marvels come of abandoning scientific research for mythology.[2]

The Reverend Monahan's expression of sheer incredulity—on display here in his profuse deployment of exclamation marks!—on encountering the Tamil antediluvium is not surprising, given that this has been historicism's response to whatever transgresses the range of mundane facts. But Monahan's critique is also revealing of the struggles undertaken by the emergent new discipline of history, as it was then practiced in Tamil India, to rationalize the received memories and reconstructed traditions of the Tamil past, which it disenchants or disavows for the sake of the historicist project of demonstrating what *actually* happened.

Since Monahan's time, others who have written against the reality of Kumarikkaṇṭam in the interest of what actually happened, fall into three camps. To recall, there are those who, caught between a disenchanted historicism and Tamil devotion's enchanted pull, attempt to historicize *kaṭalkōḷ*, the shifting of the Pandyan capitals, the creation and loss of the three *caṅkams*, and so on, by applying the scalpel of historical reason to cut the fancy out of fact in order to salvage some semblance of what actually happened. These are reluctant fellow travelers, their labors of loss shot through with skepticism but also struggling—as Tamil speakers devoted to their language and its tradition—to figure out if some land might indeed have been lost to the ocean or if some possible territorial connections might have existed with other places whose surviving memory, encoded in legend and transmitted over the ages, might have given rise to this fantastic imagination. Then there are those who establish a strict distinction between the scientific truth of Lemuria—the fact that there might indeed have been in distant geological time a continent in the Indian Ocean which subsequently submerged, drifted away, or otherwise disappeared—and the Tamil mythology of a Kumarināṭu swept away by *kaṭalkoḷ*. The latter, in the inimitable words of Nilakanta Sastri, "is all bosh," or in the quick dismissal of another, "a Dravidian dreamland." These critics have relentlessly subjected the enchanted labors of loss around the Tamil antediluvium to the unforgiving scrutiny of historical realism and rational positivism and, not surprisingly, found it utterly wanting. How could Tamil speakers, let alone human beings, have lived in the age of the dinosaurs, walking the earth with these primeval reptiles, they ask? How could a landmass have extended all the way from Kanyakumari to Antarctica, and no one else but ancient Tamil poets lived to tell about it? If all life was lost in catastrophic *kaṭalkōḷ*, how did the Pandyan king alone survive to start learning and civilization anew, they wonder?[3] Still others defer a final judgement on such questions until the ocean depths south of Kanyakumari are plumbed and the truth finally revealed by the new technologies of underwater archaeology. Until that happens, however, the imagination about a drowned Kumarināṭu remains just that, "mere imagining."[4]

All three positions share with each other their refusal to succumb to the freedom of living outside what Eliade calls the terror of history, or what occultists refer to as the tyranny of the sciences. Out of this emerges their realist and historicist assessment that the imagination of a drowned homeland, of lost capitals and vanished literary academies, and of catastrophic *kaṭalkoḷ* is filled with "falsehood, absurdities, untruths and inconsistencies." In other words, it is all "fantastic" and "fabulous." I, too, have argued for the fantastic and fabulous place-making of Kumarināṭu. Where I radically part company with these critics and naysayers is in my counterhistoricist and unrealist contention (based on my interrogation of the disenchantment of my discipline and drawing upon the oppositional and subversive semantics of the category of the fabulous) that the imagining of the Tamil prelapsarium has not been fabulous enough, and that its utopian possibilities have been left stunted and atrophied. Formulaic assertions and assurances about good kings, well-fed poets, and a happy populace, there are aplenty in these labors of loss. But why don't Tamil devotees go out on a limb and offer us detailed accounts of what everyday life would have been like in the promised land of Kumarināṭu for the average man, woman, and child? What does a society look like in which were absent the insidious caste discriminations and the ignoble disempowerment of women and the underclasses that are such ubiquitous features of the Tamil country through much of its long history? What is the modern Tamil place-maker's vision of a just world where there is no poverty or hunger, where everyone is educated, where everyone leads a life of peace and well-being? Many years ago, Paul Bloomfield insisted that "the problem of Utopia is about a sort of place," the sort of good life one leads there, and about "what it would be like to live in a world where these principles were acted upon." It is also about getting "as clear a notion as possible of what perfectibility consists in."[5] From this point of view, Tamil labors of loss miss an important opportunity to systematically contribute to the global imagination of the good society (however flawed that enterprise might be, in some assessments), or to generate a clear blueprint of a Tamil utopia that could provide a sustainable agenda for the establishment of a just and good present and future.

Is the refusal of Tamil's devotees to let their imaginations soar to utopian heights a symptom of the poverty of utopian thought in the subcontinent more generally? Does it suggest the impossibility of imagining an ideal state of plenitude and perfection in a colonial condition of subjection and misery? Or does it point to the pressure to think realistically and disenchantedly in the face of the widespread modernist condemnation of the fabulous and fantastic native mind? These are all possibilities that I have barely explored here. What I have instead argued is that, like their critics, Tamil labors of loss, too, ultimately fall victim to the terror of history and the tyranny of the physical sciences, their attempt to flirt with fantasy notwithstanding. This is not to say that it is impossible to think fantastically in Tamil India; all those who watch and appreciate Tamil cinema will concur. But it is to say that Tamil devotion has not been able to do so unabashedly and unashamedly, free of the shackles of history or other attendant disciplines, when it comes to the matter of Kumarināṭu. The magic of make-believe is compromised, again and again, by the imperative to demonstrate what *actually happened* in this most foundational of all moments for the Tamil language and its community of speakers. A prelapsarian place-world like Kumarikkaṇṭam, an elsewhere which radically transgresses the usual range of facts—beyond the known, beyond the accepted, beyond belief—deserves to be imagined into place by an imagination that gives full rein to the fantastic and to the fabulous instead of being caught up in the game of history with its harsh truth-demands. Yet Tamil's devotees cannot afford to do so as colonial and postcolonial subjects committed to demonstrating the historical reality of their antediluvian past, and to proving the primordial, singular, and exceptional presence of their language and its speakers on the subcontinent from the very beginning of time—that all of this *actually* happened. Herein lies the tragedy of their labors, as a consequence of which the Tamil prelapsarium, instead of appearing as a magical and wondrous moment out of history and beyond its disenchanted reach, seems incongruous and impoverished, meeting the demands of neither history nor fantasy.

LEMURIA AND ITS LABORS OF LOSS

But this is not my only point of departure from critics who have taken this fabulous imagination to task. As importantly, I have been concerned with interrogating the preoccupation with loss that generates place-worlds like Lemuria in the first place. And I have been interested in demonstrating that Tamil devotion's preoccupation with a lost time and a drowned past is shared by others in different parts of the modern world who are far removed from its passions or politics. Indeed, if I were to don my historicist hat and speak chronologically, the Tamil fascination with Lemuria surfaces much after the metropolitan paleo-scientist's or the Euro-American occultist's. Tamil devotion both learns and unlearns from them as it transforms the West's Lemuria into an intimate Kumarināṭu, even as the density of its preoccupation with this lost world is another revealing reminder of the hybridization of many a metropolitan thought and practice as these travel to their colonial and postcolonial address.

I have sought to understand this cross-cultural fascination with the vanished Lemuria through the analytic of labors of loss, a concept I borrow from Georges Bataille through Michel de Certeau to track those interpretive acts and narrative moves through which something is declared lost, and then recovered through the knowledge practices of modernity, the very act of recovery and naming constituting the original loss. Without these labors, there would be no apprehension of loss, no mobilization of imagination or practice around recovery and restoration. While the founding premise of this book is that a place-world like Lemuria does not exist prior to or independent of the labors of loss which imagine it into being and ensure its continued circulation, I have also insisted that there is no simple or singular way in which it is lost. In other words, loss, too, has its own poetics and politics which manifest themselves in the multiple ways in which Lemuria has been configured in the century and more of its presence in the imaginings of its diverse place-makers.

The earliest of these are metropolitan men of science—mostly natural historians, biologists, geologists, paleogeographers, prehistorians, and ethnologists—involved in solving some of the most pressing questions of their time, including the deep history of the earth and the formation of its continents and oceans, the geographical distribution of flora and fauna, and where and when the first human appeared. I have argued that Lemuria first surfaced in 1864 in this context, although it appears eccentrically—in speculative conjectures, scattered footnotes, an off-the-cuff remark here, a dissenting note there. Nevertheless, by virtue of the fact that it was witness to and flickered in and out of some of the most urgent debates of the sciences of the day, it has been ensured a quiet but continuing afterlife that I have characterized as off-modern. At the same time, Lemuria's eccentricity and off-modernity is also revealing of the general preoccupation of these sciences with the vanished worlds of lost times in the earth's primeval past. Rooted though this preoccupation might be, like in all good science, in dispassionate interest (or passionate disinterest), one cannot but miss the wonder and awe that accompanies every new discovery of a vanished land, an extinct species, or a missing link that helps to complete the human stock of knowledge about how we have come to be where we are today on this planet. Hence, my conclusion that these sciences of the earth's deep past—of its lost time and of its succession of vanished worlds and disappeared life-forms—are also involved in labors of loss, carried out though these might be in an apparently disenchanted mode. Lemuria's place-makers in the metropolitan paleo-scientific community—some of the most prominent scientists of their time—appear disenchanted only in comparison to others who labor over this lost place-world under the sign of transparent enchantment.

Of these, the earliest to do so are esotericists on either side of the Atlantic from the late 1870s. Occult labors of loss around Lemuria, enduring to this day in the metropole, are of a piece with modern esotericism's fascination with secret, hidden, buried, and astral places where the unsundered unity of matter and spirit might still be found in these times which are imagined as so utterly removed from god. The occult business is about re-divinizing such a world, and Lemuria is useful for this project because science itself had revealed its paleo-existence even while voiding it of spirit. Lemuria becomes one of the sites where the battle against Darwinian evolution is waged by Theosophy as it is transformed from Sclater's lemur habitat into a staging ground for demonstrating that man did not evolve from matter—or even more ignominiously, from the anthropoid ape—but from spirit as it manifested itself in the Third Root-Race. Beyond the late-nineteenth-century Theosophical moment, Lemuria's place in occultism changes as its American practitioners recast it as a drowned Pacific paradise whose enlightened survivors may still be found secreted away in the hidden reaches of California, and as its New Age exponents transform it into an astral abode of lost wisdom and well-being which holds out a beacon of hope for a disenchanted mankind. Although Lemuria itself might not have a singular identity, nor do Lemurians, occult labors of loss are united by the singular concern with re-divinizing an Earth that had become a-theized through the runaway triumph of a positivist materialist science that had reduced god's creation to matter, and matter alone. This project would be launched from secret archives stored in lost continents like Lemuria, to which only the occultist has access through her clairvoyant powers and her ability to commune with vanished racial memories. Esotericism's ancient penchant for the hidden and the secret finds a new perch on submerged continents and drowned landbridges revealed by modernity's own prized earth sciences.

Today, I have suggested, Lemuria is no more than a resonant place-name in occult labors of loss, with few tangible connections to the continent's ancestry in the speculations of Victorian natural history or in the spiritual and epistemological struggles against Darwinism. As a resonant name, it conjures up visions of peace and harmony, wisdom and well-being, to which humanity can aspire if it opens itself to the rule of spirit and consciousness. In the words of Diandra, a New Age healer:

> When you think of Lemuria recognize that you are going very deep into the consciousness of the human soul as it existed eons ago. Tonight we would speak to you about the time you knew as Lemuria. You will notice that we're not saying the land nor the place but the Time of Lemuria. Lemuria is the most ancient of times in this particular known expression of your consciousness. The Wondrous Time of Lemuria.[6]

No longer a placeable place, Lemuria is now, in one of its many New Age incarnations, a state of mind.

Turning to India, the metropole's Lemuria, in its reincarnation as Kumarināṭu or Kumarikkaṇṭam, is today very much a part of the symbolic paraphernalia of late Tamil modernity, although it would be fair to say that, as with many symbols, a majority of Tamil speakers have little idea what it is, really, notwithstanding a century and more of dense labors of loss around it undertaken by Tamil's devotees, by the school system, and most recently, by the state. In other words, here, too, where Sclater's vanished place-world has attracted the most intensive and variegated attention, it remains an eccentric, off-modern, even oppositional, presence, kept alive through Tamil pedagogy and Tamil intellectual activity in a society where these are themselves fairly endangered and beleaguered enterprises. And this, in turn, informs my primary argument, that the arrival here of Lemuria from the metropole in the closing years of the nineteenth century provides an occasion for those devoted to Tamil to reflect upon irrevocable loss—of Tamil words and verses, and of the original Tamil language; of Tamil antiquity and purity, sovereignty and unity; and not least, of the ancestral Tamil home-place. Living out their own lives as colonized subjects in the shadow of despair and decline, Tamil place-makers have displaced these sentiments onto a complex set of labors of loss in which their beloved language and land appear whole, pure, glorious, and perfect—once upon a time, elsewhere. Yet, yearning for the drowned Kumarināṭu also allows them to provoke their fellow speakers to love and be loyal to that which was *not* catastrophically lost once the ancestral homeland submerged. Today's Tamilnadu and the language spoken within its boundaries, shrunken though they might be in comparison to their former, antediluvian reach, are precious remnants of what once was—elsewhere—in a prelapsarian moment of plenitude and perfection. Hence my proposition that the politics of loss works to establish a productive, even pleasurable, tension between lived and lost homelands in the Tamil country. The two are intimately linked precisely because the lived is imagined as a valuable fragment—a memorative sign—of the former whole that is forever lost.

In a provocative essay, David Shulman has demonstrated the pervasiveness of the motif of the loss and recovery of precious and sacred knowledge across the various religious and literary practices of premodern India, Tamil included: "Knowledge that counts (sacred knowledge of one kind or another) is never simply there, in pristine wholeness; rather, it has undergone processes of becoming lost, defiled, buried, divided, forgotten or mutilated, and—only after this stage—recovered or restored."[7] Apprehension of loss initiates acts of recovery and reconstitution, although the whole is never fully restored. In a manner reminiscent of Bataille's observation that "sacred things are constituted through an operation of loss," Shulman as well suggests that the Indic myths he analyzes "develop in a characteristic, recurrent pattern of three stages: primeval completeness, loss or partial destruction, and reconstitution through the recovery of the essential fragment which must serve in lieu of the whole."[8] Tellingly, he insists that "it is the second stage, that of loss, which guarantees the text's sacred character."[9] "The poems *must* be lost before their sacred character can be affirmed through the process of their recovery."[10] Hence, the importance of the mythic preoccupation with loss and recovery, he concludes.

While conceding the enduring hold of such archetypes, my own concern has been with establishing the *modernity* of the labors of loss over Kumarināṭu, in which the new disciplines of the metropole—geology, ethnology, history, cartography—are mobilized to salvage and rehabilitate the Tamil past's antediluvial memories and to find a novel home for them in the earth sciences' Lemuria.[11] In this process, that which is cryptic, enigmatic, even fleeting in premodern Tamil poetry now assumes a new coherence as, underwritten by the modern sciences, a prelapsarian moment is inserted into the history of the Tamil people, their language, and their land, and reemerges as the *Tamil* antediluvium. In other words, for Tamil placemakers, and in textbooks for school and college students that they have produced, and in the documentary film that they have persuaded their government to finance, Tamil and its culture now have an established antediluvial beginning, the world's most ancient at that. The foundational moment of all of Tamil history and culture is buried in the lost Kumarināṭu. A normatively disenchanted history in which Tamil and its people seem like pale shadows of the more glorious lives and times of Sanskrit, English, or Hindi speakers is displaced by an enchanted narrative that is infinitely more satisfying and self-empowering. Satisfying because in contrast to their current state of decrepitude and despair, it shows the origins of the Tamil language and its land in a moment of promise, plenitude, and perfection; self-empowering because this can be—and is—offered to the speakers of the language today as a moment around which to rally and to realize their (lost) potential as a community.

Paradoxically, therefore, it is in a part of the world far removed from the by-lanes of Victorian science where it first surfaced that Lemuria becomes an intimate part of a people's past and their politics, however off-modern, eccentric, and oppositional its presence. Here, it is no longer a remote paleo abstraction, nor a distant occult world, but the prelapsarian homeplace of the Tamil community, the birthplace of all that is truly remarkable about its culture and civilization. This, then, is also the logic of the fascination with this vanished paleo-world among Tamil place-makers, a logic that is rooted in their labors of loss that are passionate, nostalgic, and bittersweet, "bitter because lost, all the more sweet for being lost."[12] Nostalgia—this much-maligned sentiment that has provoked so many to write so caustically against it—finds a fertile home in the modern Tamil country but operates contrarily in its labors of loss over Lemuria. On the one hand, to recall, it undoubtedly works to ensnare many a Tamil devotee in melancholic fables of the archaic, *retrospectively* yearning after that which catastrophically vanished once upon a time, long long ago. On the other hand, as I have also noted, nostalgia operates *prospectively* as well, to renew their commitment to that which survived into the present. Caught between these contrary pulls of retrospective and prospective nostalgia,[13] and in the context of the refusal (or is it failure?) of Tamil labors of loss to reach their utopian heights, the drowned Kumarināṭu has come to be frozen in antediluvial time, an object of endlessly repetitive and formulaic place-making whose potential to provoke a sustainable life of plenitude and perfection in the present or future is limited at best. Nostalgia as well, like the fabulous and the utopian, has been fatally subverted in Tamil India, so that instead of helping to engineer a just and good life in the here-and-now, as it well might have, Kumarikkaṇṭam instead lingers on merely as a ghostly reminder of all that modern Tamil speakers have lost in the deep time of their antediluvian past.

BACK TO THE INDIAN OCEAN

In conclusion, I take us back to the Indian Ocean, where it all started, in the 1860s, when Philip Lutley Sclater, ruminating over the eccentric fauna of the island of Madagascar and its relationship to the surrounding waters, summoned into existence the lost continent of Lemuria which spanned its length. Deemed one of the three largest bodies of water that make up the earth's surface today, the Indian Ocean has been variously described as "the forlorn ocean," "the neglected ocean," "the largest unknown," and so on.[14] These phrases are meant to recall the fact that, in contrast to the Atlantic and the Pacific, not much was known about what went on in the Indian Ocean through much of the past two centuries. It remained an unfathomed, mysterious place, providing a fertile ground for lost place-worlds like Lemuria to flourish in its depths.

This is not to say that metropolitan science did not attempt to crack its mystery. The 1872–74 *Challenger* expedition, which did so much to bring the wonders of the Atlantic underworld to light, also explored the southern edges of the Indian Ocean.[15] A few decades later, the British Admiralty dispatched H.M.S. *Sealark* in 1904 to undertake sounding and dredging activities in the Chagos archipelago, the Seychelles, and neighboring islands. Announcing the expedition, the *Geographical Journal* noted that "the work of the expedition will be of special interest in view of the former land connection which *almost certainly* existed between India, Madagascar, and South Africa."[16] And, indeed, a subsequent report on its findings sparked off a lively discussion in the influential Royal Geographical Society in London, where J. Stanley Gardiner, the noted Cambridge biologist who accompanied the *Sealark* on its mission, insisted in June 1906 that "the land connection of South Africa and Madagascar with India *can scarcely be disputed,* though its duration and the changes which have taken place in it may legitimately be discussed."[17] This statement, in turn, provoked the zoologist Hans Gadow, who was present at his talk, to insist that all those naysayers who had denied the existence of Lemuria had been proven wrong by the findings of the *Sealark,* and that the land connection between Africa, Madagascar, and India "must have lasted right into the middle of the Tertiary period."[18]

A few decades later, the joint Anglo-Egyptian John Murray Expedition of 1933–34 added to the growing metropolitan knowledge of the submarine world of the Indian Ocean, or at least of its western half around what is known as the Carlsberg Ridge. In the first public announcement of the expedition that appeared in London in August 1932, the correspondent for the *Times* wrote that its main objectives "will be the topography of the ocean bottom by echo soundings, to discover whether there are any traces of the continental land areas that are supposed to have stretched westwards from India and to have formed the hypothetical continent of "Lemuria," and also to ascertain whether there are ridges and peaks in this semi-enclosed ocean."[19] This announcement resulted "in several subsequent newspaper articles stressing that the Expedition was searching for the lost continent."[20] On behalf of the expedition, J. Stanley Gardiner, its secretary, responded that

> all geographers agree that India was once joined to Africa as a part of the Gondwana continent. The mechanism by which separation has been effected was, according to the older school of thought, by the foundering of the land which once lay between. The newer school suggest that there has been a drifting apart of the Indian peninsula from the African continent without the intervention of any substantial amount of land that has since sunk.[21]

Soon after the completion of the expedition, however, one of the first official statements on its findings, published in the *Geological Magazine* in

1937, concluded, "There is little or no indication that any older continental mass or land isthmus such as the hypothetical continent of Gondwanaland or the isthmus of Lemuria, ever existed except in the granite mass of the Seychelles and perhaps the corresponding granites of Socotra and the Kuria Muria Islands."[22]

By the time of the crucial multinational, multiship International Indian Ocean Expedition undertaken between 1959 and 1965 to make "a combined assault on the largest unknown area on earth,"[23] few in the metropolitan scientific community believed that they would find a drowned Lemuria in its depths, although the mapping of an elaborate system of submarine mountains, ridges, and valleys sparked off a fresh round of labors of loss in the Tamil country over a submerged Kumarināṭu. That Tamil placemakers alone maintain such a conviction has been belied in recent oceanographic work carried out in 1998–99 around the Kerguelen Plateau in the southernmost reaches of the Indian Ocean. These findings have rekindled speculations in the West about a "lost continent" in the Indian Ocean.[24]

But it is not just among scientists working on (and in) the depths of the Indian Ocean in recent times that Lemuria has surfaced, if only in passing, eccentrically. It is also strategically invoked by a motley group of individuals whom I think of as off-modern friends of this forlorn ocean. For instance, in an essay published in the London-based *Geographical Magazine* in 1957 and revealing captioned "The Peaks of Lemuria," Sir Hilary Blood, formerly of the Ceylon Civil Service and later Governor and Commander-in-Chief of the British colony of Mauritius between 1949 and 1954, wrote evocatively of the lands under his charge, in terms that recall nineteenth-century island primitivism:

> It is said that they are the unsubmerged hills of a sunken continent—all that remains above the water-line of the Peaks of Lemuria. Most of them are uninhabited, unexplored, and undeveloped, visions of tropical beauty, pocket-handkerchief paradises, protected in their unspoiled loveliness by the sweep of the ocean, known to bird and fish but not to man, inviolate save by sun and storm, existing for the glory of God who cannot but look with pleasure on these gems of his creation.[25]

These "pocket-handkerchief paradises," the remnants of the former Lemuria, were also the subject of a 1961 monograph published soon after by Oxford University Press called *Limuria: The Lesser Dependencies of Mauritius.*[26] This book, meant as a general introduction to British colonial possessions in the western Indian Ocean, similarly invoked Lemuria to convey a sense of the remoteness of these islands from contemporary human civilization, as well as the uniqueness of their floral and faunal life. More recently, Richard Edlis has published a book called *Peaks of Limuria: The Story of Diego Garcia* (1993), in which the author—who had been Commissioner for the British Indian Ocean Territory (consisting of the Chagos archipelago of fifty or so islands of which Diego Garcia is the largest) from 1988 to 1991—notes that the islands under his stewardship "comprise all that remains above sea-level of huge underwater mountains of volcanic origin which rear dramatically from the ocean bed 10,000 feet or more below. The romantic appellation for these islands is the Peaks of Limuria."[27] This "Indian Ocean Atlantis," he suggests, is "a paradise largely untouched by human hand."[28] This is what makes the presence of a large U.S. naval base in Diego Garcia so threatening. Hoping that the Chagos archipelago would soon be designated "an international environmental reservation with proper protection from the impact of human short-sightedness and greed," Edlis concludes, "otherwise, it might be better if Diego Garcia and its fellow Peaks of Limuria were to slip once more under the waters of the Indian Ocean."[29]

Transformed thus through such off-modern writings into an untouched Indian Ocean paradise, Sclater's ancestral habitat of the endangered lemur today stands on the verge of being appropriated into the regimes of global tourism that seek to promote the remote islands that might be its remnants as a haven for those moderns seeking to get away from it all for a while. It is only fitting that I give the last word to a new place-maker to join the ranks, the publicist for a hotel that calls itself "Lemuria," located in Seychelles, who tells prospective visitors in words that uncannily resonate with a century and more of the variegated labors of loss around this vanished land carried out across the globe:

> Lemuria, the inspiration for the name of the hotel, was according to legend a lost continent similar to Atlantis. Geographers looked at the Indian Ocean and saw a line of small islands stretching from the neighbourhood of western India through Seychelles to Cosmoledo and east to the Chagos. They guessed that these islands traced the outline of a drowned continent and called it Lemuria. Almost unwittingly, they uncovered elements of truth. . . . If there is such a thing as a lost paradise, it is Seychelles. Lemuria is a living legend and it is fitting that its name should be preserved by one of the finest hotels in the world, in arguably the most beautiful hotel site, where paradise may be rediscovered.[30]

NOTES

CHAPTER 1

1. Wright 1947, 15. The epigraph is from Wright's 1946 Presidential Address to the Association of American Geographers, in which he called upon his colleagues to explore "geographical truths and beliefs as these find and have found literary and artistic expression" (15). Wright's address has encouraged several scholars to think "humanistically" about geography in its presumption that "the uniqueness and specificity of a place flows from the experiences that individuals and groups associate with it" (Young 2001, 681). As Terence Young observes, humanistic geographers are "sensitive to the role of the subject in the creation and meaning of symbols and the normative significance of place" (Ibid., 681). In turn, this line of thought has been informed by Heidegger's concept of "dwelling," and Bachelard's "poetics of space" (Casey 1997). My own thoughts on place-making have been informed by this tradition of scholarship.

2. See, e.g., Ashe 1992, 8–13; Cohen 1969, 26–42; Cornell 1978, 212–16; Kolosimo 1973, 53–90; Stemman 1976, 64–95; Wellard 1975, 57–70. For works in other European languages that feature Lemuria, see, e.g., Kondratov 1967; Lugo 1978; Vincent 1969.

3. Quoted in Williams 1991, 12.

4. See, e.g., de Camp 1970; de Camp and Ley 1970; Ellis 1998; Feder 1990; Gardner 1957; Godwin 1972.

5. Williams 1991, 7.

6. See, e.g., Stiebing 1984; Vitaliano 1973; Wauchope 1962; Williams 1991.

7. Wauchope 1962, 135.

8. Ibid., 134.

9. Ibid.

10. Foucault 1980, 81–82.

11. Quoted in Bann 1989, 246.

12. Basso 1996, 6. The scholarship on the relationship between "space" and "place" is vast. In the humanist geography tradition, which has been much more cen-

trally concerned with place, "'space' is more abstract than 'place.' What begins as undifferentiated space becomes place as we get to know it better and endow it with value. . . . Place is a special kind of object. It is a concretion of value . . . an object in which one can dwell. . . . When space feels thoroughly familiar to us, it has become place" (Tuan 1977, 6, 12, 73). For Edward Relph, place is "constructed in our memories and affections, time deepened and memory-qualified, a *here* from which to discover the world, a *there* to which we can return" (quoted in Read 1996, 111). In Edward Casey's phenomenological understanding, space is "the encompassing volumetric void in which things (including human beings) are positioned" and place is "the immediate environment of my lived body—an arena of action that is at once physical and historical, social and cultural" (Casey 2001, 683).

13. Basso 1996, 5.

14. Ibid. See also Casey's discussion of "place memory" in Casey 1987a, 181–215.

15. Basso 1996, 7.

16. Ibid., 32.

17. I borrow the term "conflicted intimacy" from Richard Terdiman, who also writes that "the discourses of a society are structured in a shifting, multiform network of linked assertions and subversions, of normalized and heterodox speech. The linkage is essential, and its character is complex. . . . Counter-discourses are always interlocked with the domination they contest" (Terdiman 1985, 16).

18. The scholarly literature on lost places is limited. Peter Bishop's analysis of Tibet as a "site of contending fantasies" for European travelers is very suggestive in its exploration of the Himalayan kingdom as "a place of loss, of self-discovery, of transcendence, of ennui" (Bishop 1989, 7–8). Peter Read examines the many kinds of lost—and "wrecked" and "dead"—places in Australia that have become the subject of memorializing (Read 1996). See also John Chavez's historical study of the meaning and memory of Azatlan as a "lost homeland" to Mexican Americans (Chavez 1984).

19. Carter 1988, xxii. Although he does not deploy the notion of spatial history, Peter Bishop makes much the same observation when he writes that "places are produced by a dialogue between cultural fantasy-making and geographical landscape" (Bishop 1989, 9).

20. Carter 1988, 351. Carter's invitation to do spatial history anticipates postmodern geographer Edward Soja's complaint that "for at least the past century, time and history have occupied a privileged position in the practical and theoretical consciousness of Western Marxism and critical social science. Understanding how history is made has been the primary source of emancipatory insight and practical political consciousness. Today, however, it may be space, rather than time, that hides consequences from us, the making of geography more than the making of history that provides the most revealing tactical and theoretical world" (Soja 1989, 1). See also Foucault's charge that for generations space was treated as "the dead, the fixed, the undialectical, the immobile. Time, on the contrary, was richness, fecundity, life, dialectic" (Foucault 1980, 70). Elsewhere, commenting on the "obsession" of the nineteenth century with history, Foucault writes, "The present epoch is that of space. . . . We are at a moment when our experience of the world is less that of a long life developing through time than that of a network that connects points and inter-

sects with its own skein" (Foucault 1986, 22). As Soja reminds us, Foucault's "spatializations were not anti-history; they were the making of history entwined with the social production of space" (Soja 1989, 18).

21. Casey 1997, xi.

22. Yaeger 1996, 4.

23. Bataille 1985, 119. While Bataille himself uses the phrase "une opération de perte," I follow Michel de Certeau's translation of it as "a labor of loss" (de Certeau 1988, 324). Bataille's essay focuses on apparently "unproductive expenditures" around mourning, the construction of sumptuary monuments, potlach, and so on, even as it provokes one to think about the productivity of the principle of loss.

24. Freud famously discusses mourning as a response to a "known loss," as opposed to melancholia, the "inner labor" around "unknown loss" (Freud 1957). For a detailed treatment of psychoanalysis as a disciplinary practice that centrally concerns itself with the experience of loss, see Homans 1989.

25. For recent attempts to usefully engage the conditions under which loss is invoked, and the politics of its deployment, see Boym 2001; Harootunian 1999; Kirsch 1997; Ivy 1995; and Olalquiaga 1998.

26. Skaria 1999, 298. Although he appears to go along with this dominant diagnosis, Skaria does also suggest that "Evocations of loss . . . also make possible a politics of hope" and might underwrite a "counter-aesthetic of modernity" (ibid., 298–300).

27. Nora 1989, 8 (emphasis mine).

28. Conley 1988, viii.

29. de Certeau 1988, 47.

30. de Certeau 1988, 46. For a similar argument regarding the grounding of art history in the preoccupation with loss, see Davis 1998, 50.

31. Phillips 1985, 70.

32. de Certeau 1988, 324–25.

33. Ibid., 325.

34. Olalquiaga 1998, 11–95.

35. Ibid., 116.

36. Ibid., 129.

37. Quoted in Ibid., 129–30.

38. Ibid, 126.

39. Homans 2000, 225.

40. Baum 1970, 153.

41. Cascardi 1992, 120. See also Ferguson 2000, 155–65; Owen, forthcoming.

42. Owen 2001, 72. See also Latour 1993, especially 114–16.

43. Owen 2001, 88.

44. There is a large literature on this, but see, especially, Gould 1965; Gould 1984; Hooykaas 1963; and Huggett 1989.

45. Gould 1984, 27.

46. Quoted in Ibid., 30.

47. Hallam 1989, 30.

48. Washington 1995, 53.

49. Hallam 1989, 139. See also Ramaswamy 1999, fig. 1, 101–2.

50. On the controversy surrounding Velikovsky, see Stiebing 1984, 57–80; and Sullivan 1974, 26–33.

51. Gould 1965, 223.

52. Jackson 1981, 17. On fantasy as "the game of the impossible" that is "based on and controlled by an overt violation of what is generally accepted as possibility," see W. R. Irwin's *The Game of the Impossible,* in which fantasy is defined "as the narrative result of transforming the condition contrary to fact into 'fact' itself" (Irwin 1976, 4).

53. Armitt 1996, 4.

54. The marvelous is "anything outside the normal space-time continuum of the everyday world. . . . The marvelous element which lies at the heart of all fantasy is composed of what can never exist in the world of empirical experience" (Swinfen 1984, 5).

55. Jackson 1981, 17–18. See also Fred Weinstein's suggestion that fantasy or wishful thinking in modern times seeks to "repair reality in imagination": "Normatively perceived injustice or loss, or the failure or absence of a standpoint that can make sense especially of unexpected events, invite fantasy thinking, which has always provided the occasion for some individuals to explore, elaborate, refine, and finally, promote an alternative standpoint, reorienting people to the social world, often in unexpected ways" (Weinstein 1990, 95).

56. Jackson 1981, 19.

57. I am inspired to do this by a number of scholars who have suggested that the fantastic operates subversively in realist modernity. "The modern fantastic, the form of literary fantasy within the secularized culture produced by capitalism, is a subversive literature. It exists alongside the 'real', on either side of the dominant cultural axis, as a muted presence, a silenced imaginary other" (Jackson 1981, 180). Similarly, Lucie Armitt suggests that the fantastic is "a form of writing which is about opening up subversive spaces within the mainstream" (Armitt 1996, 3). Fred Weinstein observes that "the need to repair a reality that always appears in some measure as failed, insufficient, or threatening to someone (or some group) is an important source of reparative forms of thinking as well as of competing and multiple perspectives" (Weinstein 1990, 96).

58. Bayly 1996, 311–12. I thank Nicholas Dirks for reminding me of this.

59. Ibid., 309.

60. Murdoch 1885, 153. See also Ramaswamy 2000, 594–96.

61. Conrad 1926, 2–3, 14.

62. I borrow the concept of "planetary consciousness" from Mary Pratt's important discussion of nineteenth-century European travel-writing, which (re)cast the earth as a secular subject of discourse knowable in its entirety through modern science (Pratt 1992, 9, 15–37).

63. Boym 2001, xvi–xvii.

64. Ibid., xvi–xvii, 30–31.

65. I borrow this formulation from Dipesh Chakrabarty, who uses it to characterize postcolonial thought's relationship to European thought (Chakrabarty 2000).

66. Weeks and James 1995, 9. For a discussion of "lost continents and golden ages" as classic "eccentric geography" and "eccentric mythopoesis," see ibid., 112–35.

67. My reading of Berman 1981 and Griffin 1990, and their calls for the "reenchantment" of the physical sciences, also provoke my comments here.

68. See, e.g., Chakrabarty 2000; Nandy 1983; Lal 2000.

CHAPTER 2

1. Charles Lyell, quoted in Gardiner 1881, 241.
2. Gould 1987, 3–4.
3. Wood 1985, 7–8.
4. Quoted in Eiseley 1958, 73.
5. Quoted in Hooykaas 1963, 19.
6. Eiseley 1958, 73.
7. Winchester 2001.
8. Eiseley 1958, 83.
9. See, especially, Gohau 1991, 58–60, 92; Rossi 1984, 14–15.
10. Quoted in Gillispie 1951, 99–100.
11. Eiseley 1958, 93.
12. Ibid., 267–68.
13. Richards 1993, 6–7.
14. Ibid., 39.
15. Of landed gentry background, Sclater studied natural history in the 1840s at Oxford but went on to practice law for a few years. He was elected a Fellow of the Zoological Society of London in 1850, and in 1858 cofounded *Ibis,* the influential journal of the British Ornithologists' Union, also serving as its editor for a number of years. His 1858-paper, "On the General Geographic Distribution of the Members of the Class Aves," divided the world into six distinctive regions on the basis of their bird life. Later extended to cover all animal life, the paper continues to be read by zoogeographers. A prolific writer, Sclater authored more than a thousand books, articles, scientific papers, and short notes. There is hardly any scholarship on Sclater or his contributions, but see Browne Goode (1896), who described him as "one of the best known of living zoologists," but who makes no reference to his naming of Lemuria. Occasionally, Sclater and his labors of loss over Lemuria appear in present-day scholarly works on the history of geological thought or of zoogeography (Marvin 1971, 54–55; Wood 1985, 40–41), but less carefully researched works abound in many faulty identifications (e.g., Ellis 1998, 65; Steibing 1984, 35; Wauchope 1962, 38). So, Sprague de Camp, in his spirited and much-invoked critique of lost continents, writes that "the idea [of Lemuria] had originated with the Austrian geologist Melchior Neumayr, while the name had been invented by the British zoologist Philip Sclater and popularized by the German Ernst Haeckel" (de Camp 1970, 52). James Wellard writes that the name Lemuria "was first suggested in 1855 [*sic*] by the English zoologist Philip Lutley Sclater who postulated an area in the Pacific [*sic*] as the aboriginal home of the lemurs—a region to which, he suggested, they have an atavistic urge to return. Sclater's theory appears to have been suggested by the contemporary scientific belief that the lemming of Norway in their quadrennial suicide migration were attempting to return to their ancestral home on Atlantis. And no sooner had Sclater proposed an aboriginal habitat for the lemurs than geologists came forward prepared to support his contention, whence Lemuria was adopted as the name for an Atlantis of the Pacific" (Wellard 1975, 58–59). John Clute and John Grant do not even mention Sclater when they note that "Lemuria was originally a scientific hypothesis created to explain the sporadic distribution of lemurs in Africa, Madagascar and India. In the 1870s the German biologist Ernst Haeckel speculated

that a landbridge or small continent had once connected these regions" (Clute and Grant 1997, 574). For a similar characterization, see Hammer 2001, 100–101. Geoffrey Ashe incorrectly identifies Sclater as a geologist, but correctly identifies his authorship of Lemuria (Ashe 1992, 8). In one scholarly study, Sclater's name is even misspelled as Philip L. Schattler (Melton, Clark, and Kelly 1990, 182–83).

16. Sclater 1864. European interest in the island dates to the mid-seventeenth century, from which time many scientists have remained fascinated with Madagascar, probably because, as Alfred R. Wallace observed, it "preserve[s] to us the record of a by-gone world" (Wallace 1880, 383). Recently, Gillian Feeley-Harnik notes that up until today the island continues to be described in scholarly and popular writings in overdetermined terms such as "a lost continent," a "world out of time," "at the end of the earth," a "Living Eden," or "the Earthly Paradise of the Fall" (Feeley-Harnik 2001). See, e.g., the newly published *The Eighth Continent: Life, Death, and Discovery in the Lost World of Madagascar*, which has a rather dismissive but mostly accurate discussion of the place of Lemuria in the island's complex relationship to Western science (Tyson 2000, 34–39).

17. Robert Wood describes the lemur as "primitive monkeys, whose tree-bound nocturnal habits and ghostly demeanor had merited their Latin name for the souls of the departed" (Wood 1985, 41). Although frequently described as monkey-like, the lemur belongs to the Primate suborder Strepsirhini, which also includes the African bush baby and the Asian loris. The European "discovery" of the lemur dates to the 1650s, and between that time and Sclater's labors of loss around Lemuria there was considerable documentation of the various subspecies, anatomy, and so on of this intriguing primate. Fossils of lemur-like creatures were discovered from Eocene strata in Europe in the 1820s and in North America in the 1870s, but today the lemur is native only to Madagascar and some adjacent islands, with some related species to be found in India (Hill 1953, 264–69). The earliest maps showing the geographical spread of the lemur were published in the 1840s (Robinson 1982, 102, fig. 47; see also Murray 1866, map no. XI). Alfred R. Wallace's classic *The Geographical Distribution of Animals* includes a beautiful illustration of the lemur (Wallace 1876, 1: plate 6). To the best of my knowledge, none of the current scholarly literature on the lemur mentions Lemuria, although one of its endangered species is named, appropriately enough, "Sclater's lemur" (*Lemur macaco flavifrous*). For a website which draws attention to the plight of this threatened animal, see http://www.geocities.com/Sharna_Kestral/contents.html. At the time I looked at this site in October 2001, its creator was writing a novel for young adults entitled "The Quest for Lemuria," about a young girl who, with her lemur friends, sets out "for the lost continent of Lemuria, a place where the lemurs hope to seek sanctuary from extinction."

18. Sclater 1864, 215–19.

19. Ibid., 219.

20. Wallace 1860, 177.

21. Wood 1860.

22. Hartlaub 1877, 334. Hartlaub himself went on to insist that "this fourth continent of Isidore Geoffrey is Sclater's "LEMURIA"—that sunken land which, containing parts of Africa, must have extended far eastwards over Southern India and Ceylon, and the highest points of which we recognize in the volcanic peaks of Bour-

bon and Mauritius, and in the central range of Madagascar itself—the last resorts of the mostly extinct Lemurine race which formerly peopled it" (ibid.)

23. Barrow 2003, 99–100, 121.

24. Carter and Malouf 1989, 173.

25. Carter 1988, 137.

26. Sclater 1874, 80–81; 1875, 12–13.

27. See, especially, Browne 1996; Gohau 1991.

28. There is a vast literature on this subject, but see, especially, Hooykaas 1963; LeGrand 1988; Marvin 1973; Wood 1985.

29. Quoted in LeGrand 1988, 21.

30. Quoted in Gardiner 1881, 241.

31. The recourse to faunal landbridges as a zoogeographical explanation may be traced back to the later seventeenth century when attempts were made to understand the dispersal of animal life to the newly discovered Americas. Athanasius Kircher even drew a map in 1675 showing landbridges extending from Brazil to Greenland, and from southern California to Chile (Wood 1985, 37). For useful discussions of the modern zoogeographer's preoccupation with landbridges as faunal highways, see Browne 1983, 196–202; Fichman 1977; and Marvin 1973.

32. L. F. de Beaufort, quoted in Wegener 1966, 6 (emphasis mine). For its advocates, as Wegener (a major critic of the theory of landbridges as faunal highways) was to note, "It is probably not an exaggeration to say that if we do not accept the idea of such former land connections, the whole evolution of life on earth and the affinities of present-day organisms occurring even on widely separated continents must remain an insoluble riddle" (ibid.).

33. Browne 1983, 196–202; Fichman 1977.

34. Murray 1866, 29–38.

35. Quoted in Browne 1983, 200–201. Ten years earlier, Darwin told Dalton Hooker, "I must maintain that I have never let down or upheaved our mother-earth's surface, for the sake of explaining any one phenomenon, and I trust I have very seldom done so without some distinct evidence. So I must still think it a bold step (perhaps a very true one) to sink into the depths of the ocean, within the period of an existing species, so large a tract of surface" (quoted in Browne 1983, 200). More than half a century later, a similar complaint continued to be made by geologists like A. P. Coleman, who noted in his Presidential Address to the Geological Society of America in 1916 that some of his peers "display great recklessness in rearranging land and sea [for] the convenience of a running bird, or of a marsupial afraid to wet his feet" (quoted in Menard 1986, 19).

36. Tyson 2000, 33.

37. Murray 1866, 29.

38. Blanford 1875, 535–36.

39. Neumayr 1887, 336.

40. Oldham 1894.

41. E.g., Blanford 1875; Blanford 1890; Oldham 1894.

42. Huxley 1870.

43. Blanford 1890. William Thomas Blanford established his professional reputation working for the newly established Geological Survey of India from 1854 to 1881. Like his brother Henry Francis, he did extensive fieldwork all over the sub-

continent, and was coauthor of the first *Manual of the Geology of India* (1879). Returning to England in 1881, he held several key positions in various scientific associations and was elected president of the Geological Society (1888–90).

44. Arldt 1917, 162–72, 278–81, 663–71.

45. Blanford 1890; Oldham 1894.

46. Blyth 1871; Hartlaub 1877.

47. Haeckel 1876a, 361.

48. The eccentric visibility of Lemuria in paleogeographical thinking in these years is also apparent in the frequency with which it is noticed in contemporary surveys of the earth's paleo-zoogeography (see, e.g., Gadow 1909, 333; Heilprin 1894, 86–87, 402; Johnston 1884, 9, 511).

49. Sclater 1875, 12–13.

50. Neumayr 1887, 336.

51. Eduard Suess (1813–1914) has been described by Robert Wood as "the greatest geological syncretist of the late nineteenth century" (Wood 1985, 10). His four-volume synthesis on the history of the earth, *Das Antlitz der Erde,* took thirty-five years to complete and was published in its German original between 1885 and 1909. Suess was appointed Professor of Geology at the University of Vienna in 1857, and many of Europe's leading geologists and paleogeographers were either trained by him or influenced by his work. *Das Antlitz der Erde* was first translated into French (1897–1911) and then into English in 1904.

52. Suess 1904, 596 (emphasis mine). Suess's Gondwanaland was based on the stratigraphic term "Gondwana" coined by the British colonial geologist H. B. Medlicott in 1872 to describe Permian coal-bearing formations in central India. Suess suggested that this particular formation, with its characteristic fossil fern Glossopteris, may be found in other continents of the southern hemisphere which were united during the Mesozoic in one territorial entity called Gondwanaland. Over the four volumes of *Das Antlitz der Erde,* Gondwanaland's makeup varied, but its core was constituted by southern America, Africa, and peninsular India. For discussions of the origins and etymology of "Gondwanaland," see, especially, Sorkhabi 1996.

53. This happened in spite of the important geologist R. D. Oldham's warning against confounding the two (Oldham 1894, 174). For more on Oldham, see chapter 6.

54. Oldroyd 1996, 258.

55. A man of relatively humble background, Wallace's reputation as one of the foremost natural historians of the nineteenth century was established, not through fancy degrees or important positions in metropolitan universities but by his numerous publications based on long voyages to the Amazon (1848–52) and the Malay Archipelago (1854–62). As early as 1855, he had independently worked out the principles of evolution through natural selection that Charles Darwin elucidated in *The Origin of Species* in 1859. After his permanent return to England in 1862, he became a leading member of some of the most prominent scientific societies of his times and counted among his friends Darwin, Huxley, Lyell, and Spencer.

56. Fichman 1977, 51.

57. Wallace 1876, 1:76.

58. Ibid., 1:76–77. A zoologist whose work was much quoted by Charles Darwin, Edward Blyth (1810–73) began his professional career as Curator of the Museum of

the Asiatic Society of Bengal in Calcutta and wrote extensively on the fauna and flora of the subcontinent until his return to England in 1862. For Blyth's affirmation of Lemuria, see Blyth 1871.

59. Wallace 1877, 520. A grateful Charles Darwin wrote to Wallace: "The point which has interested me most, but I do not say the most valuable point, is your protest against sinking imaginary continents in a quite reckless manner, as was started by Forbes, followed, alas, by Hooker, and caricatured by Wollaston and Murray. . . . I have lifted up my voice against the above view with no avail" (quoted in George 1964, 150).

60. Wallace 1880, 393–94.

61. Ibid., 398.

62. Ibid., 398–99.

63. Frankel 1998, 122.

64. Wood 1985, 107.

65. Frankel 1988, 122–23; Wood 1985, 108–9.

66. Quoted in Wegener 1966, 107.

67. The son of an evangelical minister, Wegener studied at Heidelberg, Innsbruck, and Berlin, graduating with a doctorate in astronomy. His reputation as a meteorologist was established in Danish expeditions to the little-explored Greenland, in which he participated in the early years of the twentieth century. He was director of the Meteorological Research Department of the Marine Observatory at Hamburg after the First World War, and in 1926 was appointed to a chair in meteorology and geophysics at the University of Graz in Austria. By his own account, the idea of continental displacement first came to him in 1910 when he saw the remarkable congruence of the coastlines of the Americas and Africa on a world map. He publicly aired his theory for the first time at a lecture in Frankurt am Main in January 1912, following this up with his monograph *Die Entstehung der Kontinente und Ozeane* in 1915. Substantially revised editions were published in 1919, 1922, and 1929, and were soon translated into French, Spanish, and Russian. The first English translation, *The Origin of Continents and Oceans*, published in 1924, was based on the third edition of the German original, published in 1922, but the more widely read second translation, published in 1966, was based on the fourth German edition, published in 1929. There are numerous scholarly treatments of Wegener, the enormous resistance to his ideas on either side of the Atlantic up until the 1950s, and the eventual acceptance of his mobilist conception of the earth's surface by the 1960s. Of these, I have found the most useful accounts in Hallam 1973; LeGrand 1988; Marvin 1973; Oldroyd 1996; and Wood 1985.

68. Wegener 1924, 2.

69. Ibid., 2–3.

70. Ibid., facing p. 64.

71. Ibid., 64 (emphasis mine).

72. Marvin 1973, 79.

73. See, e.g., Wegener 1924, 2–3, 191.

74. Wegener 1966, 107.

75. Du Toit 1937, 293–94. Robert Wood notes, however, that du Toit himself "needed to build occasional bridges to explain details of the fossil record" (Wood 1985, 108).

76. Wood 1985, 107.

77. Ibid., 43. See also Marvin 1973, 114–19.

78. A native of New York, Bailey graduated from Columbia University and joined the U.S. Geological Survey in 1880, producing in the course of his career with the Service the "Geological Map of North America" (1907–10). He also taught geology at Johns Hopkins and at Stanford University. In Willis's scheme, giant upthrusts from ocean basins can rise above sea level and form cordilleras that "link together temporarily the permanent continental lands" (Willis 1932, 919). The classic example he gave for this phenomenon was the Panama Isthmus linking the two Americas. Because such "isthmian links" are temporary, and because they are of the same material as the ocean floor, they subside when the original pressure that had caused them to erupt subsides.

79. Willis 1932, 939, plate 27.

80. Marvin 1973, 102–3. See also Oldroyd 1996, 258–61.

81. This does not mean that it utterly disappears. For instance, an Italian zoological expedition in the 1950s to the Comoro Islands set out "in search of vanished lands, or rather in search of evidence of their past emergence from the waves of the Indian Ocean" (Prosperi 1957, 13–14). Lemuria also continues to surface in the occasional scientific paper. In a 1962 publication, the botanist van Steenis, on the evidence of the distribution of certain Jurassic angiosperms, insisted that "there must have been an isthmian connection between Madagascar and Ceylon over the Seychelles–Comores bank ("Lemuria")" (van Steenis 1962, 343). For a response that supports this hypothesis, see Wild 1965. For a recent statement (accompanied by maps) regarding Eocene–Oligocene "Lemurian Stepping-Stones" in the western Indian Ocean that facilitated overland floral migrations at a time when sea levels were dropping, see http://www.mobot.org/MOBOT/Madagasc/biomad9.html.

82. Nicolaas Rupke observes that "with plate tectonics, continents altogether lost their status as separate tectonic units and instead the tectonic plate, composed of both continental and oceanic crust, took its place" (Rupke 1996, 266).

83. E.g., Marvin 1973, 54–55; Radner and Radner 1982, 89; Sullivan 1974, 6–7; Wood 1985, 40–41. Encyclopedias or dictionaries of paleogeographical terms published in the course of the last few decades make no reference to Lemuria. I thank Eileen Zorc for her research assistance in this regard.

84. van Oosterzee 1997, 43.

85. Gohau 1991, 190–91.

86. Lewis and Wigen 1997, especially, 21–46.

87. E.g., Blanford 1890, 97–99; Griesbach 1871, 68–69; Lydekker 1896, 22–24, 224–26; Medlicott and Blanford 1879, xxxix–xl.

88. Blanford 1875, 540; Huxley 1870, lvi.

89. Haeckel 1876a, 361.

90. Frequently referred to in his own lifetime as "the Darwin of Germany," Haeckel studied medicine in Berlin, Wurzburg, and Vienna, and was appointed Professor of Zoology at the University of Jena in 1860. Haeckel's conversion to Darwinism happened in 1860, when he read *The Origin of Species* at a time when it was politically and professionally problematic to profess any admiration for the work in Germany. "It profoundly moved me at the first reading," he later wrote (quoted in Bolsche 1909, 53). He subsequently dedicated his professional life to the popular-

ization of Darwin's evolutionary ideas, going even beyond his English mentor in his numerous publications on the question of man's kinship to other primates (Kelly 1981). Author of numerous scientific and popular books on the question of evolution, Haeckel's most controversial work was *Natürliche Schöpfungsgeschichte.* Published in 1868, it was an immediate bestseller and quickly transformed him "into one of the most renowned scientists and writers in Germany" (Gasman 1969, 36). Darwin himself wrote that if he had first read Haeckel's book he probably would not have bothered with his own *Descent of Man* (1871). When it was subsequently translated into English in 1876, a reviewer noted: "The *History of Creation* gives to most Englishmen the fullest opportunity they have yet enjoyed of understanding what the complete doctrine of Development, as applied to life, really means, what is really involved in it, and on what grounds of evidence its truth is maintained. In treating of the subject, Haeckel is a man absolutely without fear. Believing in no personal creator, in no human identity apart from that of the material organism of the moment, in no future life except that of the race, he is one of the few who find themselves perfectly content with this position and this prospect" (Mott 1876, 1–2). Haeckel himself referred to his theory of human evolution as "monism," and as Daniel Gasman has argued, the biologist's materialism was considerably leavened by German Romanticism and philosophical idealism, which eventually paved the way for the appropriation of several of his ideas into National Socialism (Gasman 1969). Writing in 1929, E. Nordenskiold describes Haeckel's life as "one of the most remarkable during the epoch just closed," and his *History of Creation* as "the world's chief source of knowledge of Darwinism" (Nordenskiold 1929, 505, 515).

91. See, e.g., Anon. 1897, 111–12; Murray 1866, 69.

92. So much so that Haeckel himself has been (mis)credited with authorship of Lemuria in subsequent scholarship (see, e.g., Clute and Grant 1992, 574; de Camp 1970, 52; Hammer 2001, 101; Wauchope 1962, 38). Predictably, biographical studies on Haeckel and of his long-term influence on his scientific peers do not even mention Lemuria or his labors of loss around it (see, e.g., Bolsche 1909; Gasman 1969; Kelly 1981; Nordenskiold 1929).

93. Lyell 1914, 1

94. Huxley 1902, 77.

95. Lemuria is rarely noticed in the vast scholarly literature on human antiquity and evolution, and when it is mentioned at all, it is generally in the course of a discussion of Haeckel's theories. See, e.g., Bowler 1986, 67; Reader 1981, 40; Theunissen 1989, 6.

96. For the fraught debates on this issue, see, especially, Bowler 1986 and Lewin 1987.

97. Topinard 1878, 531, 535.

98. Quatrefages de Breau 1883, 178 (emphasis mine).

99. Mathew 1889, 339–40.

100. Brinton 1901, 222–23.

101. Haeckel 1876b, 399.

102. Haeckel 1905, 634–35.

103. Quoted in Keith 1934, 1.

104. Quoted in ibid., 2.

105. Haeckel 1876b, 295.

106. Ibid., 401. In his *Descent of Man* (1871), Darwin, too, suggested that Man descended from old world monkeys (Catarrhine) whose primitive ancestors, in turn, were extinct forms of the Lemuridae, but he does not invoke Lemuria.

107. Bloch 1986, 777–78.

108. For the persistence of polygenist ideas in the latter half of the nineteenth century in Europe and the United States, even after Darwin's resounding argument that all humanity had descended from a common ancestor, see Stocking 1968. In fact, in an address to the Anthropological Society of London in 1864, Alfred Wallace had already made a similar move to Haeckel's when, according to Stocking, he observed that "all men had in fact descended from a common root. But the moment of that single ancestry lay so far in the past that by the time man's forebears had acquired the intellectual capacities which made them truly human, the various races had already been differentiated by natural section, and it might fairly be asserted 'that there were many originally distinct races of men . . .'" (Stocking 1968, 46). Haeckel's innovation lay in suggesting a shared Lemurian homeland for the time before the differentiation.

109. Bowler 1986, 67. As Bowler and others have noted, although human fossil remains were being increasingly discovered from the 1820s, it was not until the early years of the twentieth century that they became critical in debates on human origins. So, Haeckel himself argued in the 1890s that "we can only partly acknowledge the great importance which laymen and narrow specialists attach to the evidence of such 'fossil men' and 'transitional forms between the ape and man.' He who has a thorough knowledge of comparative anatomy and ontogeny, as well as of paleontology, and who is capable of an open-minded comparison of the phenomena, has no need of those fossil documents in order to accept the 'descent of man from the ape' as an historical fact" (quoted in Theunissen 1989, 13).

110. Haeckel 1876b, 326 (emphasis mine).

111. Eiseley 1958, 264.

112. Kelly 1981, 117.

113. Haeckel 1876b, 307, 310.

114. Haeckel 1876b, 326–27.

115. Quoted in Murray 1866, 64–65.

116. Ibid., 68.

117. Haeckel's Lemuria appears briefly in Fredrick Engels's 1876 manuscript entitled *The Part Played by Labor in the Transition from Ape to Man,* in which he writes, in the very opening page: "Many hundreds of thousands of years ago, during an epoch not yet definitely determinable of that period of the earth's history which geologists call the Tertiary, and most likely towards the end of it, a particularly highly-developed species of anthropoid apes lived somewhere in the tropical zone—*probably on a great continent that has now sunk to the bottom of the Indian Ocean*" (Engels 1975, 1 [emphasis mine]). Although he does not invoke Haeckel, nor name the great sunken continent as Lemuria, he does write that the highly developed species found there was "the decisive step in the transition from ape to man" (ibid., 2). As best as I have been able to ascertain, this is the only appearance in the vast writings of Engels (or Marx) of Haeckel's hypothetical cradle of the human race. Scattered references to Lemuria as birthplace of mankind, or as the homeland of "the Negritos," can also be found in numerous contemporary scholarly journals of some influ-

ence. E.g., see *Journal of the Anthropological Institute of Great Britain and Ireland* 4 (1875): 462–63; 5 (1876): 463–64; 9 (1880): 49–50; and 12 (1883): 492–93, 516–517; and *American Journal of Sociology* 8, no. 6 (1903): 734; and 9, no. 1 (1904): 90–92.

118. Bowler 1986, 68–69; Kelly 1981, 20–26.

119. I thank Frances Gouda for bringing this connection to my attention, and for guiding me to the various sources on Dubois that I use here. See also Gouda (1995, 118–56) for an analysis of Dutch ethnological practices in Indonesia in the context of which Dubois's "discovery" has to be located. I am also immensely grateful to Lisa Klopfer for help with the Dutch sources I have used in these pages.

120. Gouda 1995, 142; Leakey and Slikkerveer 1993, 73–89, 130–37; Reader 1981, 41–42. Indeed, in an 1895 letter to Haeckel, Dubois declared passionately, "I should like to tell you how happy I am to be able to express my gratitude for the influence which you, especially through your '*Schöpfungsgeschichte*' have exerted on the whole course of my life" (quoted in Theunissen 1989, 73). However, in an earlier essay in which Dubois details his reasons for why he chose the Dutch East Indies as the possible birthplace of man, he does not mention Haeckel or his theories (Dubois 1888). Bart Theunissen downplays Haeckel's influence on Dubois. While observing that, "We shall continue to be in the dark as to the driving force behind his decision," he notes that it was the discovery of a fossil ape in the Siwaliks in 1878 that compelled the Dutch doctor to go to the East Indies in the first place (Theunissen 1989, 27–37).

121. In an earlier essay Dubois had speculated (following Wallace) that there had been a land connection (*verbinding door land*) in recent geologic time linking India with the Indonesian islands. This connection is not presented as part of Lemuria, although he did speculate that "Indo-maleische," "Indomalaya" (as he refers to this connection) was the probable home of the first human (Dubois 1888, 152–61). However, note John Reader's response to the 1891 discovery of the Trinil fossils, which is overdetermined by his conviction that Dubois was obsessed with the idea of Haeckel's submerged continent: "While not exactly confirming the existence of Lemuria, this certainly affirmed the existence of a landbridge between India and Java across which animals could have mingled" (Reader 1981, 44).

122. Bart Theunissen suggests that it is possible that Dubois chose this name made popular by Haeckel in order to garner the latter's support for his fossil find (Theunissen 1989, 60).

123. Quoted in Reader 1981, 47.

124. Haeckel 1898, 26.

125. Quoted in Theunissen 1989, 58. Indeed, as early as 1888, Dubois had noted that Lemuria, which had been previously proposed as the first human birthplace, "has since been disproved by zoologists and geologists" (Dubois 1888, 163–64). Curiously, though, several scholarly studies on Eugene Dubois insist on his fascination with the lost continent (Gouda 1995, 142; Leakey and Slikkerveer 1993; Reader 1981, 37–54). Richard Leakey's glossy publication even has a chapter evocatively called "Lemuria: Journey to the Netherlands East Indies," in which he suggests that "Dubois was convinced that humankind had evolved from these early ancestors in the so-called 'paradise' of Haeckel's *Lemuria*" (Leakey and Slikkerveer 1993, 73–91, 99).

126. Keane 1899, 1.
127. Ibid., 5.
128. Ibid., 5–6
129. Keane 1896, 229.
130. Ibid., 226, 236. Years before Keane, paleogeographers had proposed that Australia might well have been part of the vanished Indo-African territorial connection (Blanford 1875, 538–40), and Suess's Gondwanaland included Australia in some of its incarnations.
131. Ibid., 231.
132. Ibid., 229.
133. Ibid., 240.
134. Keane 1908, 2–4. Similarly, the American anthropologist Daniel Brinton, who questioned Lemuria's status as the birthplace of man, acknowledged the possibility that territorial connections between the islands of the Indian Ocean might have facilitated the migration of various black races (Brinton 1901, 223).
135. Wells 1932, 98.
136. Bowler 1986, 142.
137. Quoted in ibid., 143.
138. The Anthropological Society of Australasia was founded in 1885 by Allan Carroll, who also edited *Science of Man,* the first anthropological journal to be published in Australia, from 1898. At the time that these essays were published in the early years of the twentieth century, some paleo-scientists began to explore the possibility of Australia as the site where the earliest humans might have evolved (Bowler 1986, 138).
139. Anon. 1908a, 4.
140. Anon. 1898, 91.
141. Anon. 1908b, 70.
142. Anon 1898, 91.
143. Anon 1902, 74–75.
144. Haeckel 1876b, 329–32.
145. Keane 1896, 417, 254. Well before Keane, Richard Owen and Andrew Murray had made a similar suggestion in the 1860s, when they identified the Andaman Islanders as members of an "old race" who had inhabited a now-submerged continent in the Indian Ocean (Murray 1866, 56–76). In Keane's formulation, in contrast to Haeckel's, Dravidians were a Caucasian people who entered India much later from the Northwest.
146. Bryant 2001; Ramaswamy 2001a.
147. Mazumdar 1912a, 488.
148. Ibid., 489.
149. Ibid., 490–91.
150. Mazumdar 1912b, 12.
151. Ibid., 13.
152. Even before it appeared in Mazumdar's labors of loss, Lemuria surfaced eccentrically in colonial census reports in the course of brief discussions on the original homeland of India's "aboriginal" populations (e.g., Baines 1893, 123; Stuart 1893, 211–12; Risley and Gait 1903, 508). The occasional schoolbook also speculates in a similar vein. E.g., a textbook translated into Tamil in 1936 and published

by Longmans informs the school child that "In the Andaman Islands as well as in African forests lived a primitive uncivilized folk who are called Negrito. Animal bones as well as other artifacts have been discovered in southern India, in eastern Africa, and in the islands of the Indian Ocean. Because of this similarity, it has been suggested that countless years ago, the west coast of India was joined to Africa, and its east coast to Malaysia and to Australia. India's Negrito population is similar to the Negroes of Africa and could have migrated into India via this land connection" (H. C. 1936, 12–13).

153. I thank Gregory Possehl for bringing Roy's work to my attention.

154. Roy 1912, 3.

155. Ibid., 10–17.

156. Roy 1966, 45. The Tamil historian V. R. Ramachandra Dikshitar, whose own views on Lemuria as a birthplace of mankind are rather inconsistent, dismissed Roy's theory in 1947 as "ingenious" and "difficult to accept" (Ramachandra Dikshitar 1971, 15, 80).

157. Srinivasachari 1938, 37. Correspondingly (and in opposition to a majority of Tamil's devotees), Srinivasachari questioned the possibility that the Dravidians—who he insisted came into India from the Mediterranean—could have had their former homeland in Lemuria (ibid., 39). As early as 1914, a fellow Brahman, the litterateur M. Srinivasa Aiyangar, had already made a similar suggestion when he wrote that the "aboriginal inhabitants" of India were "the Nagas" who belonged to "the Negrito race," and who had migrated to the subcontinent "from the south when it was connected by land with Australia." Although he did not know when this happened, Srinivasa Aiyangar observed that the Nagas were eventually assimilated into the Dravidian populace that had migrated into India from somewhere in Asia Minor (Srinivasa Aiyangar 1914, 277–79).

158. See Bryant 2001; Leopold 1970.

159. Das 1921, xii.

160. Ibid., 97–98 (emphasis mine).

161. The map is entitled "Rig-Vedic India (Sapta-Sindhavah & The Deccan): A geological study in the light of Rig-Vedic evidence."

162. Das 1921, 102–3. A few decades later, Jwala Prasad Singhal, a former college principle in Moradabad, published a slightly modified version of Das's theory, complete with maps (Singhal 1963; Singhal [1968?]). I thank Andrew Grout for bringing these references to my attention.

163. Senghor 1977, 2–3.

CHAPTER 3

1. Weber 1946, 142.

2. There are almost as many definitions of "the occult" as there are varieties of esoteric practices. In *The Theosophical Glossary* (1892), the occult is defined as "the science of the secrets of nature—physical and psychic, mental and spiritual" (Blavatsky 1978, 237). One of Blavatsky's successors, Charles Leadbeater, characterizes it as "the science of the hidden" (Leadbeater 1974, 3). More recently, historian Bruce Campbell defines the occultist as "one who operates outside established religion, and has a concern for theories and practices based on esoteric knowledge.

Occultism often includes the study of writings felt to contain secrets known to ancient civilizations but subsequently forgotten" (Campbell 1980, 10). The sociologist Edward Tiryakian writes: "By 'occult,' I understand intentional practices, techniques, or procedures which (a) draw upon hidden and concealed forces in nature or the cosmos that cannot be measured or recognized by the instruments of modern science, and (b) which has as their desired or intended consequences empirical results, such as obtaining knowledge of the empirical course of events or altering them from what they would have been without this intervention" (Tiryakian 1974, 265). My own understanding of the occult is derived from these formulations.

3. Brief discussions of the recurrence of "lost continents" in modern occultism may be found in Ashe 1992, 8–15; Hanegraaff 1996, 309–11; Hammer 2001, 99–108; Jenkins 2000, 83–84; and Melton et al. 1990, 258–61. None of them theorize, however, about the occult preoccupation with lost places.

4. See, e.g., de Camp 1970, 54–129; Ellis 1998, 65–70; Gardner 1957, 165–72; Godwin 1972, 170–82; Wauchope 1962, 28–49, 118–33.

5. Rossi 1984, ix.

6. Rudwick 1986, 311–12.

7. Eiseley 1958, 81.

8. Rossi 1984, ix. See also Koyre 1957.

9. Rossi 1984, 120.

10. Georg 1931, xiv.

11. Owen 2001, 88.

12. Browning 1924, 5.

13. A native of Ukraine, Blavatsky experimented with automatic writing as a teenager, and also began traveling around Europe and Egypt learning about the occult world in its myriad forms. She moved to New York in 1873 and, partnering up with the American Henry S. Olcott (1832–97), founded the Theosophical Society in 1875. She relocated to Bombay with Olcott in 1878, and subsequently to Madras in 1882, where the headquarters for the Theosophical Society was established in Adayar. Following numerous scandals there, she departed from India in 1887 and finally settled in London until she died. A prolific writer, Blavatsky authored numerous books, including some that have become classics in occult circles. *Isis Unveiled* established her credentials as an occultist of enduring significance. Among other things, its importance lies in its attempt to clear the ground for the "idea of a wisdom-religion, a universal and ageless occult knowledge derived from a civilization in which there was a unity between science and religion" (Campbell 1980, 35–36). For a recent critique of Blavatsky's significance, see Washington 1995. For a feminist assessment, see Basham 1992. The best scholarly treatment of the Theosophical Society remains Bruce Campbell's, which defines Theosophy as "a new movement [that] claimed to transcend the cleave between science and religion by a return to the concerns of an ancient wisdom-tradition, long forgotten" (Campbell 1980, 29). The Theosophical Society itself sees its mission as the restoration "of the world to the Science of the Spirit" (see editorial statement of *The Theosophist*).

14. Blavatsky 1931, 120.

15. Ibid., 592.

16. Ibid., 575.

17. Ibid., 589.

18. Blavatsky 1880, 160. For an immediate critical response, see Biswas 1880.

19. Melton 1988.

20. The "Mahatma letters" was a compendium of missives that were received in the early 1880s by the Anglo-Indian journalist and Theosophist Alfred P. Sinnett (1840–1921). The mysterious manner in which the letters reached Sinnett (through Blavatsky's occult mediations) added to their notoriety, as did their occult authorship by two apocryphal individuals, Koot Hoomi ("K.H.") and Morya ("M."), who Blavatsky referred to as Brothers, Masters, or Mahatmas. On the Theosophical concept of the Mahatma, see Campbell 1980, 53–58. For a recent provocative assessment of Blavatsky's authorship of the Mahatma letters, see Viswanathan 2000.

21. Barker 1923, 151–52.

22. The *Secret Doctrine* was originally formulated in India, written while Blavatsky traveled across Europe, and eventually completed—with considerable help from some of her followers—while she was in residence in England. This truly transnational text is remarkable on several levels, not least in its rehabilitation of "Archaic teachings," which, it repeatedly insisted, had already anticipated what modern science was only then discovering about the earth's past, subjects like submerged continents and drowned races.

23. See, e.g., Barborka 1979; Besant 1904; Besant and Leadbeater 1922; Browning 1924; de Purucker 1979; Georg 1931; Wakelam 1979; Wright 1894.

24. Wallace's pervasive interest in occultism has been well documented (see, e.g., Oppenheim 1985, 296–325). See also note 58.

25. Williams 1991, 135–40. Ignatius Donnelly (1831–1901) was a Minnesota politician and Congressman whose book "was the major source for speculation on [Atlantis] for more than a century within the occult community" (Melton et al. 1990, 153). It is also possible that Donnelly himself had heard of Lemuria from Blavatsky (1880).

26. Blavatsky 1931, 594–95. See also Blavatsky 1928, 233–34. Louis Jacolliot (1837–90) was a colonial administrator based in India in the 1860s. His *Histoire des vierges: Les peuples et les continents disparus* (1879) appears to have been widely read in occult circles (de Camp 1970, 58–60).

27. Author of numerous esoteric works, Harris (1823–1906) was born in England but lived in New York, where he was a spiritualist and Swedenborgian minister before he moved to California in 1875 (Jenkins 2000, 39). "His writings comprised a synthesis of the occultisms of India and the West, with multiple planes of existence, a fantastic account of life on other worlds, and in the many heavens and hells surrounding these worlds" (de Camp 1970, 58).

28. Newbrough (1828–91) was a New York spiritualist and "author" of *Oahspe.* One of the first modern works produced through channeling, between 1881 and 1882, it claims that the human race originated on the Pacific continent of Pan and traces its history over the past 78,000 years. Pan was destroyed by a catastrophic flood (Melton et al. 1990, 331). Newbrough's Pan resurfaces as a subterranean Lemuria in Richard Shaver's *I Remember Lemuria* (1948), of which I write later in this chapter.

29. For Theosophy, man "is not necessarily of just the form you now see. 'Man' means that being in whom Spirit and Matter have joined hands. . . . [Man] is not limited simply to ourselves, one puny race of the vast human Hierarchy" (Besant 1904, 22–23).

30. Not surprisingly, the Theosophical appropriation of Sclater's Lemuria has been disparaged by historians of science. See, e.g., Marvin 1973, 54; and Radner and Radner 1982, 89.

31. Blavatsky 1928, 160.

32. Ibid., 8

33. It is telling that Blavatsky named the continent that housed her Third Root-Race as Lemuria, for it had other archaic and esoteric names, "which varied with the language of the nation which mentioned it in its annals and scriptures." Because these varied names might cause confusion, she decided to adopt a name that would be "more familiar to the cultured reader" (Blavatsky 1928, 5–6). In a later text, the choice of the name Lemuria is explained thus: "A modern term first used by some naturalists, and now adopted by Theosophists, to indicate a continent that, according to the Secret Doctrine of the East, preceded Atlantis. Its Eastern name would not reveal much to European ears" (Blavatsky 1978, 187). Blavatsky's most well-known successor, Annie Besant, frequently referred to Lemuria by its "Eastern" name as "Shālmali," possibly in deference to her largely Hindu audience in colonial India, where she was based for much of her occult life (Besant 1904, 74–75).

34. Blavatsky 1928, 275. See also Scott-Elliott 1904, 1.

35. It is a measure of Theosophy's conflicted intimacy with geology that it's the latter's secular periodization of Earth history that was adopted by Blavatsky and her followers. However, as Blavatsky was quick to note, Charles Lyell might have "happily invented" the terms Eocene, Miocene, and Pliocene, but he left the duration of these periods "to the speculations of specialists, [and] the greatest confusion and perplexity are the results of that happy thought. It seems like a hopeless task to succeed in quoting a single set of figures from one work, without the risk of finding it contradicted by the same author in an earlier or a subsequent volume " (Blavatsky 1928, 9).

36. Geoffrey Ashe identifies Scott-Elliot as an English merchant and amateur anthropologist whose works on Lemuria and Atlantis "won the guarded approval of the Theosophical Society" (Ashe 1992, 10).

37. Scott-Elliot 1904, 17–19. For the Victorian fascination with the newly discovered dinosaur, "dinomania," see Cadbury 2000. The Theosophical innovation lay in suggesting that man, too, was around when dinosaurs roamed the earth. This suggestion was taken on board by both post-Theosophical occultists as well as some Tamil devotees.

38. Blavatsky 1928, 6–7. In Greek antiquity, Hyperboreans inhabited Hyperborea, "the land beyond the North Wind" at the northern edge of the world. They led pure and pious lives beyond the reach of the mere mortals of the oikoumene. For nineteenth-century Europe's fascination with them, see Godwin 1993. Curiously, before they make it into Blavatsky's place-making, Hyperboreans appear in Ernst Haeckel's 1870 map, which illustrates the diffusion of the twelve races of man from Lemuria (see Fig. 3).

39. Wright 1894, 137. Given the lack of consensus in the paleo-sciences about the age of the earth or of humanity, Blavatsky preferred the certitude of her Archaic Records in dating the birth of mankind to eighteen million years ago, polemically insisting, "If Modern Science is unable to estimate the date of so comparatively

recent an era as the Glacial Epoch, it can hardly impeach the Esoteric Chronology of Race-Periods and Geological Ages" (Blavatsky 1928, 823). Besant later commented on the "great obscurity" surrounding the dates assigned to various events in Theosophy's version of human evolution, noting, however, that since she herself had "no faculty for fixing ancient dates," she had followed Blavatsky in this regard as in others (Besant 1904, 118–19).

40. Scott-Elliot 1904, 23–24. See also Besant 1904, 79–80; Wright 1894, 137.

41. Blavatsky 1928, 181.

42. Scott-Elliot 1904, 24.

43. Ibid., 34. See also Blavatsky 1928, 181, 208. Theosophists were not alone in speculating thus about other planets being inhabited. Scientists, theologians, and others in Victorian Britain also debated about what this might mean for the Christian belief that the earth is "in a unique and special manner, the field of God's Providence and Government" (Gillispie 1951, 205–6).

44. Scott-Elliot 1904, 38. See also Blavatsky 1928, 285.

45. Blavatsky 1928, 331.

46. Ibid., 330–31.

47. Georg 1931, 96–97.

48. Vitaliano 1973, 11, 184–87.

49. Besant 1904, 74–75.

50. Scott-Elliot 1954, 21.

51. Blavatsky 1928, 230.

52. Ibid., 275.

53. Scott-Elliot 1904, 1.

54. Blavatsky 1928, 348.

55. Barker 1923, 156.

56. Besant and Leadbeater 1922, 447. Melton notes that Besant and Leadbeater together "filled in the gaps of the outline of philosophical thought left by Blavatsky. . . . Leadbeater convinced Besant that she could, like him, see clairvoyantly the realities of the occult (hidden, invisible) world, and their books report the results of their investigations of this world" (Melton et al. 1990, 67). As president of the Theosophical Society from 1907 to 1933, Besant's influence on Theosophy was next only to Blavatsky's (Campbell 1988, 113–31; Viswanathan 1998).

57. Blavatsky 1928, 7.

58. Ibid. (emphasis mine). Although Blavatsky's credibility in this regard is rightfully very much in doubt, she might well have received a "private letter" from Wallace, who was a fellow traveler in Euro-American spiritualist and occult circles. But there is little to indicate that he was drawn to Theosophy. On the contrary, he wrote to a friend in April 1897 that "I have tried several Reincarnation and Theosophical books, but cannot read them or take any interest in them" (quoted in Oppenheim 1985, 321). Although several scholars have observed that Wallace dabbled in Theosophy, Janet Oppenheim concludes that "Wallace's views on progress after death were very close to Blavatsky's, . . . [but] little else about Theosophy would have been likely to hold Wallace's interest for long" (Oppenheim 1985, 470).

59. Besant 1904, 74.

60. Sinnett 1885, 280.

61. Browning 1924, 5. See also Georg 1931, xvi.

62. Blavatsky 1928, 348, 826. See also Blavatsky 1880, 279; Besant and Leadbeater 1922, ii; Wright 1894, 137.

63. Besant 1904, 74.

64. Blavatsky 1928, 180–81.

65. Blavatsky 1928, 160. Elsewhere, she shudders, "Our Divine Races are shown to be the descendants of Catarrhine apes, and our ancestor, a piece of sea-slime" (ibid., 276). At least one latter-day occultist shared the Theosophical contempt for material science's ascription of a simian origin to man. Thus, the Anglo-American James Churchward declared trenchantly in 1924, "In their anxiety to sustain their monkey theories, scientists have tried to prove that man did not appear upon the face of the earth until the early Pleistocene Time, but a pin-prick can dissipate this scientific bubble" (Churchward 1994a, 43).

66. Scott-Elliot 1904, 2.

67. Quoted in Lewin 1987, 59.

68. Blavatsky 1928, 274. See also Besant 1904, 114; Scott-Elliot 1904, 29–30; Sinnett 1885, 83.

69. Blavatsky 1928, 341. Elsewhere, she writes, "Africa, as a continent, was never part and parcel of either Lemuria or Atlantis" (ibid., 275). Here, as in other regards, Blavatsky contradicts herself, for elsewhere in the *Secret Doctrine* her Lemuria did include "at least portions of what is now Africa" (ibid., 7).

70. Sinnett 1885, 75–76. Since the mid-nineteenth century, esotericism has "encompassed belief in spiritually evolved beings in other worlds than our own" (Hammer 2001, 54). See also note 43.

71. Sinnett 1885, 75.

72. Blavatsky 1928, 200.

73. Washington 1995, 52.

74. Blavatsky 1928, 343.

75. Barker 1923, 151.

76. Besant 1904, 114 (emphasis mine).

77. Barker 1923, 154. See also Besant 1904, 144; Blavatsky 1928, 824–25.

78. Wright 1894, 126.

79. Scott-Elliot 1904, 24.

80. See Campbell 1980, 65.

81. For a nuanced argument regarding the colonial context of this evolutionary vision, see Viswanathan 1998.

82. See, for instance, Barborka 1979 and Wakelam 1975, including the latter's statement, "Let us conclude that the discoveries of physical science have not contradicted the occult traditional sources in any major sense" (74).

83. Charles Vivian's *City of Wonder* (1922) is the story of the quest for the Atlantean city of Kir-Asa, which could be reached through a valley "where ghosts chase women," and which was inhabited by "astral" survivors from Lemuria. As one of the protagonists of the novel explains, "On that first continent [Lemuria] . . . there were reptiles and low animal forms of life, but man did not evolve out of them as Darwin and his school of thought claimed. Man is a spirit—man came to earth as a spirit. . . . Anyhow, think it over—it's a more reasonable theory of evolution than Darwin's" (Vivian 1973, 73–75). In Carter's *Thongor and the Wizard of Lemuria* (1969), Lemuria appears as a primeval land teeming with thick jungles, fierce drag-

ons, and "savage beast-men." The author had clearly read both Blavatsky and Scott-Elliot, whose names appear in his acknowledgements.

84. Goodricke-Clarke 1985, 22, 52–55, 101–2.

85. In this context, note also the eccentric appearance of Lemuria in the Cosmic Ice Theory of the Viennese mining engineer Hans Hoerbiger, a theory which was quite fashionable in Nazi Germany (Ashe 1992, 14; Gardner 1957, 37–39).

86. An Austrian Catholic by birth, Steiner began his professional life as an editor of the philosophical works of Goethe, of whom he was a great admirer. He flirted with Theosophy after he moved to Berlin in 1897, but by 1906 began to create his own occult synthesis that he called Anthroposophy, which was a blend of Theosophical, Rosicrucian, and mystic Christian notions leavened by Goethe's Romantic mysticism. The Anthroposophical Society was founded in 1912 and today has numerous branches in Europe, Britain, and the United States. Peter Washington observes that Steiner became the adult that Wordsworth dreamed of becoming, "one who never lost the sense that there is beyond—and yet somehow immanent in—the visible world a celestial realm accessible to the spiritual eye" (Washington 1995, 146). Rejecting Kant's contention that we cannot experience things-in-themselves, Steiner's Anthroposophy was an attempt to explore the relationship between phenomena and noumena. For recent assessments of Steiner, see Campbell 1980, 155–58; and Washington 1995, 145–53, 248–53. Neither attempts, however, to interrogate Steiner's place-making of Lemuria (or Atlantis).

87. Steiner 1911, 171. Geoffrey Ashe likens the Akashic Records to "a kind of collective memory-bank on the astral plane, which the initiate can tap to discover facts about the past that have escaped documentation" (Ashe 1992, 11). Peter Washington describes them as an "astral library of everything that has ever happened in collective spiritual history" (Washington 1995, 120).

88. Steiner 1911, 100–101.

89. Born Carl Louis van Grasshoff in Germany, Heindel was a professional engineer before he came under Theosophy's influence in 1903 when he moved to California, even becoming vice-president of the Lodge in 1904–5. In the course of a lecture tour in Germany in 1907, he was contacted by an Elder Brother of the Rosicrucian Order, and he insisted that his own 1911 work was basically an exposition of the Brother's message. Along with his wife, Heindel established several centers of the Rosicrucian Fellowship, headquartered in Oceanside, California (Campbell 1980, 160–61; Melton et al. 1990, 208, 395).

90. Heindel 1911, 275.

91. Ibid., 281

92. Ibid., 289. See also his statement, "The Negroes and the savage races with curly hair are the last remnants of the Lemurians" (ibid., 304).

93. Ibid., 291.

94. Ibid., 306.

95. See, e.g., Bailey 1922, 138–39. Born in Manchester, Bailey was raised in the Church of England. After a brief stint in India working for the YWCA (1898–1907), she moved to the United States with her husband. She was introduced to Theosophy in 1915 when she moved to California, and she became an active member of the Society. Around 1919 she insisted that an Adept called Djwhal Khul (or "the Tibetan") had contacted her, and she became a channel for his message, which

found expression in more than a score books. The mission of the Arcane School is to implement the program of the Great White Brotherhood and to work toward the establishment of the New Age (Melton et al. 1990, 56–57).

96. Besant and Leadbeater 1922, ii.

97. Ibid., iv.

98. Donnelly 1882, 32. As I noted earlier, it is possible that Donnelly was inspired to suggest this after reading Blavatsky's 1880 essay, "A Land of Mystery," in which she wrote of "an immense submerged Pacific continent," attested to by "the most ancient traditions of various and widely-separated peoples—legends in India, in ancient Greece, Madagascar, Sumatra, Java, and all the principle isles of Polynesia, as well as those of both Americas" (Blavatsky 1880, 279).

99. Kyle 1993, 96–97.

100. Jenkins 2000, 78.

101. Quoted in ibid., 7.

102. Quoted in Wauchope 1962, 36–37. Lemuria's relationship to Atlantis, with which it is frequently twinned, varies in post-Theosophical occultism. In the psychic visions of Edgar Cayce (1877–1945), arguably the most famous of American occultists of the first half of the twentieth century, Lemuria is a Pacific continent that disappeared "in a series of cataclysms lasting 200,000 years," before the emergence of an Atlantean civilization (Johnson 1998, 64). For New Ageist Michael Baran as well, Lemuria preceded Atlantis by a few centuries; its "loftier" inhabitants had a tense relationship with Atlanteans. The Biblical fratricidal confrontation between Abel and Cain "represents the ancient power struggle between Atlantis and Lemuria" (Baran 1981, 41). In David Manley's fictional *Aros of Atlantis*, Lemurians and Atlanteans destroy themselves in a catastrophic "thermonuclear war" in 23,638 B.C. (Manley 1972, xiii–xvii). The coupling of Lemuria and Atlantis is another reason for the former's visibility in American occultism, in which Atlantis has always had an important place. This coupling is, of course, a modern innovation, since Plato's Atlantis story had no place for Lemuria.

103. Jenkins 2000, 8.

104. Taffinder 1908, 165. The essay was published in a periodical called the *Overland Monthly*, whose editor insisted that the article "is splendidly written, of absorbing interest, fantastically beautiful, and yet impossible of acceptance as a fact" (quoted in Taffinder 1908, 163).

105. Taffinder 1908, 163.

106. Ibid.

107. Ibid., 165.

108. Oliver names Lemuria also as Lemurious or Lemuros, and identifies it as "a continent of which Australia is the largest remnant to-day" (Oliver 1905, vii). Phylos offered Oliver fabulous glimpses of life on Lemuria, as well as of its destruction, as for the Theosophists, by volcanic eruptions and fire "from out of the inter-planetary depths" (ibid., 351–52, 408). For brief discussions of Oliver's place in American occultism, see Ashe 1992, 45; and Zanger 1992, 98.

109. For details on AMORC, founded in 1915 and still in existence, see Melton et al. 1990, 16–17.

110. An early chapter of the book was published in 1925 in the Rosicrucian periodical *The Mystic Triangle*. Gratified by the overwhelming response he received,

Cervé wrote his monograph. A short black-and-white film on Lemuria (based on the book) was also produced circa 1935 by AMORC. Unfortunately, the film is no longer available in any archive I have consulted, including the Rosicrucian Order's.

111. Cervé 1997, 13, 49.

112. A 1938 advertisement for the book carried the following announcement:

> In the depths of the Pacific, shrouded in darkness, lies a vast continent. Where once the great edifices reached skyward and multitudes went their way is now naught but the ceaseless motion of the sea. Centuries before the early men of Europe or Africa found the glorious spark of fire or shaped stones into crude implements, the Lemurians had attained an exalted culture. . . . Has the learning of this early civilization been completely lost? Was their strange knowledge submerged with the land upon which they dwelt? Whence came these people? And were they all destroyed? Science today is proving the physical existence of the continent, and down through the ages there has come the tale of a strange people who live today and have preserved the mystical knowledge of Lemuria. (Reprinted in Zanger 1992, 99)

113. Marty 1991, 258. Marty also writes of the "instability of American existence in the Depression era . . . the desire of many religious citizens to experiment, to reach out into experiences borrowed from elsewhere, or from nowhere, which is what Utopia means" (ibid., 259). See also Jenkins 2000, 149–64.

114. Cervé 1997, 19–20.

115. Steibing 1984, 33–34.

116. Cervé 1997, 49.

117. Ibid., 73.

118. A year before Cervé published his work, Muriel Bruce had written a novel in which its principal protagonist, a British archaeologist announced that "Mu, the Motherland, was probably the Pacific continent, Lemuria, which we have always believed to be much older than Atlantis" (Bruce 1930, 273).

119. Wauchope 1962, 44–49.

120. Churchward 1994a, Preface. This first book was part of a trilogy published subsequently in 1931 and 1933.

121. Churchward 1994b, 15.

122. Churchward 1994a, 1.

123. Ibid., 32. Here, Churchward—and others who followed in his footsteps—was obviously capitalizing on the primitivist identification of these islands as tropical Arcadias, untouched by the corruptions of modern civilization and its materialist excesses (Smith 1985).

124. A 1938 advertisement for the book told its readers:

> Alive Today? Majestic Mount Shasta, covered with eternal snow and surveying the great Pacific, harbors strange clues of an unknown people. Tradition and fact unite to tell a weird saga of a tribe reputed to be the descendants of lost Lemuria, who fled to safety, and who dwell in the mountain fastness of Mount Shasta. What are their mystical practices? Do they account for the eerie lights seen far upward toward the summit? Do they practice rituals which had their inception centuries ago? Why are they cloistered from the world? Are they

> masters of nature's laws not yet known to men of today? No other book so thoroughly explains the scientific, mystical, and spiritual achievements of the ancient Lemurians and the remnant of their descendants existing today as does this one. (Reprinted in Zanger 1992, 99)

In fact, the 1960 edition of the book identified Lemurians as "The Mystery People of Mt. Shasta" on its title page. A recent edition (1997) features on its cover a painting ("Awakened Golden Solar Angel," 1990) of Queen Califa with Mount Shasta in the background by the New Age artist and Shasta resident Cheryl Yambrach Rose. "Not many people know that California is named for this Lemurian Empress of the Golden Isle," Rose tells us in a web interview (http://iasos.com/artists/cherrose/). See also Ashe 1992, 13.

125. Cervé 1997, 39–40.

126. Ibid., 147.

127. Ibid., 165.

128. The Theosophist Charles Leadbeater had already identified "Lower California" as the land that "in scenery and climate approaches most nearly to our ideal of Paradise" (Leadbeater 1931, 9–10). In 1897 the American section of the Theosophical Society was relocated to Point Loma in southern California, where it flourished until 1942. Theosophy played the most important role in assuring the nexus between California and the occult that has been a hallmark of that region through the course of the twentieth century. For California's special place in American mysticism and occultism, see St. Clair 1972; Jenkins 2000, 90–93; and McWilliams 1986. The latter argues that migration flows and the resulting population explosion account for California's penchant for the occult. "Migration severs old ties, undermines ancient allegiances. It creates the social fluidity out of which new cultic movements arise" (ibid., 291). Philip Jenkins also usefully notes that

> California law made it very easy to establish a new religious body. The process was basically open to anyone who could lay their hands on a small filing fee and produce a couple of witnesses. [Also,] Californian real estate was very cheap by eastern standards, so that a modest investment could produce an imposing temple or sanctuary with substantial grounds. Religious entrepreneurs found in the West the means, motive, and opportunity to form new sects. Furthermore, new groups faced little opposition from established churches, which throughout the twentieth century were weaker in the western states than anywhere in the nation. (Jenkins 2000, 89)

129. Polk 1991, 13.

130. On the occult's fascination with Mount Shasta, see McGillivray 1985; Melton et al. 1990, 260–61, 292–93; and Zanger 1992, 89–105.

131. Cervé 1997, 178.

132. Ibid., 185–86.

133. Lanser 1932, 16.

134. Michael Zanger reports that in the 1930s the National Forest Service as well the local chamber of commerce began receiving letters from elsewhere in the United States as well as abroad asking about Lemurians. Summers saw the mountain slopes alive with seekers looking for evidence of the lost Lemurians, and tourist pro-

moters began to offer highly priced tours to Mount Shasta (Zanger 1992, 100). In his *Occult America* (1972), John Godwin writes about a group from the American Midwest called "the Celestials," which annually visited Mount Shasta in the hope that they would meet Lemurians (Godwin 1972, 181–82).

135. King 1934, vii. Ballard (who published his channeled visions under the name Godfré Ray King) was a former gold-mine promoter who, along with his wife Edna (who had been a clerk in an occult book shop), started the "I AM" movement around 1930 in Chicago. Although elements from Rosicrucianism, Christian Science, and other countercultural religions entered the I AM message, it was dominantly Theosophical in its thrust, representing Theosophy's "greatest popular diffusion" (Heelas 1996, 45).

136. King 1934, 89–99. For a contemporary observer's evaluation of the brief popularity of the movement, see Braden 1949, 257–307. Whitehead argues that "I AM" was a typical example of a Depression-era cult—"frenzied, short-lived, and legally suspect"—which "did considerable damage to the Occultist's public image" (Whitehead 1974, 558–59). See also Jenkins 2000, 97–99, 149–55.

137. Melton 1988, 133–34. Although the movement basically petered out in the 1940s, a few of the faithful continue to hang around Mount Shasta and continue to publish about Lemuria (e.g., Schroeder 1984).

138. Jungclaus 1990.

139. Walton 1985, cover page.

140. I thank Sandy Freitag for bringing this to my attention.

141. Spence attended the University of Edinburgh before serving as editor for various newsmagazines, including *The Scotsman*, the *Edinburgh Magazine*, and the *British Weekly* (Ellis 1998, 274–75). See also de Camp 1970, 91–98.

142. Spence 1931, 17.

143. Ibid., 218.

144. Ibid., 91.

145. Ibid., 219.

146. Ellis 1998, 44. Indeed, Spence concludes his preface to *The Problem of Lemuria* on the following suggestive note: "The proof that a native white race once dwelt in the Pacific area and that its vestiges are still to be found there is, I am convinced, of the highest moment to the whole study of a difficult question" (Spence 1933, 8).

147. Churchward 1994a, 25. See also Brown 1924 and Hayes 1972.

148. Antoine Faivre writes, "Along with the smoking factory chimneys came both the literature of the fantastic and the new phenomenon of spiritualism. These two possess a common characteristic: each takes the real world in its most concrete form as its point of departure, and then postulates the existence of another, supernatural world, separated from the first by a more or less impermeable partition. Fantasy literature then plays upon the effect of surprise that is provided by the irruption of the supernatural into the daily life, which it describes in a realistic fashion. Spiritualism . . . follows the inverse procedure, teaching how to pass from this world of the living to the world of the dead, through séances of spirit rappings and table tippings" (Faivre 1987, 38). See also Jackson 1981, 25–26.

149. Clareson 1977; Vasbinder 1982.

150. Clute and Grant 1997, 593.

151. Scott 1898.

152. Clute and Grant 1997, 575, 639, 988.

153. Bruce 1930, 274–76. A similar story is Owen Rutter's *The Monster of Mu* (1932), whose hero is an ethnologist who goes off in search of Mu, rumored to be inhabited by white priests and brown pygmies. In Nelson Bond's *Exiles of Time* (1940), a skeptical archaeologist named Lance is converted into a believer after traveling back through time to prehistoric Mu. "His scientifically trained mind discovered the solution to many mysteries that had evaded the researches of man. According to the stodgy Twentieth-century textbooks, man of this period was supposed to have been a crude savage living in caves, huddling over a tiny fire of twigs. That this was erroneous Lance knew now. The Murian [*sic*] civilization was not like that at all. It was . . . well, it was curiously paradoxical. It ran to extremes. . . . He was constantly being forced to recognize here a superiority to Twentieth-Century existence, there a primitive trait. For example, the Murians had motor-driven means of transportation. . . . These strange people built—or evidently had built in the past—gigantic buildings towering to the sky, vast amphitheatres, great monuments of intricate designing" (Bond 1940, 73–74).

154. See, e.g., Roberts 1942. On the popularity of the nether reaches of Earth as a setting for fantasies about dragons, dwarfs, lost races, and so on, see Kafton-Minkel 1989.

155. Carter 1966, 1969.

156. See, e.g., Manley 1972.

157. In this regard, see Vladimir Holan's *Lemuria* (1934–38). Written by one of the most well-known Czech poets of the twentieth century and centering on the lives of three individuals, it explores the possibilities of escape to another land and time out of the present (Holan 1940). "There is nothing about Lemuria in the book, and of course, everything in the book is Lemuria" (Rudolf Mrazek, personal communication). It is perhaps no coincidence that such a book was written and published in the crucial years when Czechoslovakia was invaded by Nazi Germany and incorporated into the Third Reich. I am immensely grateful to my colleague Rudolf Mrazek for discussing this novel with me.

158. Shaver 1948, 1–2. Years later, he changed his position slightly when he observed that it was Palmer who had given an occult slant to his own labors of loss (Cohen 1969, 41).

159. Godwin 1972, 173–74. See also Kafton-Minkel 1989, 133–53.

160. Godwin 1993, 103–4.

161. I thank Margaret Grafeld (Office of Information and Privacy Coordination) and Leo Dillon (Office of the Geographer and Global Issues, Bureau of Intelligence and Research) of the State Department for their assistance in securing the file under the Freedom of Information Act. See also Owen 1988.

162. Mott first appears in the file in May 1934, when he petitioned to have a postal service set up for his recently established "Principality of Atlantis." He also attempted unsuccessfully to enter the United States in April 1936 with a passport issued by the Principality.

163. Meeker first surfaces in the file in 1947, when she wrote to the White House seeking information on "the Atlantis-Lemurian Government." She introduced herself as "a real estate broker of California and a Mining Executive of the U.S.A. and anywhere for that matter."

164. Cervé 1997, 103–4.
165. Ibid., 73.
166. Ibid., 59–68; Churchward 1994a, 6.
167. Cervé 1997, 131.
168. Churchward, quoted in Jenkins 2000, 84.
169. Cervé 1997, 104.
170. Churchward 1994a, 30–32.

171. Although the term "New Age" has been questioned by some, I follow Philip Lucas in my understanding that its four distinguishing characteristics are "a belief in an imminent planetary spiritual transformation that will occur at the level of human consciousness; an ethic of self-empowerment and self-healing as a prerequisite to the healing of society; a desire to reconcile religion and science in a higher synthesis that enhances the human condition both materially and spiritually; [and] strong eclecticism in its embrace of healing therapies, spiritual practices, and millennial beliefs" (quoted in Johnson 1998, 33). I agree with Paul Heelas, who notes that "the New Age remains under-theorized" (Heelas 1996, 8).

172. See also Charles Leadbeater's detailed blueprint of the Utopia created by the Sixth Root-Race (Leadbeater 1931).

173. Lemurian Fellowship 1939, 27–28.

174. My analysis here differs from that of Olav Hammer, who contends that with plate tectonics and marine cartography lost lands like Atlantis have been marginalized, and that "explicit references to mythical continents in New Age books have become rare" (Hammer 2001, 108). The numerous, ever-proliferating New Age sites for Lemuria and Atlantis on the Internet alone lead me to question this.

175. I borrow the phrase "resonant place-name" from Clute and Grant 1997, 575.

176. For example, when you walk through the doorway of the Lemuria Writing Retreat, "you enter another world. . . . Instead of turning automatically towards some printed books, newspapers or human specialists, here in Lemuria you do not have to accept the dictums, principles, theories, ideas, personal conceptions or beliefs of supposed authorities. Here you can write for the sheer sake of writing. You do not have to have the goal of publication in mind. There are no judgements or critiques." Instead, as in Lemuria, writers are encouraged to use techniques of meditation and concentration to improve their craft; they can draw upon their "sixth sense," as Lemurians did, to unleash their creative potential. "In Lemuria it is possible to concentrate your attention and wait for an impression. Here it is possible to communicate with animals and trees in their own language. In Lemuria you are close to the creative force that poets have called divine" (http://www.dailywriting.net/MistyIsle.htm).

177. Lemurian Fellowship 1939, 27.

178. Ibid., 7–8.

179. Robert Stelle learned about Lemuria when the Elders of the lost continent—the Lemurian Brotherhood—channeled their wisdom to him. His *The Sun Rises* (1952) takes the form of a revelation of memories of a "co-operative" life on Lemuria 78,000 years ago, when Lemurians had adhered to Natural Laws governing human conduct and had "succeeded in building a life far superior to the insecure discontented existence we lead today" (Stelle 1952, unpaginated epilogue). See also Melton 1988, 725–26.

180. Lemurian Fellowship 1939, 26.

181. Ibid., 28.

182. Lemurian Fellowship 1945, 13.

183. Lemurian Fellowship 1939, 24. A later text of the Lemurian Fellowship notes that the Elders built a vast underground temple in Tibet where they stored their priceless wisdom for future use (Lemurian Fellowship 1945, 23).

184. Lemurian Fellowship 1939, 46–50.

185. Zitko 1941, n.p; see also pp. 1–2.

186. Melton 1988, 729.

187. Kueshana 1953. For one reader's appreciation of Kueshana's work and the journal called *The Lemurian Builder* published by the Stelle Group, see Childress 1988.

188. Subramuniyaswami 1998, n.p. *The Lemurian Scrolls* describes Lemuria as "the first continent on Earth to be inhabited by humans (Lemurians). Lemuria has been revealed by scripture and explained and made popular by numerous clairvoyants and mystics over the past hundred years as a highly advanced civilization with amazing technologies, maintaining the love of nature and appreciating its many gifts (Subramuniyaswami 1998, 278). For more on Subramaniyaswami, an American who insists he was initiated into the arcane regime of Siddha yoga in the 1940s in Sri Lanka, and who in 1979 founded the magazine *Hinduism Today* (which is read widely in the Hindu-Indian diaspora), see the website entitled "Come to Lemuria!" (http://www.hindu.org/ha/ls/index.html). Many thanks to Indira Peterson for alerting me to this site.

189. Subramuniyaswami 1998, 331.

190. Kyle 1995, 182.

191. Hanegraff 1996, 310–11; Melton et al. 1990, 253–54, 261.

192. For others, see, e.g., http://www.spiritweb.org/Spirit/atlantis-mu-lemuria.html and http://www.lemuria.net/article-lemuria-a-reflection.html.

193. E.g., Junglaus 1990. See also http://chat.lazaris.com/publibrary/publemuria.cfm and numerous other New Age healing sites on the web.

194. Lal 2000, 66.

195. Oppenheim 1985, 160.

196. Ibid., 2.

197. Steiner 1911, 2.

198. Ibid., 5–6.

199. Nandy 1995, 45–46. Nandy's provocative essay makes this plea because millions of people (in places like India, but also elsewhere) living outside disciplinary history have been "disenfranchised and oppressed" and dismissed for "their inadequate knowledge of history." For these people, the ahistorical mode of being is now confined "for private or secret use or for use as forms of fantasy useful in the creative arts." In his reading, modern historical consciousness is "very nearly a totalizing one. . . . Once you own history, it also begins to own you. You can, if you are an artist or a mystic, occasionally break the shackles of history in your own creative or meditative moments. The best you can do, by way of exercising your autonomy, is to live outside history for short spans of time" (ibid., 45). For an earlier discussion regarding "the terror of history" which confronts modern man committed to making himself within and through history, see Eliade 1959, 141–62. Although Eliade does not share Nandy's postcolonial agenda of critical traditionalism and radical social criti-

cism of (colonial) modernity's excesses, there are interesting parallels between the two thinkers in their critique of historicism and their defense of practices that have actively chosen to disregard history. In Eliade's diagnosis, however, archaic society's disregard for history stems from its penchant for the cyclical, the repetitive, and its conformity to archetypes. Nandy does not make such Orientalist claims. For a productive critique of Nandy's position, see Chakrabarty 2000, 247–49.

200. Steiner 1911, 1.

201. Browning 1924, 4.

202. Barker 1923, 151.

203. Georg 1931, xiv–xvi.

204. Thus Eugen Georg quotes from Darwin's *Origin of the Species*: "I look at the geological record as a history of the world imperfectly kept, and written in a changing dialect; of this history, we possess the last volume alone, relating only to two or three countries. Of this volume, only here and there a short chapter has been preserved; and of each page, only here and there a few lines." Georg then triumphally retorts:

> Fossil discoveries are accidental and give a bare morsel of processes that evolved through eons of time. Conclusions based on them must be incomplete. Furthermore, there has never been a systematic and reasonably satisfactory search for human fossils, and five (or seven) continents of the earth are, from this point of view, practically terra incognita. Who can predict what systematic excavations of the Atlantean, Lemurian, Arctic, and Antarctic continental remnants . . . we might find to prove the existence of man in the Tertiary or earlier periods! (Georg 1931, 155–56)

205. The Theosophists were the first to link Tibet to Lemuria when Blavatsky suggested that the Book of Dzyan was in the custody of the Brotherhood of Mahatmas who were resident there, but other occultists also follow suit. The classic works on the Western fascination with Tibet are Bishop 1989 and Lopez 1998. For a provocative discussion of Tibet's place in the modern West as "a sanitarium for the recuperation of an exhaustive knowledge that was always in danger of entropy, loss, or destruction," see Richards 1993, 11–44.

206. For a recent evaluation of modern esotericism's complex relation to science, see Hammer 2001, 202–328.

207. Blavatsky 1928, 206.

208. Sinnett 1954, iii.

209. Ibid., vii.

210. Ibid., viii.

211. Steiner 1911, 167; Sinnett 1954, ix.

212. Steiner 1911, 168.

213. Ibid., 6. For a critique of "psychic archaeology," offered very much from the perspective of professional science, see Williams 1991, 286–304.

214. Blavatsky 1928, 340.

215. Paul Brunton (1898–1981) was a British writer and journalist who first traveled to India in 1931. His first meeting with Ramana Maharshi is documented in *A Search in Secret India* (1934), among the first works to present the Hindu mystic to the West.

216. Brunton 1971, 11–14.
217. Ibid., 19–21.

CHAPTER 4

1. Mutthuvirasami Naidu, 1954–55, 456.
2. Ramaswamy 1997. This is not to say that all Tamil devotees necessarily participate with the same enthusiasm in labors of loss around Lemuria; indeed, a few have been quite critical of this enterprise in the name of Tamil itself. But it is to say that Lemuria's place-makers in the Tamil country are invariably devotees of Tamil. Labors of loss over Lemuria first surfaced within what I have identified as Tamil devotion's classicist imaginary, which was preoccupied with establishing the timeless antiquity and primordiality of Tamil but soon spread to the religious, the Indianist, and Dravidianist discourses on the language as well.
3. I borrow the notion of commemorative density from *Recovered Roots*, Yael Zerubavel's imaginative study of Zionist and Israeli nationalism, in which she notes that certain moments from a nation's past command an intensity of attention while others fall into oblivion. "In this process, such moments are elevated beyond their immediate historical context into symbolic texts that serve as paradigms for understanding other developments in the group's existence" (Zerubavel 1995, 8–9).
4. Ramaswamy 1997, especially, 179–242.
5. Directed by P. Nilakantan (who had had a long career in commercial Tamil films), the script for the film was written by R. Mathivanan, a Tamil scholar best known for his etymological work on Tamil who has also written extensively on Lemuria (see, e.g., Mathivanan 1977; Ramachandran and Mathivanan 1991). Produced at a cost of about 12,00,000 rupees, the documentary has since its original screening been shown on television in India, as well as at several other international Tamil conferences. At least one viewer was inspired to his own labors of loss over the hapless homeland (Ganapathy 1982, 1984). Another viewer wrote approvingly that the film would remind his fellow Tamils of their glorious lost past (Singaravelan 1981, 47–52). For a less enthusiastic appreciation, see Ilantiraiyan 1981, 75–77. In demonstrating the antediluvian origins and history of Tamil, the documentary remarkably combines a mobilist conception of the creation of the earth's continents based on drift theory with an earlier, fixist notion based on submerged continents, thus representing the complex hybridization at their postcolonial address of a century of fraught metropolitan debates over the history of Earth and the creation of its landmasses. Not least, the film is noteworthy for graphically presenting the catastrophic loss of Lemuria to the ravages of the ocean. The documentary was not the only presentation on Lemuria at the conference. A Tamil translation of the Russian Alexander Kondratov's *The Riddle of the Three Oceans* (1974), one of the most quoted books in Tamil labors of loss, was also released to mark the occasion (Kondratov 1995), and several featured speakers traced the origin of Tamil to the lost continent (see, e.g., Devaneyan 1981; Mahalingam 1981a; Pillay 1981b). Indeed, K. K. Pillay (1905–1981), an influential historian who once headed the history department at the University of Madras, was moved to declare, "It is no chauvinism to state that the ancestors of the Tamils were among the earliest people of the world. This view is based by certain oceanographers and ethnologists on the so-called Lemurian the-

ory" (Pillay 1981b, 36). Earlier, he would have been more skeptical of such a stance (see, e.g., Pillay 1957b; Pillay 1975, 41–55).

6. Published in Calcutta, the textbook was specifically written with the Indian student in mind and was "the only popular description that has yet appeared of the geology and climate of India" (Blanford 1873, Preface). An education report of 1878 noted that the book was taught in the Bengal Presidency and the Central Provinces, and that it was also recommended for schools in Madras, especially translated into local languages (Government of India 1878, 72–74). The textbook was revised in 1874 and 1878, but I have seen no Indian-language edition. Like his older brother William Thomas, Henry Francis started his professional career with the Geological Survey of India, working in that important institution from 1851 to 1862, when he retired to join the faculty of Calcutta's Presidency College. In 1872 he was appointed meteorological reporter to the government of Bengal. His most extensive statement on the paleo land connection between Africa and India may be found in Blanford 1875.

7. Blanford 1873, 119–20.

8. Medlicott and Blanford 1879, xxxiv–xxxv, liii–lxviii, 297. The manual was also used as a college textbook in geology courses at the University of Madras in the 1890s.

9. See also Oldham 1893, 209–14, 248–53, 490–95; and Oldham 1894. A thoroughly revised edition of GSI's official synthesis was completed in 1939. It provisionally adopted the continental displacement theory, and Lemuria's disappearance was explained in accordance with the mobilist theories of Wegener and du Toit (Pascoe 1950–76, 53, 1075–1108, 1330–32). There is no scholarly analysis of when the geological establishment in India was converted to a mobilist view of earth history. While several scholars from the 1960s discuss the formation of India in terms of continental displacement theory (see, e.g., Krishnan 1982), others are more ambiguous (e.g., Dey 1968; Mehdiratta 1954; Ray 1963).

10. Wadia 1919, 110.

11. See, e.g., Balfour 1976, 699; Holderness 1911, 23–24; Holdich 1905, 7–9; Holland 1909, 85–87; Masani 1946, 6; and Mookerji 1936, 7–9.

12. See, e.g., Anon. 1916, 2–3; H.C. 1936, 12–13; Fox 1938, 161, 291–92; Marsden 1909, 4–5; Smith 1883; Wadia 1919; and Wadia 1939.

13. Forster 1984, 135–36. Forster (1879–1970) first visited India for a few months in 1912–13 when he traveled across the country extensively, and then again in 1921–22 when he worked as Private Secretary to the Maharaja of Dewas. Although this passage has been much commented upon by Forster scholars as an instance of the colonial novelist's penchant for a timeless India, the invocation of the prehistoric land connection between the subcontinent and Africa that subsequently submerged has not drawn any attention.

14. Geology was a valuable knowledge form in official British India, particularly because its findings were essential to the economic interests of the colonial state. There are few studies of the pedagogical dissemination of modern geological knowledge among Indians, although it appears that here, as in other regards, they had to face considerable racial prejudice in colonial India (Grout 1995).

15. Holderness 1911, 23. See also Medlicott and Blanford 1879, 291, 297; Oldham 1894.

16. Wadia 1919, 4.

17. Such is the authoritative status accorded to Maclean's *Manual* in Tamil labors of loss that its author is accorded the rank of "scientist" and is even frequently referred to as "Doctor."

18. Maclean 1987, 33–34.

19. Ibid., 33.

20. Ibid., 111.

21. Elmore 1915, 9–10; Risley 1908, 46–47; Risley 1909, 298–99; Thurston 1899; Thurston and Rangachari 1987, xx–xxxvi. The authoritative *Linguistic Survey of India* (1903–28) barely hints at Lemuria in its volume on the Dravidian and Munda languages: "With regard to the Dravidas, some authorities believe *that they arrived in India from the south,* while others suppose them to have entered it from the north-west" (Grierson 1906, 5 [emphasis mine]). Sten Konow, the Sanskritist who edited this volume doubted that Dravidians were the original inhabitants of the subcontinent, noting in a letter to George Grierson, the director of the project, that, "I do not think that the present distribution of Mundas and Dravidas affords any clue as to which of them first came to India. But I think that such a clue is furnished by the fact that the Mundas are connected with those tribes which must be supposed to be the oldest inhabitants of Further India" (India Office Library, Grierson Papers, S/1/1/7, letter dated March 4, 1904). In response to this claim, Edward Gait, the then-Census Commissioner retorted, "The Dravidians (the race) are I believe generally regarded as the oldest inhabitants of India: they are allied to the African Negro and probably came from the south by way of the submerged continent of Lemuria" (India Office Library, Grierson Papers, S/1/1/7, letter dated March 17, 1904). Grierson himself was remarkably reticent on this issue, only noting in passing after commenting on the profusion of languages in the subcontinent that, "Over all, there broods the glamour of eastern mystery. Through all of them we hear the inarticulate murmur of past ages, of ages when the Aryans wandered with their herds across the steppes of Central Asia, when the Indo-Chinese had not yet issued from their home on the Tang-tse-Kiang, and perhaps when there existed the Lemurian continent where now sweep the restless waves of the Indian ocean" (Grierson 1903, 342).

22. Baines 1893, 123; Stuart 1893, 211–12; Risley and Gait 1903, 508.

23. E.g., Anon., 1916, 1–6; Arunachalam and Raghavan 1967, 10; Marsden 1909, 22–25; Marsden 1917, 7; Marsden 1930, 13; Muruku Sundaram 1966, 10–11; Rangaswami Aiyangar 1910, vii, 7–8; Ratnam Pillai 1922, 9; Seetharaman 1962, 8–9; Shanmukhasundaram 1939, 9–10; Srinivasan 1949, 2–3; Sundaravarada Acariyar 1924, 7; Tiruvenkadatayyangar 1948, 8; Tamilnadu Textbook Society 1977, 16.

24. Tamilnadu Textbook Society 1981a, 20. See also Tamilnadu Textbook Society 1981b, 20. A college history textbook in Tamil by K. K. Pillay notes in a slightly different vein, "Some geographers and ethnologists maintain that man first appeared on Earth in southern India. Material evidence in support of this theory has not yet been found. Even so, as far as the Indian subcontinent is concerned, man first appeared in southern India. There is little doubt of this" (Pillay 1981a, 27; see also Government of Tamilnadu 1975, 114–18).

25. Ramaswamy 1997, 39–41.

26. Nallasami Pillai 1898, 112 (emphasis mine). From an affluent Tiruchirapalli family, Nallasami Pillai was educated at the prestigious Presidency College in

Madras, from where he graduated with a Bachelor's degree in 1883. He went on study law and practiced as a lawyer in the Madras High Court from 1887. In 1893 he joined the colonial district administration for the next two decades. His twin devotion to Tamil and to the cause of Saiva philosophy led him to found (and edit) in 1897 the short-lived but influential bilingual periodical called *The Light of Truth or Siddhanta Deepika,* which provided one of the earliest publication forums for Tamil devotional ideas. In fact, just a few months before Nallasami Pillai published his 1898 editorial, where he hinted about Lemuria, the *Siddhanta Deepika* carried a short announcement remarking that a speaker at the meeting of the British Association in Toronto had raised the possibility of "southern India being the cradle of the human race" (*The Light of Truth or Siddhanta Deepika* 1897, 1, no. 4: 96). In 1895, two years before this announcement, V. Kanakasabhai Pillai (1855–1906) started to publish a series of articles in the *Madras Review* in which he flagged references in both ancient Tamil poems and in the Buddhist canon to an antediluvian land that had extended south of Cape Kumari. These essays were subsequently republished in 1904 in a book widely quoted by Tamil's devotees entitled *The Tamils Eighteen Hundred Years Ago* (Kanakasabhai 1966, especially 21–22). Kanakasabhai's family hailed from Jaffna, although he himself worked in the postal service in the Madras Presidency after securing his Bachelor's degree. Although not a professional historian, his work shows the influence of historicist ideas in both its methodology as well as approach to sources, chronology, and causal reasoning (note, however, that a colonial commentator, in a lecture for the Presidency College Historical Association in Madras in 1896, dismissed Kanakasabhai's novel attempt as yet another instance of the "comparative worthlessness of Hindu history") (Sturge 1897, 18).

27. Nallasami Pillai 1898, 112.

28. Maraimalai Adigal 1948, xxix–xxxi. Among the most versatile and tireless of Tamil's devotees, Maraimalai Adigal (who discarded his given Sanskrit name of Vedachalam in favor of a pure Tamil one around 1916) was a prolific writer, speaker-reformer on matters ranging from the reform of Hinduism to vegetarianism, and founder of numerous societies dedicated to the twin causes of Tamil and the Saiva religion. His devotion to Tamil was fiercely contestatory in its opposition to Sanskrit, Aryanism, and Brahmanism, all of which he held responsible for the state of decrepitude in which he found his beloved language and its speakers. His labors of loss around Lemuria were merely one among his numerous interests (Ramaswamy 1997, 144–54, 215–19).

29. Maraimalai Adigal 1948, xxxiii.

30. Basso 1996, 75–76.

31. Like Basso, Carter notes, "By the act of place-naming, space is transformed symbolically into a place, that is, a space with a history" (Carter 1988, xxiv). Elsewhere, he writes that naming brings into being a living space, rendering it habitable, "a place that could be communicated, a place where communication could occur" (ibid., 144).

32. Kanakasabhai 1966, 21; Nallasami Pillai 1898, 112; Savariroyan 1901a, 80; Tirumalaikolunthu Pillay 1900, 14.

33. Suryanarayana Sastri 1903, 6.

34. Although the term "Lemuria" is clearly a product of Victorian modernity, a recent place-maker insists that it appears in a medieval inscription in the Tamil coun-

try (Kumari Maintan 2001, 39). Thanks to Theodore Baskaran for bringing this to my attention.

35. Suryanarayana Sastri 1903, 8. A year earlier, in 1902, the Jaffna-born Mutthuthambi Pillai had referred in his *Tamil Classical Dictionary* to the postdiluvian territory that remained after the loss of land to the ocean as Kumarināṭu (Mutthuthambi Pillai 1902, 90), but it is Suryanarayana Sastri who made the explicit equation between Lemuria and Kumarināṭu that was so consequential for the rest of the century. A Smarta Brahman by birth, Suryanarayan (frequently referred to by his fellow devotees as Paritimāl Kaḷaiñar, a pure Tamil pseudonym he adopted briefly) was a Tamil pandit at the famous Madras Christian College from the 1890s until his early death. He was renowned for his mastery of literary Tamil, and also for his attempts to introduce innovative ideas from English literature into Tamil prose, plays, and poetry (Ramaswamy 1997, 12–13, 199–200).

36. The earliest documented evidence I have found for the use of the name "Kumarikkaṇṭam" for the antediluvian Tamil homeland is in a public lecture given at the University of Madras in 1926 (Purnalingam Pillai 1985, xi). Although its Tamil place-makers insist that "Kumarikkaṇṭam" is a prediluvian "Tamil" name, it is a derivative of the Sanskrit place-name "Kumārikā-khaṇḍa" (and its variant, "Kumari Dvipa"), which is imagined in the Sanskritic *Purāṇas* as the southernmost province of "India," with numerous Saiva shrines and Brahman settlements (on this, see Ramaswamy 2000, 582).

37. The Theosophical Society was headquartered in the Madras suburb of Adayar in 1882–83, where it continues to exist today. The Society was heavily patronized by the English-educated elite of Madras from the closing decades of the nineteenth century, especially under the leadership of Annie Besant. Besant's *The Pedigree of Man* (1904) was originally delivered as lectures to a Madras audience in December 1903 (for its rather unflattering portrait of the Lemurian, see Besant 1904, 79–81). However, in spite of Besant's influence in the intellectual and political life of the Presidency in the first fifteen years or so of the twentieth century, she is rarely if ever quoted by Tamil place-makers, nor indeed are other Theosophists. The one important exception is Scott-Elliot—a rather marginal figure in metropolitan Theosophy, as we have seen—whose Theosophical affiliations are rarely mentioned in Tamil labors of loss (he is frequently placed in the company of scientists such as Sclater and Haeckel). Even in such invocations, it is Scott-Elliot's fabulous chronologies (which credit Lemuria with a deep geological presence) and his views on the catastrophic disappearance of the continent through a series of ocean floods, that are embraced, not his portrayal of the simian-like Lemurian. Given the unflattering image of Lemurian life in Theosophical place-making, it is perhaps predictable that this particular strand of metropolitan place-making is only ambivalently embraced, the Theosophists' influence over Madras society notwithstanding. It is important to also note that the Theosophists' close association with a Brahmanic elite in the Presidency also made their views less acceptable to a large number of Tamil devotees, whose own views on Tamil and its past were increasingly fueled by the non-Brahman and anti-Brahman movements (Ramaswamy 1997, 27–28).

38. Some even suggested that their Kumarikkaṇṭam was distinct from the Lemuria of Western scientists (see, e.g., Mutthuthambipillai 1906–7, 196; and, more recently, Sathur Sekharan et al. 1994, 1–2, 70–71).

39. Kalyanasundaram 1959, 106. On Kalyanasundaram's place in Tamil devotion, see Ramaswamy 1997, 173–74.

40. Savariroyan 1907b, 63. See also Arunachalam 1944, 2; and Kandiah Pillai 1934, 7–8. Although these men insist that Kumarikkaṇṭam was named as such because it was ruled by women, in most Tamil labors of loss the pre-lapsarian Tamil homeland was ruled by kings belonging to the Pandyan dynasty.

41. Savariroyan 1907b, 61. Savariroyan also suggested that matrilineality in Kerala today is the only post-lapsarian remnant of the power and respect that women formerly commanded in Kumarikkaṇṭam (ibid., 62–63). Savariroyan (or Savariraya Pillai), a Christian devotee of Tamil, began his professional career as a Tamil schoolteacher in Tutukudi in 1880, and then moved to Tiruchirapalli, where he taught in a local college. His fame in the world of Tamil devotion rests on his editorship of the *Tamilian Antiquary*, a short-lived but important bilingual journal that he founded in 1907.

42. Shulman 1980, 144–56.

43. Kandiah Pillai 1945, 3, 7.

44. Shulman 1980, 148.

45. Kandiah Pillai 1945, 7.

46. Anbazhagan 1975, 9.

47. E.g., Arunachalam 1944; Subramanian 1939, 36–37.

48. Ramaswamy 1997, 114–21.

49. Ramaswamy 2000, 583.

50. Suryanarayana Sastri 1903, 8.

51. Ramaswamy 1997, 154–61.

52. Anbu Ganapathy 1951, 4.

53. I borrow the term "geo-body" from Thongchai Winichakul's study (1994) of the mapping of Siam, in which he uses it to refer to the entirely novel way of conceptualizing territory in modernity as a bounded, nameable entity with a singular biography and history, which is enabled by the discourses of scientific geography and cartography.

54. Somasundara Bharati 1912, 24. This much-quoted essay was also subsequently translated into Tamil and published in the important literary journal *Centamiḻ* (vol. 12 [1913–14]: 30–38), and also reprinted in its English version in *The Light of Truth or Siddhanta Deepika* (vol. 14 [1913]: 1–24). Somasundara Bharati's text had its origins in an essay he submitted for his Master's degree in Tamil from the University of Madras. As I suggest later, Tamil pedagogy is the primary institutional site for the production of Tamil labors of loss over Lemuria. Unlike most Tamil devotees, Somasundara Bharati led a life of comparative ease and affluence as a lawyer. Nevertheless, he combined his legal career with devotion to Tamil, and in 1933 was appointed Chair of the Tamil Department at the newly founded Annamalai University in Chidambaram. Fondly referred to in Tamil devotional circles as Nāvalar, "the eloquent," his writings on Tamil are a classic example of the manner in which historicist ideas are subverted by fabulist notions of loss, as I discuss later (see also Ramaswamy 1997, 212–13).

55. E.g., Savoriroyan 1907a, 1–2.

56. Christian Jacob quoted in Conley 1996, 8.

57. Carter and Malouf 1989, 173.

58. Nallasami Pillai 1898, 112.

59. Somasundara Bharati 1912, 24.

60. E.g., Appadurai 1941, 2–3; Aravaanan 1984; Ganapathy 1984, ix; Kandiah Pillai 1945, 1–3; Raghava Ayyangar 1979, 25–30; Somasundara Bharati 1967, 103, 116; Velan 1995, 109; Vimalanandam 1987, 1669–71. For college history textbooks in Tamil that circulate this notion, see Government of Tamilnadu 1975, 115–18; Pillay 1981a, 28–30; and Swaminathan 1996, 23–24.

61. Somasundara Bharati 1967, 116.

62. Appadurai 1941, Publisher's Preface; Devaneyan 1940, 58–59; Kandiah Pillai 1934, 7–8; Mathivanan 1983, 630.

63. For theories of these alternate homelands, see Devasikhamani 1919; Ponnambalam Pillai 1914; Savariroyan 1900–1901; Savoriroyan 1907a; and Srinivasa Aiyangar 1914.

64. Tamby Pillai 1907, 41 (emphasis mine).

65. Sesha Iyengar 1933, 26. Sesha Iyengar taught in the history department of the well-known Pachaiyappa's College in Madras and was also a member of the Historical Records Commission.

66. Somasundara Bharati 1912, 22.

67. Purnalingam Pillai 1945, 18; Sesha Iyengar 1933, 26; Srinivasa Pillai 1965, 3–5; and Tamby Pillai 1907, 43; among others.

68. Even though some conceded that it might be far-fetched to insist that mankind first emerged on Lemuria, given that geologists claimed that it had submerged before the appearance of humans on Earth, they insisted that Dravidians (and hence Tamil speakers) are the most ancient peoples of the subcontinent (see, e.g., Government of Tamilnadu 1975, 124, and Srinivasa Iyengar 1985, 12–13).

69. Bryant 2001, 108–39.

70. Sesha Iyengar 1933, 56–58.

71. From the 1870s, several European ethnologists, especially those with a connection to British India, favored India as "the cradle of mankind" (e.g., Evans 1875, lxxxvi; 1898, 15; Falconer 1868, 578–79). Even Haeckel considered this a possibility (Haeckel 1876a, 399). Such metropolitan speculations came to the attention of Tamil's devotees by the 1890s (*The Light of Truth or Siddhanta Deepika* 1, no. 4 [1897]: 96).

72. Manickam Nayakkar 1985, 77–78. Unlike a majority of Tamil's devotees, Manickam was not a professional Tamil scholar, but an electrical engineer whose devotional interests led him to explore the "mystical" aspects of his beloved language.

73. Abraham Pandither 1984, 88–90. A pioneer in the movement to restore Tamil musical forms, practices, and songs to their former prominence, Abraham Pandither, a well-known Christian devotee of Tamil, began his professional career as a schoolteacher in Dindukkal. He is most remembered for his magnum opus on the history and practice of Tamil music, *Karunāmirthasāgaram* (1917), as well as for organizing several conferences to spread awareness of Tamil song. For his efforts, he was awarded the coveted title of "Rao Sahib" by the colonial government.

74. Kandiah Pillai 1934, 24.

75. Stocking 1987.

76. Suryanarayana Sastri 1903, 8.

77. Savoriroyan 1907a, 2.

78. Sivagnana Yogi 1913, 3–4. See also Mahalingam and Sathur Sekharan 1991, 103.

79. Kandiah Pillai 1945, 6.

80. Gopala Krishnan 1956, 1–3.

81. E.g., Devaneyan 1940; Purnalingam Pillai 1925.

82. See, especially, Devaneyan 1992.

83. Bayly 1995, 168.

84. E.g., Caldwell 1856.

85. As others have noted, utopia means the good place that is no place. See also the characterization of utopias as "impossible societies" (Irwin 1976, 109–13).

86. Basso 1996, 6.

87. Scholes 1979, 206. See also Rosemary Jackson's comment: "The fantastic . . . must be so close to the real that you almost have to believe in it. . . . The fantastic cannot exist independently of that 'real' world which it seems to find so frustratingly finite" (Jackson 1981, 19–20, 27).

88. Taussig 1987, 121.

89. Basso 1996, 6 (emphasis mine).

90. Sami 1955, 27.

91. Ibid., 24.

92. Suryanarayana Sastri 1903, 94–102. As early as 1887 C. W. Damodaram Pillai (1832–1901)—who published many a forgotten Tamil literary work—wrote poignantly of the lost texts of which he had heard when he was young and that were no longer available in his time. These had been primarily lost to the flood that had seized the second *caṅkam* at Kapāṭapuram, but other forces, including gross negligence, had been at work as well (Damodaram Pillai 1971, 15, 44–69). The reminiscences of his colleague U. V. Swaminatha Aiyar (1855–1942) are also written in the shadow of loss (see Ramaswamy 1997, 208–12).

93. Abraham Pandither 1984, 20. For an academic discussion of these works, see Zvelebil 1973, 124–35.

94. Chidambaranar 1938, 8; Deveneyan 1940, 181–86.

95. Rasanayagam 1984, 181.

96. So much so that even contemporary scholars (in India and abroad) who work on ancient Tamil literature write in the shadow of loss (see, e.g., Ramanujan 1985, x–xiv; Venkatasami 1983; Zvelebil 1992). For a provocative analysis of the widespread use of the motif of loss and recovery of sacred texts in premodern India, see Shulman 1988b.

97. Ramaswamy 1997, 34–46, 208–12.

98. Somasundara Bharati 1967, 96.

99. Despite my best efforts, I have been unable to locate this text. Many scholars I spoke to in Tamilnadu knew about it and even believed in its existence but had never seen the publication themselves. The late P. Shankaralingam, former librarian of the Roja Muthiah Research Library in Chennai, told me that he remembered that his Tamil teacher in Tirunelveli used to talk about it, but even this well-stocked archive does not own a copy of this elusive book. Some verses from the work are published in Kandiah Pillai 1947, 28.

100. Aiyan Aarithan 1904–5, 283.

101. Chidambaranar 1938, 8–9. See also a similar statement in Kandiah Pillai's *Namatu Nāṭu*, "Our Land," intended for young children (Kandiah Pillai 1945, 6–7).

102. Vaiyapuri Pillai 1957, 676. For another similar dismissal, see Subramania Aiyar 1970, 57–60.

103. Trevor-Roper 1983.

104. Nallasami Pillai 1898, 112.

105. Kandiah Pillai 1945, 6–7.

106. Savariroyan 1907a, 2 (emphasis mine). See also Tamby Pillai 1907, 42–43.

107. On the complex place of Sanskrit in the Tamil life-world, see Ramaswamy 1997, 34–46, 46–47, 149–50.

108. Mathivanan 1983, 630.

109. Sami 1955, 1, 11.

110. Nambi Arooran 1980, 70–110.

111. Purnalingam Pillai 1925, ii. Purnalingam Pillai's ancestors had been Tirunelveli landowners for several generations, while he himself was among the first in his family to get a Western-style education, first at Hindu College in Tirunelveli, and then at Madras Christian College, from where he graduated with a Bachelor's degree in English and Philosophy in 1886. A lifelong professional academic, he held various teaching and administrative positions across the Presidency. However, his biographers tell us that his first commitment was to Tamil. Author of several books which demonstrated, among other things, how much Tamil had suffered over the centuries since its lofty antediluvian beginnings, he was also proprietor and editor of the Tamil journal *Ñāṉapōtiṉi*.

112. Suryanarayana Sastri 1903, 94.

113. Neduncheliyan 1953, 16.

114. For a detailed study, see Ramaswamy 2001a.

115. Purnalingam Pillai 1945, 18, 26–27.

116. *Tamilnadu Legislative Assembly Debates* 32 (1971): 521.

117. Neduncheliyan 1953, 5–6.

118. Ibid., 6–7.

119. Rajamanikkam 1944, Preface. A native of Tanjavur, Rajamanikkam held various positions, teaching Tamil first in high schools in Madras city and then in colleges in Madras and Madurai. He reached the peak of his career when he taught Tamil at the University of Madras from 1959 until his passing in 1967.

120. Rajamanikkam 1944, 66–67.

121. E.g., Kasinathan 1993. At a seminar-conference held in Chennai in 1994 on Kumarikkaṇṭam, Dr. Kasinathan, the then-Director of the Tamilnadu State Department of Archaeology, spoke of the possibilities of using marine archaeology to uncover the secrets of Sclater's vanished continent (Sathur Sekharan et. al. 1994, 17–21). In an interview I had with the Dr. Kasinathan in March 1997, he informed me that he had recently submitted a proposal to the Tamilnadu government to put in a request to UNESCO for funds for this project. Just a few weeks before this interview, the state's education minister, K. Anbazhagan (who we have already encountered in these pages), had called upon the central government to support the development of marine archaeology in a valedictory address he gave to the first International Conference of Marine Archaeology, held in Chennai in February 1997. The promise of marine archaeology for Tamilnadu lay in the fact that so much

of its past heritage—as embodied, for example, in the lost city of Poompuhar—was buried in the ocean. Poompuhar, he claimed, was a residue of the larger Lemuria landmass (*Hindu* [Madras], February 2, 1997, p. 4). See also "Governor for Law to Protect Underwater Heritage Sites," *Hindu* (Madras), February 23, 1997, p. 4.

122. Ramaswamy 1997.

123. E.g., Chidamabaranar 1938, 24; Maraimalai Adigal 1948; Suryanarayana Sastri 1903, 11–12.

124. Ramaswamy 1997, 77–78. For a fascinating *longue durée* study of Western fantasies around "the dream of a perfect language," including the search for "a unique mother tongue" from which all languages descended, see Eco 1995.

125. More so than most devotees of Tamil, Devaneyan led a life of loss and deprivation that in his writings is so clearly displaced onto his beloved Tamil. From an undercaste Tirunelveli family converted to Christianity, Devaneyan struggled to get an education even while holding down several petty jobs, finally earning a Master's degree in Tamil from the University of Madras in 1944. He writes that when he first submitted his thesis on Kumarināṭu as the homeland of the Dravidians to the University for a Master of Oriental Languages degree in 1938, it was rejected (Devaneyan 1940, 43). This did not stop him from going on to publish prolifically on this subject, a life of stark poverty and hardship notwithstanding (see also Ramaswamy 1997, 213–15).

126. R. Mathivanan, an ardent follower of Devaneyan and a scriptwriter for the 1981 government film, told me that he had included scenes in the film in which the antediluvial inhabitants of Kumarināṭu are shown speaking "Kumari Tamiḻ," the prelapsarian language of the lost homeland. These scenes, however, were cut on the editing table (interview with R. Mathivanan, Chennai, March 1997).

127. Abraham Pandither 1984, 61.

128. Somasundara Bharati hints at the existence of "non-Tamil races" to the south of Kumarināṭu (1912, 23), as does S. K. Ganapathy (1984, 6). In the 1981 documentary, the Pandyan kings of Kumarikkaṇṭam are shown undertaking "sea voyages to distant lands where they spread Tamil culture, civilization, and fine arts," and "foreign merchants visited Kumarināṭu, not only to buy pearls, ivory and the like, but also to acquire works of Tamil literature whose fame had crossed the ocean." The Pandyan kings are also shown receiving envoys from distant lands.

129. E.g., Devaneyan 1940; and Devaneyan 1966.

130. See, e.g., Abraham Pandither 1984, 25; and Srinivasa Pillai 1965, 2–3.

131. For S. Padmanabhan, a retired bank official–turned–amateur historian, and the General Secretary of the Kanyakumari Historical and Cultural Resource Center (founded in 1991–92), the Tamil spoken in Kanyakumari still carries traces of the antediluvian tongue (Padmanabhan 1971, 5–6). The irony in this claim is that other Tamil speakers would claim that the Kanyakumari dialect is contaminated by the neighboring Malayalam and hence is an impure version of chaste Tamil (*centamiḻ*).

132. See, in this regard, Appadurai 1941, Publisher's Preface; and Purnalingam Pillai 1925, iii.

133. Nedunceliyan 1953, 15.

134. Suryanarayana Sastri 1903, 6. See also Abraham Pandither 1984, 25; Kandiah Pillai 1934, 29–30; and Somasundara Bharati 1912, 23–24. For college text-

books that voice a similar contention, see Government of Tamilnadu 1975, 118; and Pillay 1981a, 29–33.

135. Purnalingam Pillai 1945, 5. Since at least the 1850s Dravidians were caricatured in colonial sociology as servile laborers who constituted the dark global underbelly of the British empire (Caldwell 1856, 5; Risley 1909, 298).

136. Purnalingam Pillai 1985, 9. Even some history textbooks prescribed for colleges in the state insist on this (see, e.g., Government of Tamilnadu 1975, 114–18).

137. Devaneyan 1940, 55. For the gendered politics and poetics of Tamil devotion, see my extended analysis in Ramaswamy 1997.

138. Maraimalai Adigal 1948, xxxx–xxxxi (emphasis mine).

139. E.g., Abraham Pandither 1984, 42; Ganapathy 1984, 5.

140. Appadurai 1941, 36–38; Kalyanasundaram 1940, 24; Manickam Nayakkar 1985, 79–80.

141. E.g., Purnalingam Pillai 1945, 23–26.

142. Government of Tamilnadu 1975, 127. See also Pillay 1981a, 32–34.

143. E.g., Purnalingam Pillai 1945, 4; Sami 1955, 26.

144. Aravaanan 1984, 121–22.

145. This does not mean an absence of ambivalence about the African connection in Tamil devotion. In the labors of loss, especially of upper-caste devotees of Tamil, the territorial and ethnic connection to "the dark continent" and the black diaspora is downplayed, or even ignored.

146. Kandiah Pillai 1957, 12.

147. Ibid., 12–70.

148. Kandiah Pillai 1945, n.p. See also Aravaanan 1984, 121; Mahalingam and Sathur Sekharan 1991, n.p. Similar maps of the dispersal of the Indo-Europeans out of their ur-homeland were used by those like Gustaf Kossina, who adopted the cultural-historical approach for the interpretation of prehistoric remains (Anthony 1995, 91).

149. For thoughtful comments on how diffusionism became popular from the 1880s in a climate of conservatism and pessimism which was inclined to the belief "that particular inventions were unlikely to be made more than once in human history," see Trigger 1989, 151–55.

150. The specific identification of the Pandyas as the lords of the antediluvian homeland was made as early as 1900 (Tirumalaikolunthu Pillay 1900, 14). See also Abraham Pandither 1984, 31, 51, 77; Chidambaranar 1938, 8–10; Somasundara Bharati 1967; Tamby Pillai 1907, 40.

151. The colonial documentation of the Pandyas began toward the end of the eighteenth century when Colin Mackenzie started collecting their genealogies and legendary histories. Their patronage of a literary academy based in Madura(i), referred to as "the College," was also briefly noted, as was the latter's role in the cultivation of Tamil letters. Subsequently, other colonial scholars in the nineteenth century consolidated the Pandyas' reputation as one of the premier dynasts of ancient southern India, and as patrons of Tamil literature. Importantly, however, the Pandyas are resolutely postdiluvial in colonial historiography, as are the boundaries of their kingdom, the duration of the *caṅkam* they supported, the capital city from which they ruled, and so on (see, e.g., Taylor 1835; and Wilson 1838). As William Taylor unequivocally observed, "The foundation of Madura was long posterior to the

flood; and probably not much more than 1,500 years before the Christian era. . . . We have discarded, for reasons adduced, the fabulous pretension [by the Pandyas] to ante-diluvian antiquity" (Taylor 1835, 1:135, 237). In 1885 Charles Maclean acknowledged the existence of "native" traditions that accorded an antediluvial past to the Pandyas in his *Manual*, only to disavow them (Maclean 1985, 120). In bestowing an antediluvial identity on the Pandyas, Tamil labors of loss thus vigorously contest the dominant colonial historiography in this regard as in others.

152. Purnalingam Pillai 1985, 4.

153. Kailasapathy 1968, 211. As David Shulman, following T. G. Aravamuthan, observes, "The notion of the sea lapping the feet of the king became a cliché of the [medieval Tamil] commentators" (Shulman 1988a, 312).

154. See also Ramaswamy 1998.

155. See, e.g., Devaneyan 1940, 43–50; Mathivanan 1983, 631–34; Sadashiva Pandarather 1940, 3–7. For a presentation of the life and times of these antediluvian kings as factual history in Tamil schoolbooks, see especially Mutthukrishna Nattar 1960, 4–6; Narayanasami 1964, 1–3; Venkatachalam Pillai 1951, 1–5; and Vilvapati et al. 1956, 14–16.

156. Tamil devotees were not alone in colonial India in idealizing kingship in their historical ruminations, even while in the political realm they were demanding modern democratic forms of rule, as Ronald Inden's analysis of the historian K. P. Jayaswal's scholarship also suggests (Inden 2000, 188–92).

157. Annapoorni 1935, 257.

158. Suryanarayana Sastri 1903, 62.

159. Somasundara Bharati 1912, 5.

160. Maraimalai Adigal 1948, xxxx.

161. Ramachandran and Mathivanan 1991, 95.

162. Purnalingam Pillai 1904, 4–5; Suryanarayana Sastri 1903, 61–62.

163. Abraham Pandither 1984, 55; Aiyan Aarithan 1904–5, 274; Devaneyan 1940, 50; Raghava Aiyangar 1938, 91.

164. Somasundara Bharati 1967, 122–23.

165. Subramania Sastri 1915–16, 420–21; Swaminatha Aiyar 1978, 14.

166. Appadurai 1993.

167. Ivy 1995, 10.

168. Stewart 1993, 144.

169. I am mindful here of Anthony Smith's observation that the nationalist spatial vision has to be ultimately tangible and practical: "It demands a terrain on which nations can be built" (Smith 1986, 184). Note also his follow-up observation: "But the vision also contains an unearthly note, an element of archaic mystery, particularly where the environment yields up vestiges of a pre-history that antedates the community" (ibid., 184).

170. E.g., Sethu Pillai 1925, 285–86.

171. Starobinski 1966, 93.

172. Daniels 1985, 76.

173. Boym 2001, xvi.

174. Ibid., xiii. See also David Lowenthal's comment that, "What nostalgia does require is a sense of estrangement; the object of the quest must be anachronistic" (Lowenthal 1975, 4).

175. Young 2000, 54. Many thanks to Laura Kunreuther for recommending this essay.

176. Ibid., 66.

177. Casey 1987b, 367.

178. Ibid., 381.

179. Ibid., 365.

180. Phillips 1985, 66. See also Fred Davis's statement that nostalgia summons up an "odd mix of present discontents, of yearning, of joy clouded with sadness, and of small paradises lost" (Davis 1979, 29).

181. Casey 1987b, 378–79.

182. Ibid., 379.

CHAPTER 5

1. Eduard Suess, quoted in Marvin 1972, 57.

2. I adapt this phrase from Marcia Yonemoto's rich analysis of the place of the ocean in the Tokugawa "geo-cultural imaginary" (Yonemoto 1999).

3. For an exquisite discussion of the Victorian fascination with "the world that drowned" and with lives that lived out their existence in the ocean's depths, see Olalquiaga 1998, 103–98.

4. Donnelly 1882, 48–50.

5. Barker 1923, 155.

6. Wood 1985, 122.

7. Frankel 1998, 122.

8. Hsu 1992, 52. Similarly, note Carl Sagan's authoritative statement, "While the ocean keeps many secrets, . . . there isn't a trace of oceanographic or geophysical support for Atlantis and Lemuria. As far as science can tell, they never existed" (Sagan 1996, 4). Or in Peter Tyson's recent assessment, "By the mid-1960s, evidence from oceanic rift structures, paleo-magnetism, and other sources finally gave Wegener's theory the upper hand, and the idea of lost continents was buried for good" (Tyson 2000, 38). These learned comments notwithstanding, the enduring trope of "lost continent" still occasionally crops up, even in erudite scientific tomes and on websites. For instance, an official website (that I saw on November 14, 2000) entitled "Lost Continent" details the undersea explorations of the Kerguelen Plateau in the southern Indian Ocean carried out in 1998–99 by the Institute of Geophysics (University of Texas at Austin) and the Massachusetts Institute of Technology (http://www.ig.utexas.edu/outreach). See also an editorial by BBC Online Science Editor David Whitehouse called "'Lost Continent' Discovered" (http://news.bbc.co.uk/hi/english/sci/tech/newsid_353000/353277.stm). Thanks to V. Narayanan for this reference.

9. Spence 1931, 17.

10. Joseph 1972, 2–4. This renewed interest was also provoked by the publication in 1974 of *The Riddles of Three Oceans,* an English translation of a Russian work by the amateur-scholar Alexander Kondratov, who wrote on Lemuria in the wake of the International Indian Ocean Expedition (Kondratov 1967). Drawing upon paleoscientific and Tamil labors of loss around Lemuria, Kondratov insisted on the truth about Sclater's lost paleo-world in terms that were extremely congenial to the Tamil project. I thank Irina Glushkova for her comments on Kondratov.

11. Haeckel 1876a, 360.
12. See, e.g., Murray 1866, 69.
13. Reprinted in Zanger 1992, 99.
14. Lemurian Fellowship 1935, 11.
15. I am inspired here by Peter Bishop's comment in his suggestive study of European travel writing on Tibet, that these travelers "did not first look at the country and then drift off into reverie. The reverie, the fantasy, was an integral part of the looking" (Bishop 1989, 219).
16. Mutthuvirasami Naidu 1954–55, 453. Other than identifying himself by his name and the suffix "B.A.," Mutthuvirasami Naidu offers no details about himself, and as best as I can tell, has written nothing else on Tamil or its literature. He may be seen as typical of the Tamil Everyman who was motivated to lament about Lemuria on learning of its loss. He is not alone in this regard.
17. Ibid., 456.
18. Ibid.
19. Paramasivanandam 1944, 22.
20. Ibid., 21–22.
21. Ibid., 23.
22. The choice by Sethu Pillai of Ilango as his alter ego is not fortuitous, for verses from the *Cilappatikāram* play a critical role in Tamil labors of loss.
23. Sethu Pillai 1950, 4.
24. Ibid., 6.
25. Ibid., 7. See also Sethu Pillai 1925, 285–26. So pervasive had such lamentations become that a 1963 essay by the literary critic P. Sri noted, "As soon as one mentions Kumari, waves of memories rise in the minds of some about the large territory that lay between the Tamil land and Africa, and that disappeared into the ocean" (Sri 1963, 34). He concludes, disenchantedly, that when he himself stood at land's end and looked out at the ocean, it was not clear to him what tales it had to tell, what songs it sang, or what secrets it concealed (ibid., 54).
26. Cohn 1996.
27. Purnalingam Pillai 1945, 2.
28. Savariroyan 1901a, 80.
29. Kanakasabhai Pillai 1966, 21.
30. *The Light of Truth or Siddhanta Dipika* 1910–11, 11, 88. Critical scholarship places *Iṟaiyaṉār Akapporuḷ* between the fourth and sixth centuries of the common era, while the commentary has been dated to between 700 and 1000 C.E. (Aravamuthan 1930; Zvelebil 1973). Note, in particular, Aravamuthan's historicist observation, "The commentator lets his fancy run riot and exhibits a decided preference for the marvelous over the probable" (Aravamuthan 1930, 192). For a thoughtful analysis of how this aphoristic grammar and its learned commentary are themselves "the subject of a particularly rich, medieval tale of loss and restoration," see Shulman 1988b (quote appears on p. 111).
31. Although these acts of enumeration are very much a reflection of the general penchant for numbers in the subcontinent from very ancient times, the formulaic quality of Nakkirar's is one of the grounds on which the historicity of the *caṅkam* has been doubted (Ramanujan 1985, x; Shulman 1988a, 294–95). There is extensive academic scholarship on the existence of the *caṅkam* as a literary institution, a

good deal of which is concerned, in fine historicist fashion, with sieving out the "true" and the "factual" from the "fabulous" and the "incredible" (see, e.g., Aravamuthan 1930; Graefe 1960; Ramachandra Dikshitar 1947; Scharfe 1973; Zvelebil 1973). For a more nuanced analysis of Nakkirar's narrative as an expression of a persistent archetype centered on the motif of "renewed creation that follows upon the deluge," see Shulman 1988a.

32. The *Cilappatikāram* has been dated by academic scholarship to the middle of the first millennium C.E., and Adiyarkunallar must have lived in the twelfth or thirteenth century C.E. The verses from the *Cilappatikāram* most frequently cited in Tamil labors of loss are 8:1–2 and 11:17–22.

33. Swaminatha Aiyar 1950, 228–29, 299–300.

34. *Kalittokai* 104:1–4; *Pura<u>n</u>ā<u>n</u>uru* 6:1–2, 17:1, 67:6.

35. Subramania Aiyar 1970, 23–31.

36. *Iraiya<u>n</u>ār Akapporuḷ*, with Nakkirar's commentary, was published in 1883, followed in 1887 by the publication of *Kalittokai*. Soon after, Swaminatha Aiyar published *Cilappatikāram*, with Adiyarkunallar's commentary, in 1892, and *Pu<u>r</u>anā<u>n</u>ū<u>r</u>u* in 1894 (Zvelebil 1992, 221–22).

37. Savariroyan 1901b, 198.

38. Shulman 1988a, 294–309.

39. Aravaanan 1977, 167–69; Hancock 2002, 258–61. Also, conversations with R. Mathivanan, N. Kasinathan, A. Chellaperumal, and A. Dhananjayan (March 1997).

40. E.g., Baliga 1960; Blanford 1877.

41. Tirumalaikolunthu Pillay 1900, 18.

42. There are some exceptions here. Damodaram Pillai suggested in 1887 that the two great "calamities" that had affected Tamil over the centuries were *kaṭalkōḷ* and the arrival of Muslims in the Tamil country. The latter set fire to Tamil books and destroyed them (Damodaram Pillai 1971, 44–45). This accusation was reiterated a few years later (Chelvakasavaraya Mudaliar 1904, 24), but since then, I have not seen it repeated, probably because of the growing influence of the secular Dravidian movement on Tamil devotion. For an extended discussion of how Brahmans and *kaṭalkōḷ* were responsible for the loss of Tamil books, see Devaneyan 1940. In 1955 K. P. Sami noted that the sea and Sanskrit were the twin forces that destroyed Tamil books (Sami 1955, 1).

43. Shulman 1988a, 295.

44. Ibid., 302.

45. Aravaanan 1984, 9.

46. The 1981 documentary film provides a graphic illustration of this in its attempt to combine current mobilist theories of continental drift with earlier fixist conceptions of drowned landmasses. For some recent essays that insist on the truth of a submerged Lemuria, even using drift theory to do so, see Kumari Maintan 2001; Kalcina Valuti 2001; and Poonkunran 2001.

47. Poonkunran 2001, 24.

48. Kandiah Pillai 1934, 20; Tamilnadu Textbook Society 1996b, 44; Tamilnadu Textbook Society 2001, 44.

49. Devaneyan 1940, 47–48.

50. Appadurai 1941, 34–35; Kandiah Pillai 1934, 19.

51. Dundes 1988.
52. Raghava Aiyangar 1938, 91–92.
53. Devaneyan 1940, 55; Appadurai 1941, 2–3.
54. See, e.g., Appadurai 1941, 30–58; Chidambaranar 1948, 57–62; Devaneyan 1940, 57–58; Subramania Pillai 1949, 10–11.
55. Savariroyan 1901a, 80–81.
56. E.g., Chidambaranar 1948, 163–73; Kanakasabhai 1966, 21; Mahalingam and Sathur Sekharan 1991, 44–45; Raghava Aiyangar 1938, 94–96; Raghava Ayyangar 1979, 31–49; Ramachandran and Mathivanan 1991, 1.
57. E.g., Mutthuthambipillai, 1906–7, 197–98; Subramania Pillai 1949, 10–11; 24–25. At least one critic of such attempts wrote, " 'Professor Elliot' is an unknown figure in the geological world, and what he means by 'erosions' and how he fixes their dates, will need to be critically examined, before his observations can be accepted by Hindus as, at any rate, safely and scientifically inferential" (*The Light of Truth or Siddhanta Deepika* 1910–11, 11, 89).
58. See, e.g., Aiyan Aarithan 1904–5, 278–79; Kanakasabhai Pillai 1966, 21; Raghava Ayyangar 1979, 43–45. The devotee turns to the dates 2387 B.C.E., 504 B.C.E., and 306 B.C.E., furnished by colonial historiography as the dates for the floods mentioned in the Buddhist canon (Tennent 1977, 6).
59. Purnalingam Pillai 1945, 13.
60. As Lucie Armitt notes with regard to fantasy, "Temporality itself becomes obsolete, insofar as forms like the fairy tale or the utopia contextualize narrative chronology in terms of what are effectively two competing spatial dimensions. Thus, within the outer frame of timelessness, we have an inner frame of sequentially structured time" (Armitt 1996, 5). See also Ashis Nandy's comment that "temporality cannot be allowed to determine authenticity in an ahistorical or epic culture" (Nandy 1983, 24).
61. The term "historicism" has been used in different ways in the philosophy of history. I use it here to designate the modern discipline's preoccupation with temporal continuities, progressions, and relations between things (Nora 1989, 9). I also follow Dipesh Chakrabarty in his suggestion that historicism, a nineteenth-century European conception fundamental to the idea of modernity, is premised on the assumption that "to understand anything it has to be seen both as a unity and in its historical development" (Chakrabarty 2000, 6). Mostly, the historicist impulse is to always historicize. To narrate the past as it "actually happened" was the new dogma of nineteenth-century historicism.
62. See, especially, Chakrabarty 2000; Guha 1997; Lal 1999; Lal 2000; Nandy 1995.
63. Nandy 1995, 46. Nandy's observation resonates with Pierre Nora's argument that "the pressure of a fundamentally historical sensibility" that accompanies the modern has displaced "the warmth of tradition," "the silence of custom," and "the repetition of the ancestral" (Nora 1989, 7).
64. Krishnaswami Aiyangar 1911, 321–24.
65. Srinivasa Iyengar 1982, lvi.
66. Sundaram Pillai 1895, 9–10. On Sundaram Pillai's place in Tamil devotion, see Ramaswamy 1997, 17–18, 80–81.
67. Srinivasa Iyengar 1982, lvi.

68. The classic here is Gillespie 1951.

69. For Blavatsky's discussion of the centrality of the Noachian Deluge to the antediluvial history of Earth, see Blavatsky 1928, 147–57.

70. Cohn 1996. But note also William Stiebing's comment that the Deluge "is one popular belief about man's ancient past that archaeologists and historians will probably never succeed in eradicating" (Stiebing 1984, 27).

71. For discussions of areas where the Flood continues to be a subject of concern well into the twentieth century, such as Biblical archaeology or Creationism, see Gould 1988.

72. Given the conflicted intimacy between Tamil devotion and the modern disciplinary formation of history, it goes without saying that the degree to which Tamil place-makers are historicist varies enormously, from the few who are completely in history to some who are utterly outside it, with a majority falling somewhere in between. The spread of historicist assumptions among modern Tamil intellectuals has yet to be systematically documented by scholars, but for a suggestive essay that links Brahmanical approaches to the past with (imperial) historicism, see Geetha 1993.

73. Here I follow Ann Swinfen in my understanding that pure fantasy generates an imaginative and imaginary world through faithfully observing "rules of logic and inner consistency which, although they may differ from those operating in our own world, must nevertheless be as true to themselves as their parallel operations in the normal world" (Swinfen 1984, 3). The world created by pure fantasy has to necessarily be complete, self-consistent, and uncompromised by the demands of prosaic realism.

74. There is virtually no scholarship on the pedagogic dissemination of history in the Presidency, but see some scattered comments in Sattianadhan 1894 and Pillay 1957a.

75. Nora 1989, 11.

76. *Tamilnadu Legislative Assembly Debates* 24 (1972): 76. This is not the first time that some members of the Legislative Assembly called upon the government to write the "correct" history of Tamilnadu, beginning with the antediluvial Kumarikkaṇṭam (see also *Madras Legislative Assembly Debates* 32 [1965]: 619–68, and *Tamilnadu Legislative Assembly Debates* 32 [1970]: 495–527).

77. Subrahmanian 1996, 25. Subrahmanian received his doctorate in history from Annamalai University and began his professional career at the University of Madras before moving on to teach in Madurai.

78. Sundaram Pillai 1895, 10–15. See also Nallasami Pillai's statement that "the scalpel of remorseless, historical criticism, applied by generations of European Orientalists to Sanskrit chronology, has yet to be applied to Tamil" (*The Light of Truth or Siddhanta Deepika* 1910–11, 11, 134).

79. Seshagiri Sastri 1897, iv. Seshagiri Sastri also tellingly complained about those "selfish authors who created history not from real facts but from the depth of their own imaginations" (ibid., iii).

80. My comments here are informed by my reading of Nora 1989, especially his observation, "The 'acceleration of history' . . . confronts us with the brutal realization of the difference between real memory . . . and history, which is how our hopelessly forgetful modern societies, propelled by change, organize the past. . . . This conquest and eradication of memory by history has had the effect of a revelation, as

if an ancient bond of identity had been broken and something had ended that we had experienced as self-evident—the equation of memory and history. . . . Memory and history, far from being synonymous, appear now to be in fundamental opposition" (Nora 1989, 8). I take heed, however, of those critics of Nora who question his stark opposition between "history" and "memory" (e.g., Zerubavel 1995, 4–5).

81. Srinivasa Aiyangar 1914, 201.

82. Sivaraja Pillai 1932, Appendix XII. A native of Nanjil Nadu in Travancore, Sivaraja Pillai got his Bachelor's degree from Madras Christian College. A bureaucrat in his early professional life, his literary and historical interests found expression in the *Malabar Quarterly Review,* which he founded. Between 1927 and 1934, he taught Tamil at the University of Madras, as well as participating in the Tamil Lexicon project.

83. Sivaraja Pillai 1932, Appendix XII.

84. Nora 1996, 12. See also Lal 2000, 160–61.

85. Sivaraja Pillai 1932, v. See also Nilakanta Sastri 1956, 39.

86. Seshagiri Sastri 1897, 39. For a later reiteration of the same sentiment, see Subramania Aiyar 1970, 22, 35–39.

87. Krishnaswami Aiyangar 1911, 336. A native of Kumbakonam and educated in Bangalore, where he began his early professional career as a schoolteacher, Krishnaswami Aiyangar was appointed to the newly created Chair of Indian History and Archaeology in the University of Madras in 1914, a position he held until 1929. He was editor of the *Journal of Indian History* in the mid-1920s. For a sample of Krishnaswami Aiyangar's historicist notions, see his *Ancient India* (1911, 314–29).

88. Meenakshi Mukherjee calls attention to the Bengali writer Pramatha Chaudhuri's observation that "when the British came, rhyme gave way to reason" (Mukherjee 1985, 16). See also Viswanathan 1989, 39–40, 47–48, 81–82, 110–11, 118–41. In my *Passions of the Tongue,* I consider the division of labor between poetry and prose in the cultural politics of Tamil devotion, where the former is used to constitute an intimate community of devotees united around their love for their mother/language, whereas prose is largely deployed for matters relating to language and cultural policy, for the promotion of Tamil in education, government, and public activities, and for petitioning the state (Ramaswamy 1997, 80–85). For a comparable discussion of the split between the poetic and the prosaic in early-twentieth-century Bengal, see Chakrabarty 2000, 149–79. Chakrabarty suggests that prose lends itself to historicism, whereas the poetic stands outside historical time.

89. Nilakanta Sastri 1972, 12. See also Srinivasa Iyengar 1982, 152; Subramania Aiyar 1970, 23–70.

90. Rocher 1986, 1–31, 115–31. For perceptive observations on the progressive disavowal of the *Purāṇas* by disciplinary history in colonial Bengal, see Guha 1997, 152–212.

91. Rangachariyar 1891b, 37.

92. Saravana Mutthu, 1892, 15–16. A former native of Ceylon who received his Bachelor's degree from the University of Madras, Saravana Mutthuppillai was Librarian, Presidency College, Madras, at the time of the publication of these comments. Although this is not the prevailing opinion among most scholars, Vimalanandam also credits him with the authorship of the first historical novel in Tamil in 1865 (Vimalanandam 1987, 925).

93. Seshagiri Sastri 1897, 7. See also Subramania Sastri 1915–16, especially, 417, 429–30.

94. Aravamuthan 1930, 308 (emphasis mine).

95. Saravana Mutthu 1892, 15–16.

96. Vaiyapuri Pillai 1956, 672. See also Aravamuthan 1930, 307; Pillay 1957b, 113–14.

97. Seshagiri Sastri 1897, v.

98. Chakrabarty 2000, 76.

99. Nora 1996, 7.

100. Srinivasa Iyengar 1982, 152. Srinivasa Iyengar began his professional career in 1884 as a college lecturer in Tiruchirapalli. He was appointed Reader in Indian History and Archaeology at the University of Madras in 1928, and in 1930 moved to the newly established Annamalai University as Professor of History and Politics.

101. Srinivasa Iyengar 1982, 232.

102. Ibid., 232. See also Seshagiri Sastri 1897, 8–9; Srinivasa Aiyangar 1914, 236–37, 252–53.

103. Sivaraja Pillai 1932, 25.

104. Rangachariyar 1891a, 748–49.

105. Nilakanta Sastri 1956, 87.

106. Joseph 1972, 4.

107. Nora 1989, 9.

108. Seshagiri Sastri 1897, iv–v , 8–9.

109. Krishnaswami Aiyangar 1911, 337. See also K. G. Sesha Aiyar's suggestion—in an otherwise disenchanted historicist account—that "tradition is really human testimony *regarding the lost past*; and though like all human testimony it is liable to error, it should not on that account be discarded as wholly unworthy of attention" (Sesha Aiyar 1937, 4 [emphasis mine]).

110. Pillay 1957b, 114.

111. Krishnaswami Aiyangar 1911, 336.

112. See, e.g., Pillay 1975, 41–55; Subrahmanian 1996, 10, 27.

113. Pillay 1975, 98.

114. Aravamuthan 1930, 192.

115. Nandy 1995. See also Lal 2000, 157–62.

116. Krishnaswami Aiyangar 1911, 1.

117. Quoted in Guha 1997, 153.

118. Duara 1995, 17–33.

119. *The Light of Truth or Siddhanta Deepika* 5, no. 3 (1901): 30–31. About fifty years before Vinson, Caldwell had dated the earliest works of Tamil literature to as late as the eighth or ninth centuries C.E. (Caldwell 1856, 61).

120. Savariroyan 1901a, 80.

121. Ibid., 81.

122. *The Light of Truth or Siddhanta Deepika* 5, no. 12 (1902): 193–94.

123. Savariroyan 1901b, 197–98.

124. *The Light of Truth or Siddhanta Deepika* 1, no. 3 (1897): 71.

125. Tirumalaikolunthu Pillay 1900, 18.

126. Kumari Maintan 2001, 38–39. A similar argument is made by another,

whose name (Kalcina Valuti) echoes the name of the first ruler of the antediluvial Kumarināṭu (Kalcina Valuti 2001). These essays are a response to Jayakaran 2001.

127. See, e.g., Suryanarayana Sastri 1903, 96–97.

128. Purnalingam Pillai 1945, Foreword.

129. Somasundara Bharati 1967, 94.

130. Raghava Aiyangar 1938, 78–79. A Brahman devotee of Tamil whose professional reputation was built on his editorship of the leading literary journal *Centamiḻ*, as well as his participation in the University of Madras Tamil Lexicon project, M. Raghava Aiyangar was awarded the coveted title of Rao Sahib by the colonial government in 1936. For his complex place in Tamil devotion, see Ramaswamy 1997, 195–96.

131. Savariroyan 1901b, 198.

132. See, e.g., Kumari Maintan 2001, 38–39.

133. That Tamil labors of loss are not alone in colonial (and postcolonial) South Asia in deploying science to recuperate tradition is clear from Gyan Prakash's analysis of the development of "Hindu science" in modern India (Prakash 1999).

134. Mutthuthambipillai 1906–7, 196. Mutthuthambipillai was born and educated in what was then Ceylon, although he moved to the Indian mainland in 1880, where he worked as an accountant and in the medical business for a number of years, even while participating in various literary and publishing activities. Returning to Ceylon in 1893, he published, among other things, his encyclopedia of Tamil literature, a first of its kind, called *Apitāṉa Kōcam* (1902).

135. Savariroyan 1901b, 198.

136. Suryanarayana Sastri 1903, 97.

137. Purnalingam Pillai 1904, 8–9 (emphasis mine).

138. E.g., Abraham Pandither 1984, 17, 21–24; Savariroyan 1900–1901, 104–5.

139. See, e.g., Subramania Sastri 1915–16, 429–37.

140. See, e.g., Damodaram Pillai 1971, 9–15; Suryanarayana Sastri 1903.

141. Geetha 1993, 132. See also Meenakshi Mukherjee's analysis of the inextricable entanglement of "elements of fantasy and intimations of history" in historical novels in late colonial India (Mukherjee 1985, 38–67), and Sudipta Kaviraj's discussion of "imaginary history," where he considers the deployment of the "fictive imagination" in the treatment of historical subjects in late-nineteenth-century Bengal (Kaviraj 1995).

142. Raghava Aiyangar 1938, 81–85.

143. A good example is the 1981 documentary film, with its depiction of life and events in the antediluvian Pandyan courts. See also Ramachandran and Mathivanan 1991, 15, and Sathur Sekhar et al. 1994, 72. The most striking instance of an undiluted flight of fantasy takes the form of a serialized novel originally published in the 1960s by the well-known poet Kannadasan (1927–81). Set in the antediluvian southern land, the novel is an excellent example of how geographical fantasy can offer an opportunity for exploring sexual fantasies as well (Kannadasan 1994).

144. There were undoubtedly many differences of opinion that prevailed in the metropole about the aims and parameters of academic geography, but I am principally concerned here with the translation of these into an imperial discipline that underwrote European colonial projects across the globe (see Godlewska and Smith 1994).

145. Edney 1997; Ramaswamy 2001b.
146. Heindel 1911, 275.
147. Stelle 1952.
148. E.g., Devaneyan 1940, 220–65; Sivagnana Yogi 1913, 3.
149. Ramaswamy 2000, 593–97.
150. Thongchai 1994.
151. E.g., Annapoorni 1935; Srinivasa Pillai 1965, 1–3.
152. Daniel 1992; Sivagnanam 1974, 717–74.
153. It is a measure of the geopolitical strength of the category of "India" that few Tamil place-makers call for the incorporation into the modern Tamil state of islands such as the Maldives, Sri Lanka, Java, Sumatra, and others which are also claimed by some to be surviving remnants of the former Kumari Nāṭu (see Abraham Pandither 1984, 90–91; Mutthuthambi Pillai 1906–7, 193; Sami 1955, 26). Sri Lanka, in particular, is a fascinating anomaly in this regard, given its proximity to south India and the fact that numerous Tamil devotees who hailed from the island participated in the labors of loss around Kumarināṭu. Yet, I have come across no irredentist intentions with regard to the incorporation of the island into the Tamil mainland in these labors of loss.
154. Swaminatha Aiyar 1950, 228. Many have noted the formulaic quality of these numbers which are all multiples of the number 7, which appears in other literary contexts as well (Shulman 1988a, 294).
155. This is the gloss provided in 1927 by Purnalingam Pillai (1945, 3). A decade earlier, Abraham Pandither identified these in 1917 as "the land of the Naawel tree, the land of the Peepul tree, the land of the Reed, the land of the Andil bird (the nightingale of India), the land of the Elephants, the land of the cocoanut, and the land of the Arecanut Palm" (Abraham Pandither 1984, 90).
156. This resignification began as early as 1903 (Suryanarayana Sastri 1903, 61–62).
157. On this, see, especially, Somasundara Bharati 1967.
158. Chidambaranar 1948, 150; Kandiah Pillai 1945, 2.
159. Disenchanted historiography also acknowledges that the ancient Pandyas moved their capital several times, but attributes this to local politics or shifts in river beds, not to antediluvial losses of territory to the ocean.
160. Arunachalam 1906, 75–76.
161. Murray 1866, 60. For the contrary position, that Ceylon was connected to Malaya and Sumatra rather than to India, see Tennent 1977, 11–12.
162. Tennent 1977, 5–8.
163. To say the least, if Tamil labors of loss went down this path, they would encounter the hostility of both the Indian and Sri Lankan states, and not perhaps freely circulate as they have.
164. Basso 1996, 44.
165. Arunachalam 1944; Padmanabhan 1971.
166. Nora 1989.
167. I am paraphrasing here from Nora 1989, 12, 19, and 24.
168. That Kanyakumari had served as such a *lieu de mémoire* for others in colonial India is clear from the reminiscences of Vivekananda, who meditated there on a

rock in 1892, when a moving vision of all of India's long history flashed before his eyes.

169. Basso 1996, 62.

170. After this precocious appearance, Lemuria rarely features in geography textbooks, either in English or Tamil, for their dominant concern is with describing the earth as empirically observable (for an exception, see Fox 1938, 161, a textbook meant for high-school students studying physical geography). A textbook in Tamil on the geography of Madurai district alludes to territorial loss: "What was the ancient capital of the Pandyas? The first capital of this country was on a land now submerged in the ocean. The second capital [was] Kapāṭapuram. Today, this too is under the ocean" (Ashirvatham 1931, 108). A few decades earlier, another mentions the antediluvian Pandyan capitals without, however, invoking territorial loss (Chithambera Sastriar 1883, 17–18). These books are, however, rare exceptions.

171. While most history schoolbooks, even when written by Tamil writers, discuss Lemuria in the formulaic and impersonal terms of colonial labors of loss, there are a few exceptions where Tamil place-making sentiments creep in. For example, "Ancient texts tell us that once upon a time, there were 49 territories to the south of Kumari and that these were seized by the ocean. Geologists believe that once upon a time the southern ocean adjacent to this land was once a vast territory, and that this was the birthplace of man" (Rangaswami Ayyangar 1927, 2). Or, "*Cilappatikāram*, our famous Tamil epic, tells us about the seizure by the ocean of Kumarikkaṇṭam. Scholars tell us that the first man emerged on Kumarikkaṇṭam. The evidence for this may be found in archaeological remains in Tamilnadu" (Duraikkannu Mudaliar 1958, 11–12). See also Shanmukhasundaram 1939, 9–10; Tamilnadu Textbook Society 1977, 16, 36; Tiruvenkadatayyangar 1948, 8. History schoolbooks published after the mid-1980s rarely invoke Lemuria, even in the speculative terms of colonial labors of loss.

172. See, e.g., Damodaram Pillai 1919, 103–6; Suryanarayana Sastri and Chelvakesavaraya Mudaliar 1918, 91–95.

173. The *caṅkam* narrative (as recounted in Nakkirar's commentary) appears for the first time in a Tamil-language schoolbook in 1904. The authors, however, make no connection between the *caṅkams* and Kumari Nāṭu, which is not even alluded to (Suryanarayana Sastri and Chelvakesavaraya Mudaliar 1904, 33–34). See also Kailasapillai 1930, 8–9; Mutthuswami Pillai 1934, 62; Subramania Sharma 1922, 85–87.

174. E.g., Kanakaraja Aiyar 1934, 35; Venkataramayyar 1934, 48–51.

175. E.g., see Kalyanasundaram 1959, 105–11; Venkataramayyar 1934, 48–51.

176. "The Tamil language flourished with great excellence at a remote period beyond the passage of time. It is more ancient than peoples like Babylonians, Egyptians, Greeks and others who scholars point to for their antiquity. It emerged even before the appearance of the Himalayas. It also flourished on the landmass called Lemuria which was swallowed up by the ocean" (Venkataramayyar 1934, 48). Other textbooks state that Tamil had been spoken for "thousands of years" (Mutthuswami Pillai 1934, 62; see also Meenatchisundaram 1958, 1, and Nagarajan 1961, 2–3), and that man appeared for the first time on the Tamil land called Kumarināṭu (Anbu Ganapathy 1951, 5). See also Kalyanasundaram 1959, 106; Mutthukrishna Nattar 1960, 5; Tamilnadu Textbook Society 1982, 88.

177. Gopalakrishan 1956, 1–3. See also Anbu Ganapathy 1951, 5–6; Nannan and Chandrasekharan 1958, 2–3; Venkatachalam Pillai 1951, 1–5.

178. Arunagirinathar and Veeraraghavan 1949, 2; Meenatchisundaram 1958, 2; Mutthuswami Pillai 1934, 62; Nagarajan 1961, 3; Sivasailam Pillai 1951, 17–18; Tamilnadu Textbook Society 1996a, 34; Venkatachalam Pillai 1951, 4–5; Venkataramayyar 1934, 48.

179. Kalyanasundaram 1959, 105–6.

180. Srinivasan 1956, 21–24.

181. See, e.g., Gopala Krishnan 1956, 1; 1954, 1; Nagarajan 1961, 1; Nannan and Chandrasekharan 1958, 1; Srinivasan 1956, 20.

182. Sivasailam Pillai 1951, 15. This textbook was published by the South India Saiva Siddhanta Works Publishing Society, one of the premier publishing outlets for Tamil devotional sentiments (Ramaswamy 1997, 222–24). See also Saccithanandan 1999, 52–53.

183. Kalyanasundaram 1959, 111. See also Arunagirinathar and Veeraraghavan 1949, 2; Gopala Krishnan 1956, 4; Nagarajan 1961, 4; Nannan and Chandrasekharan 1958, 3; Sivasailam Pillai 1951, 19; Srinivasan 1956, 25; Tamilnadu Textbook Society 1982, 94; Tamilnadu Textbook Society 1996b, 2; Tamilnadu Textbook Society 2001, 48; Venkatachalam 1956, 4; Venkatachalam Pillai 1951, 5; Vilvapati et al. 1956, 22–22.

184. Government of Tamilnadu 1975, 22–25, 114–28. The committee was chaired by the noted Tamil litterateur M. Varadarajan, and included among its members K. R. Hanumanthan and C. E. Ramachandran, who were historians at Presidency College, Madras, and the University of Madras, respectively. Also on the committee were the archaeologist R. Nagaswamy, and K. Appadurai (the author of the first monograph-length book in Tamil on Lemuria, published in 1941). The committee was established in 1971 and was mandated not only to rectify the traditional omission of southern India in history textbooks, but also to highlight "the great antiquity" of the Tamil country in light of recent research (*Tamilnadu Legislative Assembly Debates* 14 [1971]: 117–19; see also *Tamilnadu Legislative Assembly Debates* 24 [1972]: 273–79).

185. Government of Tamilnadu 1975, 25.

186. Ibid., 116 (emphasis mine).

187. Books by scholars who have written against the reality of Lemuria or of the three *caṅkam*s, such as K. A. Nilakanta Sastri, Seshagiri Sastri, or M. Srinivasa Aiyangar, have also frequently been prescribed in the college curriculum from the early years of the twentieth century.

188. See also Pillay 1981a, 21–49; and Swaminathan 1996, 19–40. In recent years, the DMK—one of the two Tamil nationalist parties that have held power in the state since 1967—has also incorporated the teaching of the truth about Kumarikkaṇṭam in correspondence courses it offers to its card-carrying members. I thank Mr. Sunderarajan of the Perasiryar Anbazhagan Research Library, Chennai, for this information.

CHAPTER 6

1. Jervis 1938, 9.

2. Haeckel 1870, enclosure.

3. Blakemore and Harley 1980, 17–22.

4. Harley and Woodward 1987, xvi. See also Harley's insistence that we ought to move "the reading of maps away from the canons of traditional cartographical criticism with its string of binary oppositions between maps that are 'true and false,' 'accurate and inaccurate,' 'objective and subjective,' 'literal and symbolic,' or that are based on 'scientific integrity' as opposed to 'ideological distortion'" (Harley 1988, 278).

5. Harley 1987, 2.

6. Harley 2001, 35.

7. Cosgrove 1999, 1–2.

8. "The map, as Jean Baudrillard puts it, has come to precede the territory. Rather than the map being a product of the territory, as it is usually understood, coming only after it—both temporally and conceptually—and remaining answerable to it, there has been a curious reversal. The debate is conducted in terms of the map rather than the territory itself" (King 1996, 1–2).

9. Note Blakemore and Harley's observation that "maps enable men [*sic*] to see at a glance objects and relationships which had previously been hidden from the single human perspective" (Blakemore and Harley 1980, 97).

10. Harley 2001, 14.

11. *Cartographica* 26, no. 3–4 (1989): 96. Wood and Fels similarly write that there is nothing natural about the scientific map, its claims to the contrary notwithstanding. Instead, even the scientific map is a cultural artifact which is a cumulation of choices, "every one of which reveals of value: not the world, but the slice of a piece of the world; not nature but a slant on it; not innocent, but loaded with intentions and purposes; not directly, but through a glass; not straight, but mediated by words and other signs" (Wood and Fels 1986, 65).

12. See, in this context, Jeremy Black's observation: "Mapping imaginary communities and species—Barsetshire to Middle Earth, the world of Long John Silver to that of Winnie the Pooh and Toad of Toad Hall—makes it possible to map in the absence of the politics and polemics of the real world and without the risk of complaint" (Black 1998, 101).

13. It is only since the late eighteenth century, after the 1761 invention of the chronometer, that the precise calculation of longitudes became possible. Samuel Edgerton writes that "the grid system . . . reduced the traditional heterogeneity of the world's surface to complete geometrical uniformity. . . . [It] posed . . . an immediate mathematical unity. The most far flung places could be precisely fixed in relation to one another by unchanging coordinates so that their proportionate distances, as well as their directional relationships, would be apparent" (Edgerton 1976, 113–14). Chandra Mukerji observes that "the Cartesian grid, symbolizing the profane, the contestable, the measurable and the impersonal, was the appropriate language for scientific domination of images of nature" (Mukerji 1984, 32). See also Carter 1988, 202–29.

14. Conley 1996, 2.

15. There are several studies of the world map as a genre, but see, especially, Whitfield 1994. Note, especially, his comment "that the impulse to depict the world on paper has always been associated with the desire to make some statement about the world" (ibid., 2). Whitfield also suggests that the drawing of world maps is pro-

voked by the imperative "to see an image of the entire world focused before us, clear, self-contained, comprehensible, and masterable" (ibid., 4).

16. As a scientist and man of the world in Victorian London, Sclater would have been undoubtedly familiar with Alexander K. Johnston's *Physical Atlas,* originally published in 1848, roughly based on Heinrich Berghaus's *Physikalisher Atlas,* published in German between 1845 and 1848 (Robinson 1982, 64–66, 105–6). These atlases included some of the earliest zoogeographical maps showing the worldwide distribution of various fauna, one of which mapped the spread of monkeys and lemurs (Robinson 1982, 102, fig. 47). Two years after Sclater's essay on Lemuria was published, Andrew Murray published his monumental *The Geographical Distribution of Mammals* (1866), in which Map. No. 11 illustrated the worldwide distribution of the lemur. This map, although it was published after Sclater's hypothesis on Lemuria was articulated, does not, however, show the former continent (although Murray himself, as I noted in chapter 2, was favorable to such submerged worlds). For a thoughtful argument that (biogeographical) maps were crucial to Sclater's contemporaries like Darwin and Wallace, see Camerini 1987.

17. Wegener 1924, 5. In a letter that he wrote to his father-in-law (a fellow climatologist), Wegener also noted, "You consider my primordial continent to be a figment of my imagination, but it is only a question of the interpretation of observations. I came to the idea on the grounds of matching coastlines, but the proof must come from geological observations" (quoted in Dobson 1992, 188). Critics of Wegener's drift theory, however, of whom there were plenty in the early years, suggested that he had "succumbed to the temptations of [the] map," and "had indulged in some advantageously creative cartographic generalizations" (Monmonier 1995, 164). One of them, the Yale geologist Chester Longwell, observed, "It is not improbable that gazing at the map of South America and Africa has the effect of hypnotizing the student. The coast lines appear to be such exact counterparts, even in detail—Wegener must be right!" (quoted in Monmonier 1995, 166). Mark Monmonier himself notes, "Wegener's skilled and relentless use of cartography was a significant force in the history of geology, and the success of his efforts attests to the map's rhetorical power to explain and persuade" (ibid., 169). Far before Wegener, others had similarly "succumbed to the temptations of the map," at least since 1596, when the famous Dutch mapmaker Abraham Ortelius observed the fit between the American and African coastlines on the maps of the world that had been published over the course of the sixteenth century (Romm 1994).

18. Marsden 1909, 4. By the time Marsden wrote these lines, there were numerous world geology maps in which the similarity between the rock formations of India, Madagascar, and Africa was clearly apparent (Robinson 1982, 86–100).

19. Kandiah Pillai 1945, 2. Other similar invocations of the map form by Tamil place-makers may be found in Appadurai 1941, 24; Devaneyan 1940, 50; Kandiah Pillai 1934, 4; Suryanarayana Sastri 1903, 62; and Velan 1995, 111–12.

20. Rutter 1932, 16–17. Similarly, in the 1942 novel, *New Trade Winds for the Seven Seas,* three American adventurers chance upon the lost descendants of former Lemurians living on a remote Pacific island, who take them to a temple on whose walls are drawn ancient maps which show the former continent as it had existed thousands of years ago, before it drowned. On seeing the maps, the adventurers go

in search of the lost city of Motu, Lemuria's capital in a vanished age (Roberts 1942, 128–42).

21. There are a couple exceptions in this regard. Alexander Kondratov's *The Riddles of the Three Oceans* features on its inside cover a bi-colored map which shows Lemuria. In contrast to all other maps of the lost place-world, its iconography harks back to the maps of the European age of discovery. Neptune appears on the top of the map gazing down on a map of the world that shows Lemuria as a landmass connecting Africa and India. A mermaid and a diver are drawn at the bottom, holding up the map of the world. Two ships and a submarine are part of the iconography (Kondratov 1974). Similarly, see also an untitled map which is posted on the website of a company that calls itself Lemurian Imports. This map, which also echoes in its composition antique maps, features Lemuria as a gateway through which one can access marketplaces for various esoteric commodities such as incense, tarot cards, crystal balls, and so on. These marketplaces of Lemuria bear names such as Incantia, Shadowlands, and Cosmica, which appear on the map as islands scattered among bodies of water with names such as Enchanted Strait, Floral Sea, and the Oceane Azure. Presumably, one could sail to these marketplaces in the sailing ship that also appears on the map (http://www.lemurian-imports.com).

22. See, e.g., the following publications, all in Tamil: Appadurai 1975; Chidambaranar 1948; Kandiah Pillai 1945; Kandiah Pillai 1984; Mathivanan 1977; and Puratcidasan 1995.

23. Anderson 1991, 175.

24. Smith and Godlewska 1994, 1.

25. See, especially, Godlewska and Smith 1994; Hooson 1994.

26. Stoddart 1986, 30.

27. Rudwick 1976.

28. Jay 1994.

29. Quoted in Camerini 1987, 60.

30. Mitchell 1988, 6.

31. Edney 1997.

32. Ramaswamy 2001b; 2002.

33. For some beginnings, see Brosius 1997; Dasgupta 1995; Krishna 1994; and Ramaswamy 2001b; 2002.

34. Ramaswamy 2001b; 2002.

35. Ramaswamy 2001b, 108–9.

36. Brosius 1997.

37. Gole 1989; Schwartzberg 1992.

38. Haeckel 1870.

39. Haeckel 1876b, 399. See also Haeckel 1870, 677–79.

40. Haeckel 1876b, 400 (emphasis in original). See also Haeckel 1870, 677–79.

41. On the Hyperboreans, see chapter 3, note 38.

42. Wallis and Robinson 1987, 105.

43. Ibid., 105. On the persuasiveness of such thematic maps, see Harley 2001, 8, 218.

44. Anthony 1995, 91.

45. Melchior Neumayr (1845–90), the son of a Bavarian bureaucrat, studied law

at Munich University but went on to get a doctorate in geology and paleontology from Heidelberg. In 1868 he joined the Austrian Geological Survey, but subsequently became Professor of Paleontology in Vienna. Like Haeckel, he, too, was an enthusiastic Darwinian.

46. E.g., Wood 1985, 41. Recently, however, David Oldroyd has reprinted a French map entitled "Carte du globe a l'epoque jurassique" which was originally published in 1860. It shows a paleo-supercontinent called "Americo–Africo–Australie," which he suggested dominated the surface of Earth in the Jurassic epoch (Oldroyd 1996, 258–59).

47. Marvin 1973, 55–56.

48. Oldham 1894, 175. See also Ramaswamy 1999, fig. 3.

49. Oldham 1894, 176. Richard Dixon Oldham was the son of Thomas Oldham, the first director of the Geological Survey of India. In 1893 he published a revised version of the Survey's influential *Manual of the Geology of India.* He was also president of the Geological Society of London between 1920 and 1922.

50. See, e.g., Arldt 1917, 373–401; Haug 1907; Holdich 1905; Schuchert 1931; von Ihering 1907.

51. Fortey 1981, 193.

52. Such sequential maps which track the fate of paleo-continents over time are akin to the "historical map" which Benedict Anderson flags as one of the "avatars" of modern mapmaking. "Through chronologically arranged sequences of such maps, a sort of political-biographical narrative of the realm [comes] into being, sometimes with vast historical depth" (Anderson 1991, 174–75).

53. Wegener 1924, 6. An earlier version of this illustration was included for the first time in the 3rd revised German edition of Wegener's *The Origin of Continents and Oceans* (1922). For a useful analysis of how Wegener created his map, see Marvin 1972, 72–76.

54. Wegener 1924, 6.

55. Ibid., 64–65.

56. See, e.g., Joleaud 1939, plates II, XVI, XXVI, XXVII, XLII, and LIX; Perrier 1925, 21–29; Termier 1960, xix–xxix.

57. See, e.g., Haug 1907, 111; Joleaud 1939, plate XIV; Termier 1960, xviii–xx.

58. See, e.g., Termier 1960; van Steenis 1962; Wild 1965. But in places like India the occasional geology book continued to publish maps in which Lemuria appears (see, e.g., Dey 1968, fig. 40). See also a map of the "Eocene–Oligocene Lemurian Stepping-Stones" at http://www.mobot.org/MOBOT/Madagasc/biomad9.html.

59. See, e.g., Smith and Briden 1977; Smith, Smith, and Funnell 1994.

60. Whitfield 1994, 130.

61. Quoted in van der Gracht et al. 1928, 144.

62. Willis 1932, 939, plate 27.

63. Sinnett 1954, ix.

64. Scott-Elliot 1954, 17.

65. Scott-Elliot 1904, 12.

66. Ibid., 13.

67. Ibid. (emphasis in original).

68. Ibid.

69. Churchward 1994b, 81.

70. Hayes 1972, xv.
71. Ibid., xv–xvi.
72. Scott-Elliot 1954, map no. 1.
73. Ibid., 18.
74. Ibid., 21.
75. Scott-Elliot 1904, 14.
76. Ibid., 14.
77. Ibid., 14–15.
78. Goodricke-Clarke 1985, 54, 101. See also chapter 3.
79. Browning 1924, inside cover of dust jacket.
80. Cervé 1997, 49. For Cervé's discussion of his maps, see 51–68.
81. Lemurian Fellowship 1945; Stelle 1952.
82. Churchward 1994a, 36; 1994b, 22, 24, 81, 91.
83. Lemurian Fellowship 1945; Stelle 1952. See also Lemurian Fellowship 1939.
84. Scafi 1999, 70. For a useful conceptual analysis of this shift, see also Koyre 1957.
85. Gillies 1994, 62, 172.
86. Ibid., 62.
87. Subramania Sastri 1915–16, 466–67.
88. Arasan Shanmugham Pillai 1995, 187–232; Subramania Sastri 1915–16, 465–74. Arasan Shanmughanar, a native of Madurai, was teacher to some of Tamil devotion's most well-known practitioners, such as Maraimalai Adigal, Kathiresan Chettiar, and Somasundara Bharati. His comments on the loss of Tamil land appear in the context of an extended commentary he published in 1905 on the opening verse of the *Tolkāppiyam.* Tellingly, he identified the *Tolkāppiyam* as the earliest Tamil grammar, written twelve thousand years ago, and also suggested that its author was a member of the first *caṅkam* at Te<u>n</u>maturai in the years before it was seized by the ocean (Arasan Shanmugham Pillai 1995, 187–88).
89. Subramania Sastri 1915–16, 421.
90. On the commitment to the subcontinent's peninsularity in modern Indian mapping practices, see Ramaswamy 2002.
91. Purnalingam Pillai 1945, enclosure. See also Ramaswamy 1999, fig. 11.
92. Pulavar Kulanthai 1971, n.p.
93. Mahalingam 1981b, 136–37.
94. Anon. 1995, 95.
95. Mathivanan 1977, 1.
96. Ramachandran and Mathivanan 1991, 1.
97. Thanikachalam 1992, n.p.
98. Manuel Raj 1993, xii. See also Ramaswamy 2000, fig. 3.
99. Sivasailam Pillai 1951, facing page 15.
100. Chidambaranar 1948; Kandiah Pillai 1945; and Pulavar Kulanthai 1971.
101. See, e.g., Anon. 1995; Mahalingam 1981b.
102. See maps of Lemuria in Kandiah Pillai 1945 and Manual Raj 1993.
103. Carter 1988; Harley 2001, 178–87; Mundy 1996, 135–79.
104. Harley 1992a.
105. Harley 2001, 179, 181.
106. Harley 1992a, 530.

107. It is perhaps fitting that Appadurai, perhaps most well known for this book among Tamil's devotees, was himself a native of the Kanyakumari area. Proficient in many languages, including Hindi (which he taught for several years before his conversion to Tamil devotion), Appadurai was a translator of numerous works into Tamil, as well as a participant in the University of Madras Tamil–English dictionary project. Much of his devotion to Tamil is clearly in the vein of what I have characterized as contestatory classicism, his passion for it even leading him to insist that Tamil should become the national language of independent India (Ramaswamy 1997, 43–44).

108. Appadurai 1941, 8, 56. See also Government of Tamilnadu 1975, 285–87; Manual Raj 1993.

109. Chandrababu 1996, 3.

110. Gardiner 1933.

111. Quoted in *Illustrated Weekly of India*, July 29, 1934, 44.

112. Lawrence 1999, 41; Matthews 1967.

113. Behrman 1981. See also International Indian Ocean Expedition 1975, xi–xiii.

114. Matthews 1967, 554.

115. Ibid., 567.

116. S. Padmanabhan (see chapter 4, note 131) told me in the course of a conversation (March 1997) that he had been inspired in his own labors of loss over Kumarināṭu when he saw the *National Geographic* map. See also a rough reproduction of the map in Ramachandran and Mathivanan 1991.

117. Saccithanandan 1999, 52–53. I thank Theodore Baskaran for alerting me to this publication, and Mary Rader for acquiring the atlas for me.

118. See, e.g., Mathivanan 1977, 14; Government of Tamilnadu 1975, 283. See also Mutthuswamy Aiyar and Appuswamy Aiyar 1933, 253, 329.

119. See, e.g., Government of Tamilnadu 1975, 267; Mahalingam 1981b, 165–66; Mahalingam and Sathur Sekharan 1991, 177. The 1981 documentary film includes such maps as well.

120. See, e.g., maps in Aravaanan 1984; Chandrababu 1996; Chidambaranar 1948; Manual Raj 1993; Ramachandran and Mathivanan 1991.

121. I borrow this formulation from Homi Bhabha's provocative discussion of colonial mimicry (Bhabha 1994, 85–92).

122. See, e.g., maps in Chidambaranar 1948; Mahalingam 1981b.

123. See, e.g., maps in Kandiah Pillai 1947; Mathivanan 1977; Manuel Raj 1993; Rajasimman 1944; Ramachandran and Mathivanan 1991; Thanikachalam 1992.

124. See, e.g., maps in Chidamabaranar 1948; Kandiah Pillai 1945; Mahalingam 1981b.

125. Chidambaranar 1948; Ramachandran and Mathivanan 1991.

126. This is also the first time the island of Sri Lanka appears on Tamil maps of Lemuria as a province of the former homeland. Since that time, numerous other maps show the island as an intimate part of Kumarikkaṇṭam's geography (see, e.g., maps in Kandiah Pillai 1947; Mahalingam 1981b; Pulavar Kulanthai 1971; and Sivasailam Pillai 1951).

127. Ryan 1994, 116. See also Harley's observation that the empty spaces on a map are "positive statements, and not merely passive gaps in the flow of language" (Harley 2001, 14).

128. Ryan 1994, 126–27. See also Harley 2001, 187–95.

129. Mignolo 1995, 219–309. In my analysis of Tamil maps of Lemuria, I have been influenced by Mignolo's nuanced reading of the "fractured," "subaltern" maps of Amerindians from sixteenth-century Peru. He argues that alternate territorial conceptions "cohabit" and "coexist" within the same frame, although social and political power is unequally distributed between colonizer and colonized. The very prevalence of such coexisting conceptions suggests that Western cartography was unable to suppress alternate cosmologies. While I agree with this analysis, I prefer to also see the Tamil maps of Lemuria as exhibiting more than a benign or passive "coexistence" of "native" and "Western" knowledges of the lost continent, for they also actively work to transform and hybridize the Euro-American place-worlds of the paleo-scientist and the occultist into the lived Tamil homeland. At the same time, as Mignolo rightly reminds us, unlike the modern map produced in the West which attained universal status with the global expansion of Europe, "native" cartographies, resistant and alternate though they may be, invariably remain local, subaltern forms, consumed only by other natives.

130. A rare exception are the maps published by Chidambaranar in 1948.

131. See, e.g., Appadurai 1941; Chandrababu 1996; Chidambaranar 1948.

132. See, e.g., Mathivanan 1977; Mahalingam 1981b; Thanikachalam 1992.

133. Mignolo 1995, 219–33.

134. Thongchai 1994, 56.

135. Edney 1997.

136. Harley 1988, 301.

137. See also Ramaswamy 1999.

138. Pulavar Kulanthai (1906–72), scholar and litterateur whose radical participation in Tamil devotion I have considered in *Passions of the Tongue*, is one among a handful of Lemuria's place-makers who have written in verse about the loss of this hapless land (Pulavar Kulanthai 1971, 9–71). His *Irāvaṇa Kāviyam* (an epic poem on Ravana), in which he proposed that the Sanskrit *Rāmāyaṇa*'s demon-villain was actually a Tamil hero and monarch of the ancestral Tamil homeland now lost to the ocean, was originally banned when it was first published in 1946, but subsequently felicitated (and reprinted) in 1971, after the political triumph of Tamil nationalism in 1967.

139. Other maps showing the deployment of the synoptic operation may be found in Chidambaranar 1948; Chandrababu 1996; Kandiah Pillai 1945, Manuel Raj 1993; and Thanikachalam 1992.

140. Other examples of the sequential strategy may be found in Mathivanan 1977 and Thanikachalam 1992.

141. Rajasimman 1944, 5.

142. Kandiah Pillai 1945, n.p.; 1957, 12; 1949, 6.

143. Thus, a 1984 bilingual map entitled "The Dispersal of the Negro People," shows three lines radiating out from a place called "Kumari." The lines are meant to designate the "Dravidian Blacks" (who are shown moving to India), "East Blacks" (who are shown moving to Australia), and "West Blacks" (who are shown moving to Africa) (Aravaanan 1984, 121). Similarly, Puratcidasan published a map in 1995 that graphically represented numerous lines radiating out from a place named "Kumari" to various parts of the known world (Puratcidasan 1995, 81). See also Mahalingam and Sathur Sekharan 1991.

144. I adapt this phrase from Renfrew 1987, 86.

145. Thus, one of Chidambaranar's maps dates antediluvian "Tamilnāṭu" to 20,000,000 B.C. (Chidambaranar 1948, facing page 1), whereas Mahalingam's 1981 map places it in 30,000 B.C. (fig. 9).

146. Anderson 1991, 174–75.

147. Thongchai 1994, 152.

148. Harvey 1990, 249.

149. Anderson 1991, 164.

150. Ibid., 173. On colonialism and scientific cartography, see, especially, Edney 1997; Harley 1992a; 2001, 170–95; Mignolo 1995; Mundy 1996; Ryan 1996; and Thongchai 1994.

151. Ramaswamy 1999, 121.

152. Quoted in King 1996, 1–2.

153. Thongchai 1994, 130.

154. Harley 1992a, 532.

155. Robinson and Petchenik 1986, 4.

156. Heidegger 1977. For a nuanced application to a colonial context of Heidegger's proposition that "the fundamental event of the modern age is the conquest of the world as picture," see Mitchell 1988, especially his insistence that colonial power "enframed" the colony, made it "picture-like and legible," and ready for "political and economic calculation" (33). I would add that cartography is critical to this colonial "enframing" of the colony.

157. Harvey 1990, 244–53.

158. Quoted in Gregory 1994, 65–66.

159. Bhabha 1994, 85–92.

CHAPTER 7

1. Eliade quoted in Daniels 1985, 88.

2. Monahan 1903–4, 30–31.

3. See, e.g., Joseph 1972, 2–4; Nilakanta Sastri 1956, 87; Subramania Aiyar 1970. See also Scharfe 1973. For a rare argument that a drowned Lemuria is impossible given what we know of plate tectonic theory, see Ramachandran 1988. For a recent critique of Tamil labors of loss based on the rationalities of current ocean science, see Jayakaran 2001 and 2002.

4. See, e.g., Subramania Aiyar 1970, Preface, 65–66. See also Joseph 1958, 130.

5. Bloomfield 1932, 237, 270–71.

6. http://www.spiritweb.org/Spirit/atlantis3-diandra.html.

7. Shulman 1988b, 110–11.

8. Ibid., 119.

9. Ibid.

10. Ibid., 114 (emphasis mine).

11. My analysis in this regard has been influenced by my reading of Marilyn Ivy's provocative *Discourses of the Vanishing,* in which she argues that in modern Japan one witnesses "a longing for pre-modernity, a time before the West, before the catastrophe of westernization. Yet the very search to find authentic survivals of pre-modern, pre-western Japanese authenticity is inescapably a modern endeavor, essentially

enfolded within the historical condition that it would seek to escape. Thus, that search speaks also of the denials of modernity's ruptures" (Ivy 1995, 241–42).

12. Phillips 1985, 66.

13. I borrow this distinction from Svetlana Boym's discussion of nostalgia, in which she writes, "Nostalgia is not always about the past; it can be retrospective but also prospective. Fantasies of the past determined by needs of the present have a direct impact on realities of the future. Consideration of the future makes us take *responsibility* for our nostalgic tales" (Boym 2001, xvi [emphasis mine]).

14. See, e.g., Behrman 1981; Toussaint 1967, 1–11.

15. Rice 1986, 35–37.

16. Anon. 1904, 594 (emphasis mine).

17. Gardiner 1906, 316 (emphasis mine). Gardiner's essay is accompanied by maps of the Indo-African land connection based on Melchior Neumayr's (ibid., 319, 323). For the discussion among members of the Royal Geographical Society of Gardiner's statement, see *Geographical Journal* 28, no. 4 (1906): 465–71.

18. *Geographical Journal* 28, no. 4 (1906): 469–70.

19. "Survey of Indian Ocean, A New Expedition," *Times* (London), August 2, 1932, p. 9.

20. Rice 1986, 314.

21. Gardiner 1933, 570.

22. Wiseman and Sewell 1937, 230.

23. Behrman 1981, 12.

24. See chapter 5, note 8.

25. Blood 1957, 516. For a useful discussion of the modern metropolitan fascination with tropical islands as lost Edens, see Grove 1995, 16–72. Denis Cosgrove observes that from at least the mid-eighteenth century, "the global imagination of Europeans was no longer captivated by land. It was dominated by the island, whose self-contained spatialities offered the perfect geographical template for the systematic theories of 'nature' debated by rationalist philosophers and naturalists such as Montesquieu, Rousseau, Linnaeus, and Buffon" (Cosgrove 2001, 188–89). Cosgrove also writes that "even after longitude became fixed, imaginary islands continued to appear on maps and charts. Distortions of vision at sea by mist, mirage and heat haze, not to mention human errors of observation caused by tiredness, alcohol, or simple mendacity, ensured that British Admiralty charts registered nonexistent islands in the Pacific Ocean into the era of satellite observation" (ibid., 285).

26. Scott 1961. The book was subsequently reprinted in 1974 by Greenwood Press.

27. Edlis 1993, 15.

28. Ibid., ix.

29. Ibid., 87. Edlis is also a member of a U.K.-based nonprofit organization called Friends of the Chagos, which "aims to promote conservation, scientific and historical research and to advance education concerning the Chagos Archipelago." Although the organization does not invoke Lemuria in its publicity material (http://www.ukotcf.org/members/chagos.htm), I learned about it when one of its members contacted me to learn more about Lemuria, having found out I was writing a book on the lost continent.

30. http://www.7south.net/lemuria_3.htm.

BIBLIOGRAPHY

Abraham Pandither, M. 1984. *Karunāmirthasāgaram* (A Treatise on Music or Isait-Tamil, which is one of the Main Divisions of Muttamil, or Language, Music, and Drama). Reprint ed. Delhi: Asian Educational Services.

Aiyan Aarithan. 1904–5. Iṭaiccaṅkam [The middle academy]. *Centamiḻ* 3, no. 8: 271–83.

Anbazhagan. 1975. *Tamiḻk Kaṭal Alai Ōcai* [Sounds of the waves of the Tamil ocean]. Madras: South India Saiva Siddhanta Works Publishing Society.

Anbu Ganapathy. 1951. *Teṉṟal Vācakam: Iraṇṭām Puttakam (Iraṇṭām Pārattiṟkuriyatu)* [Tamil reader for class seven]. 2nd ed. Madras: Duco Publishing House.

Anderson, Benedict. 1991. *Imagined Communities: Reflections on the Origin and Spread of Nationalism.* 2nd ed. London: Verso.

Annapoorni, K. 1935. Tamiḻnāṭum Paṇṭait Tamiḻarkaḷum [Tamilnadu and ancient Tamilians]. *Āṉantapōtiṉi* 21, no. 4: 257–62.

Anonymous. 1897. The Original Home of Mankind. *Australasian Anthropological Journal* 1, no. 5: 111–12.

———. 1898. The Lost Continent in the Indian Ocean: The First Home of Mankind. *Science of Man: Journal of the Royal Anthropological Society of Australasia,* n.s., 1, no. 4: 90–91.

———. 1902. The Migrations of Primitive Men and Blacks of Australia. *Science of Man: Journal of the Royal Anthropological Society of Australasia,* n.s., 5, no. 5: 74–76.

———. 1904. The Proposed Indian Ocean Expedition. *Geographical Journal* 24, no. 5: 593–94.

———. 1908a. The Original Home of the Earliest Men. *Science of Man: Journal of the Royal Anthropological Society of Australasia* 10, no. 1: 3–4.

———. 1908b. The Evolution of Mankind. *Science of Man: Journal of the Royal Anthropological Society of Australasia* 10, no. 5: 70–71.

———. 1916. *A History of India for Junior Students with* Introduction by L. F. Rushbrook Williams. London: Longmans, Green.

———. 1995. Kaṭal Koṇṭa Kumarik Kaṇṭattiṉ Eñjiya Pakutitāṉ Kumari Māvaṭṭam

[Kanyakumari district is the remaining portion of Kumarikkaṇṭam that was seized by the ocean]. *Āyvukaḷanciyam* 2, no. 7: 4–9.

Anthony, David W. 1995. Nazi and Eco-feminist Prehistories: Ideology and Empiricism in Indo-European Archaeology. In *Nationalism, Politics, and the Practice of Archaeology*, edited by P. L. Kohl and C. Fawcett. Cambridge: Cambridge University Press.

Appadurai, Arjun. 1993. Number in the Colonial Imagination. In *Orientalism and the Postcolonial Predicament*, edited by C. A. Breckenridge and P. v. d. Veer. Philadelphia: University of Pennsylvania Press.

Appadurai, K. 1941. *Kumarik Kaṇṭam Allatu Kaṭal Koṇṭa Teṉṉaṭu* [Kumarikkaṇṭam; Or, the southern land seized by the ocean]. Chennai: South India Saiva Siddhanta Works Publishing Society.

———. 1975. *Kumarik Kaṇṭam Allatu Kaṭal Koṇṭa Teṉṉaṭu* [Kumarikkaṇṭam; Or, the southern land seized by the ocean]. Reprint ed. Chennai: South India Saiva Siddhanta Works Publishing Society.

Arasan Shanmugham Pillai, S. 1995. *Tolkāppiya Caṉmukaviruttiyiṉ Mutalāvatu Pakutiyākiya Pāyira Virutti* [Arasan Shanmughanar's commentary on the *Tolkāppiyam*'s opening verse]. Tanjavur: Karanthai Tamil Sangam.

Aravaanan, K. P. 1984. *Tamiḻariṉ Tāyakam* [Motherland of the Tamilians]. Madras: International Institute of Tamil Studies.

Aravaanan, Thayammal. 1977. Dravidians and Africans. In *Dravidians and Africans*, edited by K. P. Aravaanan. Madras: Tamil Kootam.

Aravamuthan, T. G. 1930. The Oldest Account of the Tamil Academies. *Journal of Oriental Research* 4: 183–201, 289–317.

Arldt, Theodor. 1917. *Handbuch der Paléogéographie*. Leipzig: Palaktologie.

Armitt, Lucie. 1996. *Theorising the Fantastic*. London: Arnold.

Arunachalam, M. 1944. Kumari. *Kumari Malar (Annual Issue)* (December): 1–3.

Arunachalam, P. 1906. Sketches of Ceylon History. *The Light of Truth or Siddhanta Deepika* 7, no. 1: 70–76.

Arunachalam, P. G., and V. S. Raghavan. 1967. *Vācaṉ Varalāṟṟu Nūl (Patiṉōrām Vakuppiṟkuriyatu)* [Vasan history book for class eleven]. Madras: Shanta Publishers.

Arunagirinathar, S. S., and P. Veeraraghavan. 1949. *Putumuṟai Caṅka Vācakam (Nāṉkām Puttakam)* [New Sangam Tamil reader: Book four]. 5th rev. ed. Madras: Association Publishing House.

Ashe, Geoffrey. 1992. *Atlantis: Lost Lands, Ancient Wisdom*. London: Thames and Hudson.

Ashirvatham, Michael. 1931. *Māṉiṭa Pūmicāstiram: Maturai Jillā* [Human geography of Madura district]. Madura: Raja Press.

Bailey, Alice. 1922. *The Consciousness of the Atom*. London: Lucis Press.

Baines, J. A. 1893. *Census of India, 1891: General Report*. London: Indian Government.

Balfour, Edward, ed. 1976. *Encyclopedia Asiatica Comprising Indian Subcontinent, Eastern and Southern Asia (Commercial, Industrial, and Scientific)*. Reprint of 3rd ed. New Delhi: Cosmo Publications.

Baliga, B. S. 1960. *Madras District Gazetteers: Madurai*. Madras: Government of Madras.

Bann, Stephen. 1989. *The True Vine: On Visual Representation in the Western Tradition.* Cambridge: Cambridge University Press.
Baran, Michael. 1981. *Twilight of the Gods: Astounding, Documented Research on the Legendary Lost Continents and Human Prehistory, and their Influence on Mankind's Destinies and Modern Mysteries.* Smithtown, NY: Exposition Press.
Barborka, Geoffrey. 1979. *The Story of Human Evolution.* Madras: Theosophical Publishing House.
Barker, A. Trevor, ed. 1923. *The Mahatma Letters to A. P. Sinnet from The Mahatmas M. & K. H.* New York: Rider.
Barrow, Ian J. 2003. *Making History, Drawing Territory: British Mapping in India, c. 1756–1905.* New Delhi: Oxford University Press.
Basham, Diana. 1992. Through the Looking Glass: Madame Blavatsky and the Occult Mother. In *The Trial of Woman: Feminism and the Occult Sciences in Victorian Literature and Society.* New York: New York University Press.
Basso, Keith. 1996. *Wisdom Sits in Places: Landscape and Language among the Western Apache.* Albuquerque: University of New Mexico Press.
Bataille, Georges. 1985. The Notion of Expenditure. In *Visions of Excess: Selected Writings, 1927–1939,* edited by A. Stoekl. Minneapolis: University of Minnesota Press.
Baum, Gregory. 1970. Does the World Remain Disenchanted? *Social Research* 37, no. 2: 153–202.
Bayly, C. A. 1996. *Empire and Information: Intelligence Gathering and Social Communication in India, 1780–1870.* Cambridge: Cambridge University Press.
Bayly, Susan. 1995. Caste and 'Race' in the Colonial Ethnography of India. In *The Concept of Race in South Asia,* edited by P. Robb. Delhi: Oxford University Press.
Behrman, Daniel. 1981. *Assault on the Largest Unknown: The International Indian Ocean Expedition, 1959–65.* Paris: UNESCO Press.
Berman, Morris. 1981. *The Reenchantment of the World.* Ithaca, NY: Cornell University Press.
Besant, Annie. 1904. *The Pedigree of Man: Four Lectures Delivered at the Twenty-Eight Anniversary Meetings of the Theosophical Society, at Adyar, December, 1903.* Benares and London: Theosophical Publishing Society.
Besant, Annie, and C. W. Leadbeater. 1922. *Man: Whence, How, and Whither (A Record of Clairvoyant Investigation).* Chicago: Theosophical Press.
Bhabha, Homi. 1994. *The Location of Culture.* London: Routledge.
Bishop, Peter. 1989. *The Myth of Shangri-La: Tibet, Travel Writing, and the Western Creation of Sacred Landscape.* Berkeley: University of California Press.
Biswas, Amrita Lal. 1880. Notes on "A Land of Mystery." *The Theosophist* 1 (August): 278–79.
Black, Jeremy. 1998. *Maps and Politics.* Chicago: University of Chicago Press.
Blakemore, M. J., and J. B. Harley. 1980. *Concepts in the History of Cartography: A Review and Perspective.* Toronto: University of Toronto Press.
Blanford, Henry F. 1873. *The Rudiments of Physical Geography for the Use of Indian Schools Together with a Sketch of the Physical Structure and Climate of India, and a Glossary of Technical Terms Employed.* Calcutta: Thacker, Spick.
———. 1875. On the Age and Correlation of the Plant-bearing Series of India and

the Former Existence of an Indo-Oceanic Continent. *Quarterly Journal of the Geological Society of London* 31: 519–42.

———. 1877. Catalogue of Cyclones in the Bay of Bengal. *Journal of the Royal Asiatic Society* 46: 328–38.

Blanford, William T. 1890. Anniversary Address of the President: The Permanence of Ocean Basins. *Quarterly Journal of the Geological Society* 46: 43–110.

Blavatsky, Helena P. 1880. A Land of Mystery. *The Theosophist* 1 (August): 159–61, 170–73, 224–227, 277–79.

———. 1928. *The Secret Doctrine: The Synthesis of Science Religion and Philosophy.* Vol. 2: *Anthropogenesis.* Reprint ed. London: Theosophical Society Publishing House.

———. 1931. *Isis Unveiled: A Master-Key to the Mysteries of Ancient and Modern Science and Theology.* Vol. 1: *Science.* Reprint ed. Los Angeles: Theosophy Company.

———. 1978. *Theosophical Glossary.* Reprint ed. Bangalore: Theosophy Company (Mysore) Private Limited.

Bloch, Ernst. 1986. Eldorado and Eden: The Geographical Utopias. In *The Principle of Hope.* Cambridge, MA: Basil Blackwell.

Blood, Hilary. 1957. The Peaks of Lemuria. *Geographical Magazine* 29, no. 10: 516–22.

Bloomfield, Paul. 1932. *Imaginary Worlds or the Evolution of Utopia.* London: Hamish Hamilton.

Blyth, Edward. 1871. A Suggested New Division of the Earth into New Zoological Regions. *Nature: A Weekly Illustrated Journal of Science* 3, no. 74: 427–29.

Bolsche, Wilhelm. 1909. *Haeckel: His Life and Work.* New and revised ed. London: Watts.

Bond, Nelson S. 1940. *Exiles of Time.* New York: Paperback Library.

Bowler, Peter J. 1986. *Theories of Evolution: A Century of Debate, 1844–1944.* Baltimore: Johns Hopkins University Press.

Boym, Svetlana. 2001. *The Future of Nostalgia.* New York: Basic Books.

Braden, Charles S. 1949. *These Also Believe: A Study of Modern American Cults and Minority Religious Movements.* New York: Macmillan.

Brinton, Daniel G. 1901. *Races and Peoples: Lectures on the Science of Ethnography.* Philadelphia: David McKay.

Brosius, Christiane. 1997. Motherland in Hindutva Iconography. *The India Magazine of Her People and Culture* 17, no. 12: 22–28.

Brown Goode, G., ed. 1896. *The Published Writings of Philip Lutley Sclater, 1844–1896.* Washington, DC: Government Printing Office.

Browne, Janet. 1983. *The Secular Ark: Studies in the History of Biogeography.* New Haven, CT: Yale University Press.

———. 1996. Biogeography and Empire. In *Cultures of Natural History,* edited by N. Jardine, J. A. Secord, and E. C. Spary. Cambridge: Cambridge University Press.

Browning, K. 1924. *Lemuria and Atlantis: Two Lost Continents.* London: Theosophical Society.

Bruce, Muriel. 1930. *Mukara.* New York: Rae D. Henkle.

Brunton, Paul. 1971. *Message from Arunachala.* New York: Samuel Weiser.

Bryant, Edwin. 2001. *The Quest for the Origins of Vedic Culture: The Indo-Aryan Migration Debate.* New York: Columbia University Press.

Cadbury, Deborah. 2000. *The Dinosaur Hunters: A Story of Scientific Rivalry and the Discovery of the Prehistoric World.* London: Fourth Estate.
Caldwell, Robert. 1856. *A Comparative Grammar of the Dravidian or South-Indian Family of Languages.* London: Harrison.
Camerini, Jane. 1987. Darwin, Wallace and Maps. Ph.D. diss., University of Wisconsin, Madison.
Campbell, Bruce. 1980. *Ancient Wisdom Revived: A History of the Theosophical Movement.* Berkeley: University of California Press.
Carter, Lin. 1966. *Thongor of Lemuria.* London: Tandem.
———. 1969. *Thongor and the Wizard of Lemuria.* New York: Berkley Corporation.
Carter, Paul. 1988. *The Road to Botany Bay: An Exploration of Landscape and History.* New York: Alfred A. Knopf.
Carter, Paul, and D. Malouf. 1989. Spatial History. *Textual Practice* 3: 173–83.
Cascardi, Anthony J. 1992. *The Subject of Modernity.* Cambridge: Cambridge University Press.
Casey, Edward. 1987a. *Remembering: A Phenomenological Study.* Bloomington: Indiana University Press.
———. 1987b. The World of Nostalgia. *Man and World* 20: 367–84.
———. 1997. *The Fate of Place: A Philosophical History.* Berkeley: University of California Press.
Casey, Edward S. 2001. Between Geography and Philosophy: What Does It Mean To Be in the Place-world. *Annals of the Association of American Geographers* 91, no. 4: 683–93.
Cervé, Wishar S. 1997. *Lemuria: The Lost Continent of the Pacific.* With a Special Chapter by Dr. James D. Ward. Reprint ed. San Jose: Supreme Grand Lodge of AMORC.
Chakrabarty, Dipesh. 2000. *Provincializing Europe: Postcolonial Thought and Historical Difference.* Princeton, NJ: Princeton University Press.
Chandrababu, B. S. 1996. *The Land and People of Tamilnadu: An Ethnographic Study.* Madras: Emerald Publishers.
Chavez, John R. 1984. *The Lost Land: The Chicano Image of the Southwest.* Albuquerque: University of New Mexico Press.
Chelvakasavaraya Mudaliar, T. 1904. *Tamil: An Essay (Tamiḻ).* Madras: S.P.C.K. Press.
Chidambaranar, Thudisaikizhar A. 1938. *The Antiquity of the Tamils and Their Literature.* Coimbatore: Krishna Vilas Press.
———. 1948. *Tamiḻc Caṅkaṅkaḷiṉ Varalāṟu* [History of the Tamil academies]. Madras: South India Saiva Siddhanta Works Publishing Society.
Childress, David Hatcher. 1988. *Lost Cities of Ancient Lemuria and the Pacific.* Stelle: Adventures Unlimited Press.
Chithambera Sastriar, M. 1883. *A Geography of the Madura District Designed for the Use of Elementary Schools* [Maturajillā Pūmicāstiram]. 4th ed. Madras: Kalaratnakaram Press.
Churchward, James. 1994a. *The Lost Continent of Mu: The Motherland of Man.* Reprint ed. Albuquerque: Be, Books.
———. 1994b. *The Children of Mu.* Reprint ed. Albuquerque: Be, Books.
Clareson, Thomas D. 1977. Lost Lands, Lost Races: A Pagan Princess of Their Very Own. In *Many Futures, Many Worlds: Theme and Form in Science Fiction,* edited by T. D. Clareson. Kent, Ohio: Kent State University Press.

Clute, John, and John Grant, eds. 1997. *The Encyclopedia of Fantasy*. New York: St. Martin's Press.

Cohen, Daniel. 1969. *Mysterious Places*. New York: Dodd, Mead.

Cohn, Norman. 1996. *Noah's Flood: The Genesis Story in Western Thought*. New Haven, CT: Yale University Press.

Conley, Tom. 1988. Translator's Introduction for a Literary Historiography. In *The Writing of History*, by Michel de Certeau. New York: Columbia University Press.

———. 1996. *The Self-Made Map: Cartographic Writing in Early Modern France*. Minneapolis: University of Minnesota Press.

Conrad, Joseph. 1926. Geography and Some Explorers. In *Last Essays*. London: Dent.

Cornell, James. 1978. *Lost Lands and Forgotten People*. New York: Sterling Publishing.

Cosgrove, Denis. 1999. Introduction: Mapping Meaning. In *Mappings*, edited by D. Cosgrove. London: Reaktion Books.

———. 2001. *Apollo's Eye: A Cartographic Genealogy of the Earth in Western Imagination*. Baltimore: Johns Hopkins University Press.

Damodaram Pillai, C. W. 1919. *Tamil Seventh Reader for Form 3 and Class VIII: Ē<u>l</u>ām Vācaka Pustakam*. Madras: Macmillan.

———. 1971. *Tāmotaram: C. W. Tāmōtarappiḷḷai E<u>l</u>utiya Patippuraikaḷi<u>n</u> Tokuppu*. Prefaces written by C. W. Damodaram Pillai. Jaffna: Jaffna Cooperative Tamil Books Publication.

Daniel, D. 1992. *Travancore Tamils: Struggle for Identity, 1938–1956 (Part I)*. Madurai: Raj Publishers.

Daniels, E. B. 1985. Nostalgia: Experiencing the Elusive. In *Descriptions*, edited by D. Ihde and H. J. Silverman. Albany: State University of New York Press.

Das, Abinas Chandra. 1921. *Rig-Vedic India*. Calcutta: University of Calcutta.

Dasgupta, Keya. 1995. A City Away from Home: The Mapping of Calcutta. In *Texts of Power: Emerging Disciplines in Colonial Bengal*, edited by P. Chatterjee. Minneapolis: University of Minnesota Press.

Davis, Fred. 1979. *Yearning for Yesterday: A Sociology of Nostalgia*. New York: Free Press.

Davis, Whitney. 1998. Winckelman Divided: Mourning the Death of Art History. In *The Art of Art History: A Critical Anthology*, edited by D. Preziosi. Oxford: Oxford University Press.

de Camp, L. Sprague. 1970. *Lost Continents: The Atlantis Theme in History, Science, and Literature*. 2nd rev. ed. New York: Dover.

de Camp, L. Sprague, and Willy Ley. 1970. *Lands Beyond*. New York: Rinehart.

de Certeau, Michel. 1988. *The Writing of History*. Translated by Tom Conley. New York: Columbia University Press.

de Purucker, G. 1979. *Fundamentals of the Esoteric Philosophy: Commentary and Elucidations of H.P.B's "The Secret Doctrine."* Reprint ed. Pasadena: Theosophical University Press.

Devaneyan, G. 1940. *Oppiya<u>n</u> Mo<u>l</u>inūl (Muta<u>n</u> Maṭalam): Tirāviṭam (Muta<u>r</u>pākam): Tami<u>l</u>* [Comparative linguistics: Dravidian: Tamil]. N.p.

———. 1966. *The Primary Classical Language of the World*. Katpadi: Nesamani Publishing House.

———. 1981. Tami<u>l</u>a<u>n</u>i<u>n</u> Pi<u>r</u>antakam [The birthplace of the Tamilian]. In *Aintām*

Ulakat Tamiḻ Mānāṭu Viḻā Malar [Commemoration volume of the fifth international Tamil conference]. N.p.

———. 1992. *Paṇṭait Tamiḻar Nākarikamum Paṇpātum* [Civilization and culture of ancient Tamilians]. Reprint ed. Madras: South India Saiva Siddhanta Works Publishing Society.

Devasikhamani, S. K. 1919. *The Tamils and Their Language.* Tiruchirapalli: Young Men's Tamilian Association.

Dey, A. K. 1968. *Geology of India.* New Delhi: National Book Trust, India.

Dobson, Jerome E. 1992. Spatial Logic in Paleogeography and the Explanation of Continental Drift. *Annals of the Association of American Geographers* 82, no. 2: 187–206.

Donnelly, Ignatius. 1882. *Atlantis: The Antediluvian World.* New York: Harper and Brothers.

Duara, Prasenjit. 1995. *Rescuing History from the Nation: Questioning Narratives from China.* Chicago: University of Chicago Press.

Dubois, Eugene. 1888. Over de wenschelijkheid van een onderzoek naar de diluviale fauna van Ned. Indie, inhet bizonder van Sumatra. *Natuurkundig tijdschrift voor Nederlandsch-Indiee* 48: 148–65.

Dundes, Alan, ed. 1988. *The Flood Myth.* Berkeley: University of California Press.

Duraikkannu Mudaliar, P. 1958. *Camūka Nūl (Eṭṭām Vakuppu)* [Social studies for class eight]. Madras: Orient Longmans.

du Toit, Alexander. 1937. *Our Wandering Continents: An Hypothesis of Continental Drifting.* Edinburgh: Oliver and Boyd.

Eco, Umberto. 1995. *The Search for the Perfect Language.* Translated by James Fentress. London: Blackwell.

Edgerton, Samuel Y. 1976. *The Renaissance Rediscovery of Linear Perspective.* New York: Harper and Row.

Edlis, Richard. 1993. *Peaks of Limuria: The Story of Diego Garcia.* London: Bellew Publishing.

Edney, Matthew H. 1997. *Mapping an Empire: The Geographical Construction of British India, 1765–1843.* Chicago: University of Chicago Press.

Eiseley, Loren. 1958. *Darwin's Century: Evolution and the Men who Discovered It.* New York: Doubleday.

Eliade, Mircea. 1959. *Cosmos and History: The Myth of Eternal Return.* New York: Harper and Row.

Ellis, Richard. 1998. *Imagining Atlantis.* New York: Alfred A. Knopf.

Elmore, Wilber T. 1915. *Dravidian Gods in Modern Hinduism: A Study of the Local and Village Deities of Southern India.* Hamilton, NY: Privately printed.

Engels, Fredrick. 1975. *The Part Played by Labour in the Transition from Ape to Man.* Peking: Foreign Languages Press.

Evans, John. 1875. Anniversary Address of the President. *Quarterly Journal of the Geological Society of London* 31: xxxvii–lxxvi.

Faivre, Antoine. 1987. Occultism. In *The Encyclopedia of Religion,* edited by M. Eliade. New York: Macmillan.

Falconer, Hugh. 1868. Primeval Man and His Contemporaries. In *Paleontological Memoirs and Notes of the Late Hugh Falconer,* edited by C. Murchison. London: Robert Hardwicke.

Feder, Kenneth. 1990. *Frauds, Myths, and Mysteries: Science and Psuedoscience in Archaeology.* Mountain View, CA: Mayfield Publishing.

Feeley-Harnik, Gillian. 2001. *Ravenala Madagascariensis Sonnerat:* The Historical Ecology of a "Flagship Species" in Madagascar. *Ethnohistory* 48, no. 1–2: 31–86.

Ferguson, Harvie. 2000. *Modernity and Subjectivity: Body, Soul, Spirit.* Charlottesville: University of Virginia Press.

Fichman, Martin. 1977. Wallace: Zoogeography and the Problem of Land Bridges. *Journal of Historical Biology* 10: 45–63.

Forster, E. M. 1984. *A Passage to India.* Reprint ed. New York: Harcourt Brace.

Fortey, R. A. 1981. Maps of the Past. In *The Evolving Earth,* edited by L. R. M. Cocks. Cambridge: Cambridge University Press.

Foucault, Michel. 1980. *Power/Knowledge: Selected Interviews and Other Writings, 1972–1977.* Translated by Colin Gordon et al. New York: Pantheon.

———. 1986. Of Other Spaces. *Diacritics* 16: 22–27.

Fox, Cyril S. 1938. *Physical Geography for Indian Students (Being a Completely Revised and Enlarged Edition of Simmons and Stenhouses's "Class Book of Physical Geography").* London: Macmillan.

Frankel, Henry. 1998. Continental Drift and Plate Tectonics. In *Sciences of the Earth: An Encyclopedia of Events, People, and Phenomena,* edited by G. A. Good. New York: Garland.

Freud, Sigmund. 1957. Mourning and Melancholia. In *The Standard Edition of the Complete Psychological Works of Sigmund Freud,* edited by J. Strachey. London: Hogarth Press.

Gadow, Hans. 1909. Geographical Distribution of Animals. In *Darwin and Modern Science,* edited by A. C. Seward. Cambridge: Cambridge University Press.

Ganapathy, S. K. 1982. *Cantappāvil Tamiḻar Caritam: Mutaṟ Pakuti-Kaṭalkōḷ Carukkam* (History of Tamils in verse, pt. 1, Seizure by the ocean). Madras: Pari Nilayam.

———. 1984. *Kumarik kaṇṭat Tamiḻar* [Tamilians of Kumarikkaṇṭam]. Salem: Omkara Patippakam.

Gardiner, J. Stanley. 1906. The Indian Ocean. *The Geographical Journal* 28, no. 4: 313–32.

———. 1933. The John Murray Expedition to the Indian Ocean. *Geographical Journal* 81: 570–73.

Gardiner, J. Starkie. 1881. Subsidence and Elevation, and on the Permanence of Oceans. *Geological Magazine,* n.s., 8, no. 2: 241–45.

Gardner, Martin R. 1957. *Fads and Fallacies in the Name of Science.* New York: Dover.

Gasman, Daniel. 1969. Social Darwinism in Ernst Haeckel and the German Monist League: A Study of the Scientific Origins of National Socialism. Ph.D. diss., University of Chicago.

Geetha, V. 1993. Re-writing History in the Brahmin's Shadow: Caste and the Modern Historical Imagination. *Journal of Arts and Ideas* 25–26: 127–37.

Georg, Eugen. 1931. *The Adventure of Mankind.* Translated by Robert Bek-Gran. New York: E. P. Dutton.

George, Wilma. 1964. *Biologist Philosopher: A Study of the Life and Writings of Alfred Russel Wallace.* London: Abelard-Schuman.

Gillies, John. 1994. *Shakespeare and the Geography of Difference.* Cambridge: Cambridge University Press.

Gillispie, Charles Coulston. 1951. *Genesis and Geology: A Study in the Relations of Scientific Thought, Natural Theology, and Social Opinion in Great Britain, 1790–1850*. Cambridge, MA: Harvard University Press.

Godlewska, Anne, and Neil Smith, eds. 1994. *Geography and Empire*. Oxford: Blackwell.

Godwin, John. 1972. *Occult America*. New York: Doubleday.

Godwin, Joscelyn. 1993. *Arktos: The Polar Myth in Science, Symbolism, and Nazi Survival*. London: Thames and Hudson.

Gohau, Gabriel. 1991. *A History of Geology*. Translated by Albert V. Carozzi and Marguerite Carozzi. New Brunswick, NJ: Rutgers University Press.

Gole, Susan. 1989. *Indian Maps and Plans: From Earliest Times to the Advent of European Surveys*. New Delhi: Manohar Publications.

Goodrick-Clarke, Nicholas. 1985. *The Occult Roots of Nazism: The Ariosophists of Austria and Germany, 1890–1935*. Wellingborough, Northamptonshire: Aquarian Press.

Gopala Krishnan, T. P. 1956. *Tēcīya Tamiḻ Vācakam: Nāṉkām Puttakam (Nānkām Vakuppukkuriyatu)* [National Tamil reader for class four]. 3rd rev. ed. Madras: Vartha Publishing House.

Gouda, Frances. 1995. *Dutch Culture Overseas: Colonial Practice in the Netherlands Indies, 1900–1942*. Amsterdam: Amsterdam University Press.

Gould, Stephen Jay. 1965. Is Uniformitarianism Necessary? *The American Journal of Science* 263: 223–28.

———. 1984. Toward the Vindication of Punctuational Change. In *Catastrophes and Earth History: The New Uniformitarianism*, edited by W. A. Berggren and J. A. Van Couvering. Princeton, NJ: Princeton University Press.

———. 1987. *Time's Arrow, Time's Cycle: Myth and Metaphor in the Discovery of Geological Time*. Cambridge, MA: Harvard University Press.

———. 1988. Creationism: Genesis vs. Geology. In *The Flood Myth*, edited by A. Dundes. Berkeley: University of California Press.

Government of India. 1878. *Textbooks in Indian Schools: Report of the Committee Appointed to Examine the Textbooks in Use in Indian Schools, with Appendices*. Calcutta: Home Secretariat Press.

Government of Tamilnadu. 1975. *Tamiḻnāṭṭu Varalāṟu: Tolpaḻaṅkālam* [History of Tamilnadu: Prehistoric times]. Madras: Tamilnadu Aracu.

Graefe, W. 1960. Legends as Milestones in the History of Tamil Literature. In *Professor P. K. Gode Commemoration Volume*, edited by H. K. Hariyappa and M. M. Patkar. Poona: Oriental Book Agency.

Gregory, Derek. 1994. *Geographical Imaginations*. Cambridge, MA: Blackwell.

Grierson, George A. 1903. Language. In *Census of India, 1901: Report*. Calcutta: Office of the Superintendent of Government Printing.

———, ed. 1906. *Munda and Dravidian Languages, Linguistic Survey of India*. Vol. 4: *Munda and Dravidian Languages*. Calcutta: Office of Superintendent of Government Printing.

Griesbach, Charles L. 1871. On the Geology of Natal. *Quarterly Journal of the Geological Society of London* 27: 53–72.

Griffin, David Ray. 1990. Introduction: The Reenchantment of Science. In *The Reenchantment of Science: Postmodern Proposals*, edited by D. R. Griffin. Albany: State University of New York Press.

Grout, Andrew. 1995. Geology and India, 1770–1851: A Study in the Methods and Motivations of a Colonial Science. Ph.D. diss., University of London.

Grove, Richard. 1995. *Green Imperialism: Colonial Expansion, Tropical Island Edens and the Origins of Environmentalism, 1600–1860*. Cambridge: Cambridge University Press.

Guha, Ranajit. 1997. *Dominance without Hegemony: History and Power in Colonial India*. Cambridge, MA: Harvard University Press.

Haeckel, Ernst. 1870. *Natürliche Schöpfungsgeschichte*. 2nd ed. Berlin: Verlag von Georg Weimar.

———. 1876a. *History of Creation: On the Development of the Earth and Its Inhabitants by the Action of Natural Causes (A Popular Exposition of the Doctrine of Evolution in General, and of That of Darwin, Goethe, and Lamarck in Particular)*. Vol. 1. New York: D. Appleton.

———. 1876b. *History of Creation: On the Development of the Earth and its Inhabitants by the Action of Natural Causes (A Popular Exposition of the Doctrine of Evolution in General, and of That of Darwin, Goethe, and Lamarck in Particular)*. Vol. 2. New York: D. Appleton.

———. 1898. *The Last Link: Our Present Knowledge of the Descent of Man*. Translated by Hans Gadow. London: Adam and Charles Black.

———. 1905. *The Evolution of Man: A Popular Scientific Study*. Translated by J. McCabe. 2 vols. London: Watts.

Hallam, Anthony. 1973. *A Revolution in the Earth Sciences: From Continental Drift to Plate Tectonics*. Oxford: Clarendon Press.

———. 1989. *Great Geological Controversies*. 2nd ed. Oxford: Oxford University Press.

Hammer, Olav. 2001. *Claiming Knowledge: Strategies of Epistemology from Theosophy to the New Age*. Leiden: Brill.

Hancock, Graham. 2002. *Underworld: The Mysterious Origins of Civilization*. New York: Crown Publishers.

Hanegraaff, Wouter. 1996. *New Age Religion and Western Culture: Esotericism in the Mirror of Secular Thought*. Leiden: Brill.

Harley, J. Brian. 1987. The Map and the Development of the History of Cartography. In *Cartography in Prehistoric, Ancient, and Medieval Europe and the Mediterranean*, edited by J. B. Harley and D. Woodward. Chicago: University of Chicago Press.

———. 1988. Maps, Knowledge, and Power. In *The Iconography of Landscape: Essays on the Symbolic Representation, Design, and Use of Past Environments*, edited by D. Cosgrove and S. Daniels. Cambridge: Cambridge University Press.

———. 1992a. Rereading the Maps of the Columbian Encounter. *Annals of the Association of American Geographers* 82: 522–42.

———. 1992b. Deconstructing the Map. In *Writing Worlds: Discourse, Text, and Metaphor in the Representation of Landscape*, edited by T. J. Barnes and J. S. Duncan. London: Routledge.

———. 2001. *The New Nature of Maps: Essays in the History of Cartography*. Baltimore: Johns Hopkins University Press.

Harley, J. Brian, and David Woodward. 1987. Preface. In *Cartography in Prehistoric, Ancient, and Medieval Europe and the Mediterranean*, edited by J. B. Harley and D. Woodward. Chicago: University of Chicago Press.

Harootunian, Harry. 1999. Memory, Mourning, and National Morality: Yasukuni Shrine and the Reunion of State and Religion in Postwar Japan. In *Nation and*

Religion: Perspectives on Europe and Asia, edited by P. van der Veer and H. Lehmann. Princeton, NJ: Princeton University Press.
Hartlaub, G. 1877. General Remarks on the Avifauna of Madagascar and the Mascarene Islands. *Ibis: A Quarterly Journal of Ornithology,* 4th ser., 1: 334–36.
Harvey, David. 1990. *The Condition of Postmodernity: An Enquiry into the Origins of Cultural Change.* Cambridge: Blackwell.
Haug, Emile. 1907. *Traité de Géologie.* Vol. 2. Paris: Librarie Armand Colin.
Hayes, Christine. 1972. *Red Tree: Insight into Lost Continents, Mu and Atlantis, as Revealed to Christine Hayes.* San Antonio: Naylor.
H.C. 1936. *Intiya Carittiram: Intu, Muslim Āṭcikaḷiṉ Varalāṟu* [History of India: Hindu and Muslim rule]. Translated by S. K. Devasikhamani. Bombay: Longmans, Green.
Heelas, Paul. 1996. *The New Age Movement: The Celebration of the Self and the Sacralization of Modernity.* Oxford: Blackwell.
Heidegger, Martin. 1977. The Age of the World Picture. In *The Question Concerning Technology and Other Essays.* New York: Garland.
Heilprin, Angelo. 1894. *The Geographical and Geological Distribution of Animals.* 2nd ed. London: Kegan Paul, Trench, Trubner.
Heindel, Max. 1911. *Rosicrucian Cosmo-Conception or Mystic Christianity: An Elementary Treatise upon Man's Past Evolution, Present Constitution and Future Development.* 5th ed. Oceanside, CA: Rosicrucian Fellowship.
Hill, W. C. Osman. 1953. *Primates: Comparative Anatomy and Taxonomy (I-Strepsirhini).* New York: Interscience Publishers.
Holan, Vladimir. 1940. *Lemuria (1934–1938).* Prague: Melantrich.
Holderness, Thomas W. 1911. *Peoples and Problems of India.* London: Williams and Norgate.
Holdich, Thomas H. 1905. *India.* New York: D. Appleton and Co.
Holland, T. H. 1909. Geology. In *The Imperial Gazetteer of India: The Indian Empire.* Vol. 1, *Descriptive.* Oxford: Clarendon Press.
Homans, Peter. 1989. *The Ability to Mourn: Disillusionment and the Social Origins of Pschyoanalysis.* Chicago: University of Chicago Press.
———. 2000. Loss and Mourning in the Life and Thought of Max Weber: Toward a Theory of Symbolic Loss. In *Symbolic Loss: The Ambiguity of Mourning and Memory at Century's End,* edited by P. Homans. Charlottesville: University of Virginia Press.
Hooson, David, ed. 1994. *Geography and National Identity.* Oxford: Blackwell.
Hooykaas, R. 1963. *Natural Law and Divine Miracle: The Principle of Uniformity in Geology, Biology, and Theology.* Leiden: E. J. Brill.
Hsu, Kenneth J. 1992. *Challenger at Sea: A Ship that Revolutionized Earth Science.* Princeton, NJ: Princeton University Press.
Huggett, Richard. 1989. *Cataclysms and Earth History: The Development of Diluvialism.* Oxford: Clarendon Press.
Huxley, Thomas. 1870. Anniversary Address of the President. *Proceedings of the Geological Society of London* 1870: xxix–lxiv.
———. 1902. *Man's Place in Nature and Other Anthropological Essays.* New York: D. Appleton and Co.
Ilantiraiyan, S. 1981. *Aintāvatu Tamiḻ Mānāṭu* [Fifth Tamil conference]. New Delhi: Salai Publications.

Inden, Ronald. 2000. *Imagining India.* Reprint ed. Bloomington: Indiana University Press.

Irwin, W. R. 1976. *The Game of the Impossible: A Rhetoric of Fantasy.* Urbana: University of Illinois Press.

Ivy, Marilyn. 1995. *Discourses of the Vanishing: Modernity, Phantasm, Japan.* Chicago: University of Chicago Press.

Jackson, Rosemary. 1981. *Fantasy: The Literature of Subversion.* London: Methuen.

Jay, Martin. 1994. *Downcast Eyes: The Denigration of Vision in Twentieth-Century French Thought.* Berkeley: University of California Press.

Jayakaran, S. C. 2001. Kumarik Kaṇṭam–Lemuria Kuḻappam [The problem of Kumarikkaṇṭam–Lemuria). *Kalaccuvadu* 33: 52–57.

———. 2002. *Kumari Nilanīṭci: Kumarikkaṇṭam (Lemuria)—Oru Āyvu* [Kumari land extension: Kumarikkaṇṭam (Lemuria); An exploration]. Nagercoil: Kalachuvadu Pathippagam.

Jenkins, Philip. 2000. *Mystics and Messiahs: Cults and New Religions in American History.* New York: Oxford University Press.

Jervis, W. W. 1938. *The World in Maps: A Study in Map Evolution.* 2nd ed. New York: Oxford University Press.

Johnson, K. Paul. 1998. *Edgar Cayce in Context: The Readings, Truth, and Fiction.* Albany: State University of New York Press.

Johnston, Keith, ed. 1884. *Stanford's Compendium of Geography and Travel, based on Hellwald's "Die Erde und Inhre Volker": Africa.* Revised and Corrected by E. G. Ravenstein, with Ethnological Appendix by A. H. Keane. London: Edward Stanford.

Joleaud, Leonce. 1939. *Atlas de Paléobiogéographie.* Paris: Lechevalier.

Joseph, P. 1958. Lost Lemuria—Fresh Evidence? *Tamil Culture* 7, no. 2: 121–31.

———. 1972. *The Dravidian Problem and the South Indian Culture Complex.* Madras: Orient Longmans.

Jungclaus, David. 1990. *The Lemurian–Atlantis Vision Wheel.* Westlake Village, CA: Lost World Publications.

International Indian Ocean Expedition. 1975. *Geological–Geophysical Atlas of the Indian Ocean.* Oxford: Pergamon Press.

Irwin, W. R. 1976. *The Game of the Impossible: A Rhetoric of Fantasy.* Urbana: University of Illinois Press.

Kafton-Minkel, Walter. 1989. *Subterreanean Worlds: 100,000 Years of Dragons, Dwarfs, the Dead, Lost Races, and UFOs from Inside the Earth.* Port Townsend, WA: Loompanics Unlimited.

Kailasapathy, K. 1968. *Tamil Heroic Poetry.* London: Oxford University Press.

Kailasapillai, T. 1930. *Pālapāṭam: Iraṇṭām Puttakam* [Children's lessons: Book two]. 10th ed. Jaffna: [Navalar Printing Press].

Kalcina Valuti. 2001. Kumarik Kaṇṭa Maṟuppiṉ Piṉṉaṇi. *Oppuravu* 1: 39–41.

Kalyanasundaram, T. V. 1940. *Intiyāvum Viṭutalaiyum* [India and independence]. Madras: Sadhu Printing Press.

———. 1959. Tamiḻ Nāṭu [The Tamil land]. In *Centamiḻp Poḻil: Oṉpatām Vakuppu,* edited by P. Celvarajan. Madras: Aciriyar Nurpatippu Kazhagam.

Kanakaraja Aiyar, N., ed. 1934. *Tamiḻ Nūṟ Kōvai (Mutaṟ Pakuti)* [Garland of Tamil books, part one]. Tiruchirappalli: T. G. Gopala Pillai.

Kanakasabhai, V. 1966. *The Tamils Eighteen Hundred Years Ago*. Reprint ed. Madras: South India Saiva Siddhanta Works Publishing Society.

Kandiah Pillai, N. S. 1934. *Tamiḻakam* [The Tamil home-place]). Madras: Orrumai Office.

———. 1945. *Namatu Nāṭu* [Our nation]. Madras: South India Saiva Siddhanta Works Publishing Society.

———. 1947. *Muccaṅkam* [The three academies]. Madras: Muttamil Nilayam.

———. 1949. *Tamiḻ Intiyā* [Tamil India]. 2nd ed. Tirunelveli: South India Saiva Siddhanta Works Publishing Society.

———. 1957. *Varalāṟṟuk Kālattiṟku Muṟpaṭṭa Paḻantamiḻar* [Prehistoric ancient Tamilians]. 3rd ed. N.p.: Progressive Printers.

———. 1984. *Namatu Moḻi* [Our language]. Reprint ed. Chennai: South India Saiva Siddhanta Works Publishing Society.

Kannadasan. 1994. *Kaṭal Koṇṭa Teṉṉaṭu* [The southern land seized by the ocean]. 4th ed. Madras: Kannadasan Patippakam.

Kasinathan, Natana. 1993. Submerged Land of Tamilagam. *Kalveṭṭu* 38: 1–8.

Kaviraj, Sudipta. 1995. Imaginary History. In *The Unhappy Consciousness: Bankimchandra Chattopadhyay and the Formation of Nationalist Discourse in India*. Delhi: Oxford University Press.

Keane, A. H. 1896. *Ethnology*. 2nd rev. ed. Cambridge: Cambridge University Press.

———. 1899. *Man: Past and Present*. Cambridge: Cambridge University Press.

———. 1908. *The World's Peoples: A Popular Account of Their Bodily and Mental Characters, Beliefs, Traditions, Political and Social Institutions*. London: Hutchinson.

Keith, Arthur. 1934. *The Construction of Man's Family Tree*. London: Watts.

Kelly, Alfred. 1981. *The Descent of Darwin: The Popularization of Darwinism in Germany*. Chapel Hill: University of North Carolina Press.

King, Godfré Ray. 1934. *Unveiled Mysteries*. Chicago: Saint Germain.

King, Goeff. 1996. *Mapping Reality: An Exploration of Cultural Cartographies*. New York: St. Martin's Press.

Kirsch, Stuart. 1997. Lost Tribes: Indigenous People and the Social Imaginary. *Anthropological Quarterly* 70: 58–67.

Kolosimo, Peter. 1973. *Timeless Earth*. Translated by Paul Stevenson. London: Garnstone Press.

Kondratov, Alexander M. 1967. *Tainy Trekh Okeanov*. Leningrad: Gidrometeoizdat.

———. 1974. *The Riddles of Three Oceans*. Translated by Leonard Stoklitsky. Moscow: Progress Publishers.

———. 1995. *Intumākkaṭal Marmaṅkaḷ: Lemūriak Kaṇṭam* [The riddles of the Indian Ocean: The continent of Lemuria]. Translated by R. Parthasarathy. Reprint ed. Madras: New Century Book House.

Krishna, Sankaran. 1994. Cartographic Anxiety: Mapping the Body Politic in India. *Alternatives* 19: 507–21.

Krishnaswami Aiyangar, S. 1911. *Ancient India*. London: Luzac.

Kueshana, Eklala. 1953. *The Ultimate Frontier*. Chicago: Stelle Group.

Kumari Maintan. 2001. Kumarik Kaṇṭa Maṟuppiṉ Piṉṉaṇi. *Oppuravu* 1: 38–39.

Kyle, Richard. 1993. *The Religious Fringe: A History of Alternative Religions in America*. Downers Grove, IL: InterVarsity Press.

———. 1995. *The New Age Movement in American Culture.* Lanham, MD: University Press of America.

Lal, Vinay. 1999. History and Politics. In *India Briefing: A Transformative Fifty Years,* edited by M. Bouton and P. Oldenburg. London: M. E. Sharpe.

———, ed. 2000. *Dissenting Knowledges, Open Futures: The Multiple Selves and Strange Destinations of Ashis Nandy.* Delhi: Oxford University Press.

Lanser, Edward. 1932. A People of Mystery. *Los Angeles Times Sunday Magazine,* 22 May, pp. 4, 16.

Latour, Bruno. 1993. *We Have Never Been Modern.* Translated by Catherine Porter. New York: Harvester.

Lawrence, David M. 1999. Mountains under the Sea. *Mercator's World* 4, no. 6: 36–43.

Leadbeater, Charles W. 1931. *The Beginnings of the Sixth Root Race* [a reprint in more convenient form of the final chapters of *Man: Whence, How, and Whither*]. Madras: Theosophical Publishing House.

———. 1974. *The Hidden Side of Things.* Reprint ed. Madras: Theosophical Publishing House.

Leakey, Richard E., and L. Jan Slikkerveer. 1993. *Man-Ape Ape-Man: The Quest for Humans' Place in Nature and Dubois' "Missing Link."* Leiden: Netherlands Foundation for Kenya Wildlife Service.

LeGrand, H. E. 1988. *Drifting Continents and Shifting Theories.* Cambridge: Cambridge University Press.

Lemurian Fellowship. 1939. *Lemuria, the Incomparable, the Answer.* Milwaukee: Lemurian Press.

———. 1945. *Into the Sun: A Treatise Devoted to the Dawning of the New Order of the Ages—Novus Ordo Seclorum—and Setting Forth the Manner of the Formation and Development of the First and Greatest of All Civilizations Now in Process of Recapitulation.* Los Angeles: Lemurian Fellowship.

Leopold, Joan. 1970. The Aryan Theory of Race in India, 1870–1920. *Indian Economic and Social History Review* 7, no. 2: 271–97.

Lewin, Roger. 1987. *Bones of Contention: Controversies in the Search for Human Origins.* New York: Simon and Schuster.

Lewis, Martin M., and Karen E. Wigen. 1997. *The Myth of Continents: A Critique of Metageography.* Berkeley: University of California Press.

Lopez, Donald S. 1998. *Prisoners of Shangri-La: Tibetan Buddhism and the West.* Chicago: University of Chicago Press.

Lowenthal, David. 1975. Past Time, Present Place: Landscape and Memory. *The Geographical Review* 65, no. 1: 1–36.

Lugo, Francisco Aniceto. 1978. *Los Pueblos Maestros: Atlantida, Tiahuanaco y Lemuria.* Mexico City: Editorial Diana.

Lydekker, Richard. 1896. *A Geographical History of Mammals.* Cambridge: University Press.

Lyell, Charles. 1914. *The Geological Evidence of the Antiquity of Man.* London: J. M. Dent.

Maclean, Charles D. 1987. *Manual of the Administration of the Madras Presidency.* Vol. 1. Reprint ed. New Delhi: Asian Educational Services.

Mahalingam, N. 1981a. Kumari Kandam—The Lost Continent. In *Proceedings of the Fifth International Conference-Seminar of Tamil Studies, Madurai,* edited by M. Arunachalam. Madras: International Association of Tamil Research.

———, ed. 1981b. *Gems from the Prehistoric Past.* Madras: International Society for the Investigation of Ancient Civilization.

Mahalingam, N., and Sathur Sekaran. 1991. *Concise Form of Concept of Kumari Kandam.* Madras: International Linguistic Center.

Manickam Nayakkar, P. V. 1985. *The Tamil Alphabet and Its Mystic Aspect.* Reprint ed. New Delhi: Asian Educational Services.

Manley, David L. 1972. *Aros of Atlantis.* Philadelphia: Dorrance.

Manuel Raj, J. David. 1993. *History of Silamban Fencing: An Indian Martial Art.* Madras: privately printed.

Maraimalai Adigal. 1948. Introduction. In *India's Language Problem,* by K. Appadurai. Madras: Tamil India.

Marsden, Edmund. 1909. *History of India for Senior Classes.* Pt. 1, *The Hindu Period.* London: Macmillan.

———. 1917. *Intutēca Carittiram* [History of India for junior classes]. Rev. ed. Madras: Macmillan.

Marsden, Edmund, and Henry Sharp. 1930. *A History of India for High Schools.* London: Macmillan.

Marty, Martin. 1991. *Modern American Religion.* Vol. 2, *The Noise of Conflict, 1919–1941.* Chicago: University of Chicago Press.

Marvin, Ursula. 1973. *Continental Drift: The Evolution of a Concept.* Washington, DC: Smithsonian Institution Press.

Masani, Minoo. 1946. *Our India.* Reprint ed. Calcutta: Oxford University Press.

Mathew, Rev. John. 1889. The Australian Aborigines. *Journal and Proceedings of the Royal Society, New South Wales* 23, no. 3: 335–449.

Mathivanan, R. 1977. *Ilemūriā Mutal Arappā Varai* [From Lemuria to Harappa]. Madras: Sekhar Patippakam.

———. 1983. The Riddles of Kumarikkandam. In *Historical Heritage of the Tamils,* edited by S. V. Subramanian and K. D. Thirunavukkarasu. Madras: International Institute of Tamil Studies.

Matthews, Samuel W. 1967. Indian Ocean Unveiled on a Dramatic New Map: Science Explores the Monsoon Sea. *National Geographic Magazine* 132, no. 4: 554–75.

Mazumdar, B. C. 1912a. Man: His Origin and Original Home. *Modern Review: A Monthly Review and Miscellany* 11, no. 5: 488–92.

———. 1912b. The Dravidians of India. *Modern Review: A Monthly Review and Miscellany* 12, no. 1: 12–16.

McGillivray, Rosalyn. 1985. *From the Great Spirit to the Lemurians: The Legends of Mount Shasta.* Chico: California State University.

McWilliams, Carey. 1986. California: Mecca of the Miraculous. In *The California Dream,* edited by D. Hale and J. Eisen. New York: Macmillan.

Medlicott, H. B., and W. T. Blanford. 1879. *A Manual of the Geology of India Chiefly Compiled from the Observations of the Geological Survey.* Vol. 1, *Peninsular India.* Calcutta: Government of India.

Meenatchi Sundaram, S. 1958. *Putumuṟai Tamiḻ Vācakam (Nāṉkām Puttakam)* [New Tamil reader: Fourth book]. Erode: Sivalinga Publishing House.

Mehdriratta, R. C. 1962. *Geology of India and Pakistan.* 2nd ed. New Delhi: Atma Ram and Sons.

Melton, J. Gordon. 1988. Ancient Wisdom Family. In *Encyclopedia of American Religion.* Detroit: Gale Research.

Melton, J. Gordon, et al. 1990. *New Age Encylopedia.* Detroit: Gale Research.

Menard, H. W. 1986. *The Ocean of Truth: A Personal History of Global Tectonics.* Princeton, NJ: Princeton University Press.

Mignolo, Walter D. 1995. *The Darker Side of the Renaissance: Literacy, Territoriality, and Colonization.* Ann Arbor: University of Michigan Press.

Mitchell, Timothy. 1988. *Colonising Egypt.* Berkeley: University of California Press.

Monahan, C. H. 1903–4. A History of the Tamil Language: Review of *A History of the Tamil Language,* by V. K. Suryanarayana Sastri. *Madras Christian College Magazine* 21: 28–32.

Monmonier, Mark. 1995. Continental Drift and Geopolitics: Ideas and Evidence. In *Drawing the Line: Tales of Maps and Cartocontroversy.* New York: Henry Holt.

Montgomery, Ruth. 1976. *The World Before.* New York: Coward, McCann and Geoghegan.

Mookerji, Radha Kumud. 1936. *Hindu Civilization (From the Earliest Times to the Establishment of Maurya Empire).* London: Longmans, Green.

Mott, Albert J. 1876. *On Haeckel's "History of Creation."* Liverpool: Literary and Philosophical Society of Liverpool.

Mukerji, Chandra. 1984. Visual Language in Science and the Exercise of Power: The Case of Cartography in Early Modern Europe. *Studies in Visual Communication* 10, no. 3: 30–45.

Mukherjee, Meenakshi. 1985. *Realism and Reality.* Delhi: Oxford University Press.

Mundy, Barbara E. 1996. *The Mapping of New Spain: Indigenous Cartography and the Maps of the Relaciones Geográficas.* Chicago: University of Chicago Press.

Murdoch, John. 1885. *The Indian Teacher's Manual with Hints on the Management of Vernacular Schools.* Madras: Christian Vernacular Education Society.

———, ed. 1865. *Classified Catalogue of Tamil Printed Books with Introductory Notes.* Madras: Christian Vernacular Education Society.

Murray, Andrew. 1866. *The Geographical Distribution of Mammals.* London: Day and Son.

Muruku Sundaram, T. 1966. *Jūpiṭar Varalāṟṟu Nūl (Ēḻām Vakuppu)* [Jupiter history book for class seven]. Salem: Jupiter Publishing House.

Mutthukrishna Nattar, C. 1960. *Kaṭaiccaṅka Kālap Pāṇṭiyar* [The Pandyas of the last academy]. Madras: Palaniappa Brothers.

Mutthuswami Pillai, P. K. 1934. *Tamiḻkkalai Pāṭam: Ēḻām Pāṭap Puttakam* [Tamil lessons: Seventh textbook]. Tirunelveli: South India Saiva Siddhanta Works Publishing Society.

Mutthuswamy Aiyar, T. N. , and P. N. Appuswamy Aiyar. 1933. Intiyā Uṇṭāṉatu Eppaṭi? [How was India created?]. *Kalaimakaḷ* 3: 47–59, 142–48, 252–57, 329–341.

Mutthuthambipillai, A. 1902. *The Tamil Classical Dictionary, Apitāṉa Kōcam.* Jaffna: Navalar Press.

———. 1906–7. Kumariyāṟu [The River Kumari]. *Centamiḻ* 5, no. 4: 192–200.

Mutthuvirasami Naidu, C. 1954–55. Kumarimuṉai [Cape Kumari]. *Centamiḻc Celvi* 29, no. 10: 453–57.

Nagarajan, S. 1961. *Vaḷḷuvar Tamiḻ Vācakam: Aintām Vakuppu* [Valluvar Tamil reader for class five]. 3rd ed. Erode: Sivalinga Nurpatippu Kazhagam.

Nallasami Pillai, J. 1898. Ancient Tamil Civilization. *The Light of Truth or Siddhanta Deepika* 2, no. 5: 109–13.

Nambi Arooran, K. 1980. *Tamil Renaissance and Dravidian Nationalism, 1905–1944*. Madurai: Koodal.

Nandy, Ashis. 1983. *The Intimate Enemy: Loss and Recovery of Self under Colonialism*. New Delhi: Oxford University Press.

———. 1995. History's Forgotten Doubles. *History and Theory* 34, no. 2: 44–66.

Nannan, M., and M. R. Chandrasekharan. 1958. *Tennāṭṭu Vācakam: Nāṉkām Vakuppu* [Tennadu Tamil reader for class four]. Madras: Tennadu Patippakam.

Narayanasami, A. 1964. *Tamiḻakamum Tamiḻum* [The Tamil realm and Tamil]. 3rd rev. ed. Madras: P. G. Paul.

Nedunceliyan, R. 1953. *Maṟainta Tirāviṭam* [Lost Dravidian land]. Madras: Manram Patippakam.

Neumayr, Melchior. 1887. *Erdgeschichte* [History of the earth]. Vol. 2. Leipzig: Verlag des Bibliograph.

Nilakanta Sastri, K. A. 1956. *Historical Method in Relation to Problems of South Indian History*. Madras: University of Madras.

———. 1972. *The Sangam Age: Its Cults and Cultures*. Madras: Swathi Publications.

Nora, Pierre. 1989. Between Memory and History: *Les Lieux des Memoires: Representations* 26: 7–25.

———. 1996. General Introduction: Between Memory and History. In *Realms of Memory: Rethinking the French Past*, Vol. 1. New York: Columbia University Press.

Nordenskiold, E. 1929. Haeckel and Monism. In *The History of Biology: A Survey*. London: Kegan Paul, Trench, Trubner.

Olalquiaga, Celeste. 1998. *The Artificial Kingdom: A Treasury of the Kitsch Experience with Remarkable Objects of Art and Nature, Extraordinary Events, Eccentric Biography, and Original Theory plus Many Wonderful Illustrations Selected by the Author*. New York: Pantheon Books.

Oldham, R. D. 1893. *A Manual of the Geology of India Chiefly Compiled from the Observations of the Geological Survey by H. B. Medlicott and W. T. Blanford (Stratigraphical and Structural Geology)*. 2nd rev. ed. Calcutta: Government of India.

———. 1894. The Evolution of Indian Geography. *The Geographical Journal* 3, no. 3: 169–96.

Oldroyd, David R. 1996. *Thinking about the Earth: A History of Ideas in Geology*. Cambridge, MA: Harvard University Press.

Oliver, Fredrick Spencer. 1905. *A Dweller on Two Planets, Or The Dividing of the Way by Phylos the Thibetan*. Reprint ed. Los Angeles: Baumgardt Publishing.

Oppenheim, Jamet. 1985. *The Other World: Spiritualism and Psychical Research in England, 1850–1914*. Cambridge: Cambridge University Press.

Owen, Alex. 2001. Occultism and the "Modern" Self in Fin-de-Siēcle Britain. In *Meanings of Modernity*, edited by M. Daunton and B. Rieger. Oxford: Berg.

———.Forthcoming. *Magic and Modernity*. Chicago: University of Chicago Press.

Owen, David. 1988. Ephemeral States. *Atlantic* 261, no. 1: 14–17.

Padmanabhan, S. 1971. *The Forgotten History of the Land's End*. Nagercoil: Kumaran Pathipagam.

Paramasivanandam, A. M. 1944. Kumariyiṉ Niṉaivukaḷ [Memories of Kumari]. In *Kaṭṭuraip Pattu* [Ten essays]. Wallajahbad: Tamilkalai Patippakam.

Pascoe, Edwin H. 1950–76. *A Manual of the Geology of India and Burma Compiled from the Geological Survey of India and from Unofficial Sources by H. B. Medlicott, W. T. Blanford, V. Ball, and F. R. Mallet.* 4 vols. Delhi: Government of India.

Perrier, Edmond. 1925. *The Earth Before History: Man's Origin and the Origin of Life.* New York: Alfred A. Knopf.

Phillips, James. 1985. Distance, Absence, and Nostalgia. In *Descriptions,* edited by D. Ihde and H. J. Silverman. Albany: State University of New York Press.

Pillay, K. K. 1957a. *History of Higher Education in South India: University of Madras, 1857–1957.* 2 vols. Madras: University of Madras.

———. 1957b. Historical Ideas in Early Tamil Literature. *Tamil Culture* 6, no. 2: 113–32.

———. 1975. *A Social History of the Tamils.* Vol. 1. 2nd ed. Madras: University of Madras.

———. 1981a. *Tamiḻaka Varalāṟu: Makkaḷum Paṇpāṭum* [History of Tamilians: People and culture]. Rev. ed. Madras: Tamilnadu Textbook Society.

———. 1981b. Some Aspects of the Early History of the Tamils. In *Proceedings of the Fifth International Conference-Seminar of Tamil Studies, Madurai,* edited by M. Arunachalam. Madras: International Association of Tamil Research.

Polk, Dora Beale. 1991. *The Island of California: A History of the Myth.* Spokane: Arthur H. Clark.

Ponnambalam Pillai, T. 1914. President's Address. In *Mūṉṟāvatu Caiva Makā Caṅkap Piracaṅkaṅkaḷ* [Lectures delivered during the third Saiva conference]. Palayamkottai: Siava Sabha.

Poonkunran, M. 2001. Kumarikkaṇṭam. *Teṉmoḻi* 33, no. 10: 19–24.

Prakash, Gyan. 1999. *Another Reason: Science and the Imagination of Modern India.* Princeton, NJ: Princeton University Press.

Pratt, Mary Louise. 1992. *Imperial Eyes: Travel Writing and Transculturation.* London: Routledge.

Prosperi, Franco. 1957. *Vanished Continent: An Italian Expedition to the Comoro Islands.* Translated by David Moore. London: Hutchinson.

Pulavar Kulanthai. 1971. *Irāvaṇa Kāviyam* [epic poem on Ravana]. Reprint ed. Erode: Vela Patippakam.

Puratcidasan. 1995. *Koṭuṅkaṭal Koṇṭa Kumarikkaṇṭam* [Kumarikkaṇṭam seized by the cruel ocean]. Madras: Pasumpon Patippakam.

Purnalingam Pillai, M. S. 1904. *A Primer of Tamil Literature.* Madras: Ananda Press.

———. 1925. Foreword to *Paḻantamiḻ* [Ancient Tamil], by A. Sivaprakasar. Tirucirapuram: Saraswati Accukuttam.

———. 1945. *Tamil India.* Reprint ed. Tirunelveli: South India Saiva Siddhanta Works Publishing Society.

———. 1985. *Tamil Literature (Revised and Enlarged).* Reprint ed. Thanjavur: Tamil University.

Quatrefages de Breau, Armand de. 1883. *The Human Species.* New York: D. Appleton.

Radner, Daisie, and Michael Radner. 1982. *Science and Unreason.* Belmont, CA: Wadsworth Publishing.

Raghava Aiyangar, M. 1938. *Ārāycci Tokuti* [Research compendium]. N.p.

———. 1979. *Tamiḻ Varalāṟu* [History of Tamil]. Reprint ed. Annamalainagar: Annamalai University.

Rajamanikkam, M. 1944. *Maṟainta Nakaram Allatu Moheñco-tarō: Ciṟuvark Kuriyatu*

[The Lost City, Or Mohenjo-Daro: For children]. Madras: South India Saiva Siddhanta Works Publishing Society.

Rajasimman. 1944. Kaṭal Koṇṭa Kumari Nāṭu [Kumari Nāṭu seized by the ocean]. *Utayaṉ* 1: 4–5.

Ramachandra Dikshitar, V. R. 1947. *Studies in Tamil Literature and History.* Madras: University of Madras.

———. 1971. *Origin and Spread of the Tamils.* Reprint ed. Madras: South India Saiva Siddhanta Works Publishing Society.

Ramachandran, K. N. 1988. Intiya Tuṇaik Kaṇṭatiṉ "Payaṇam" [The journey of the Indian subcontinent]. *Tiṉamaṇi Cuṭar: Aṟiviyal Iṇaippu,* 24 September, pp. 1–2.

Ramachandran, M., and R. Mathivanan. 1991. *The Spring of the Indus Civilization.* Madras: Prasanna Pathippagam.

Ramanujan, A. K. 1985. *Poems of Love and War from the Eight Anthologies and the Ten Long Poems of Classical Tamil.* New York: Columbia University Press.

Ramaswamy, Sumathi. 1997. *Passions of the Tongue: Language Devotion in Tamil India, 1891–1970.* Berkeley: University of California Press.

———. 1998. Language of the People in the World of Gods: Ideologies of Tamil before the Nation. *Journal of Asian Studies* 57, no. 1: 66–92.

———. 1999. Catastrophic Cartographies: Mapping the Lost Continent of Lemuria. *Representations* 67: 92–129.

———. 2000. History at Land's End: Lemuria in Tamil Spatial Fables. *Journal of Asian Studies* 59, no. 3: 575–602.

———. 2001a. Remains of the Race: Archaeology, Nationalism, and the Yearning for Civilization in the Indus Valley. *Indian Economic and Social History Review* 38, no. 2: 105–45.

———. 2001b. Maps and Mother Goddesses in Modern India. *Imago Mundi* 53: 97–113.

———. 2002. Visualizing India's Geo-Body: Globes, Maps, Bodyscapes. *Contributions to Indian Sociology* 36, no. 1/2: 157–95.

Rangachariyar, M. 1891a. The Yugas: A Question of Hindu Chronology and History. *Madras Christian College Magazine* 8: 742–54, 808–15, 907–20.

———. 1891b. The Yugas: A Question of Hindu Chronology and History. *Madras Christian College Magazine* 9: 30–38, 187–98.

Rangaswami Aiyangar, K. V. 1910. *A History of India.* Pt. 1, *The Pre-Musalman Period (with Illustrations and Maps).* London: Longmans, Green.

Rangaswami Ayyangar, T. R. 1927. *Tamiḻ Nāṭu* [Tamil land]. Kumbakonam: B. N. Publishing House.

Rasanayagam, C. 1984. *Ancient Jaffna: Being a Research into the History of Jaffna from very early times to the Portug(u)ese Period.* New Delhi: Asian Educational Services.

Ratnam Pillai, G. 1922. *Junior Indian History: Book 1 for Form II (Ciṟuvarkaḷukkāṉa Intu Tēca Carittiram).* Madras: Coomaraswamy Naidu and Sons.

Ray, S. 1963. *Fifty Years of Science in India: The Progress of Geology.* Calcutta: Indian Science Congress Association.

Read, Peter. 1996. *Returning to Nothing: The Meaning of Lost Places.* Cambridge: Cambridge University Press.

Reader, John. 1981. *Missing Links: The Hunt for Earliest Man.* London: Collins.

Rice, Anthony L., ed. 1986. *Deep-Sea Challenge: The John Murray/Mabahiss Expedition to the Indian Ocean, 1933–34*. Paris: UNESCO.

Richards, Thomas. 1993. *The Imperial Archive: Knowledge and the Fantasy of Empire*. London: Verso.

Risley, Herbert H. 1908. *The People of India*. Calcutta: Thacker, Spink.

———. 1909. Ethnology and Caste. In *The Imperial Gazetteer of India: The India Empire*, Vol. 1, *Descriptive*. Oxford: Clarendon Press.

Risley, H. H., and Edward Gait. 1903. *Census of India, 1901: Report*. Calcutta: Office of the Superintendent of Government Printing.

Roberts, Alaric J. 1942. *New Trade Winds for the Seven Seas*. Santa Barbara: J. F. Rowny Press.

Robinson, Arthur. 1982. *Early Thematic Mapping in The History of Cartography*. Chicago: University of Chicago Press.

Robinson, Arthur H., and Barbara B. Petchenik. 1986. *The Nature of Maps: Essays toward Understanding Maps and Mapping*. Chicago: Chicago University Press.

Rocher, Ludo. 1986. *The Purāṇas*. Wiesbaden: Otto Harrassowitz.

Romm, James A. 1994. A New Forerunner for Continental Drift. *Nature* 367, no. 6462: 407–8.

Rossi, P. 1984. *The Dark Abyss of Time: The History of the Earth and the History of Nations from Hooke to Vico*. Chicago: University of Chicago Press.

Roy, Sarat Chandra. 1912. *The Mundas and their Country*. Bombay: Asia Publishing.

———. 1966. Presidential Address to the Seventh Oriental Congress. In *Studies in Indian Anthropology*, edited by N. K. Bose. Calcutta: Indian Studies.

Rudwick, Martin J. S. 1976. The Emergence of a Visual Language for Geological Science, 1760–1840. *History of Science* 14: 149–95.

———. 1986. The Shape and Meaning of Earth History. In *God and Nature: Historical Essays on the Encounter between Christianity and Science*, edited by D. C. Lindberg and R. L. Numbers. Berkeley: University of California Press.

Rupke, N. A. 1996. Eurocentric Ideology of Continental Drift. *History of Science* 34, no. 105: 251–72.

Rutter, Owen. 1932. *The Monster of Mu: The Thrilling Story of a Race for Treasure in the Pacific and its Tragic Ending*. London: Francis James Publishing.

Ryan, Simon. 1994. Inscribing the Emptiness: Cartography, Exploration, and the Construction of Australia. In *De-Scribing Empire*, edited by C. Tiffin and A. Lawson. London: Routledge.

Saccithanandan, M. K. 1999. *Paḷḷi Māṇavarkaḷukkāṉa Nilavarai: Ilankai* [Atlas for school students: Sri Lanka]. Chennai: Kanthazhagam.

Sadashiva Pandarathar, T. V. 1940. *Pāṇṭiyar Varalāṟu* [History of the Pandyas]. Kumbakonam: Shri Mahabharat Press.

Sagan, Carl. 1996. *The Demon Haunted World: Science as a Candle in the Dark*. New York: Random House.

Sami, K. P. 1955. *Varalāṟṟil Tamiḻakam* [The Tamil realm in history]. Kuala Lampur: Kural Manram.

Saravana Mutthu, T. 1892. *Tamiḻ Ilakkiya Ārāycci* [The study of Tamil literature: An essay]. Madras: Sri Nilayam Accu Kootam.

Sathur Sekharan, et al., eds. 1994. *Kumarikkaṇṭam Maṟṟum Cintuveḷi Māṉāṭṭu Malar*

[Conference on Lemuria and the Indus Valley]. Madras: International Linguistic Center.
Satthianadhan, Samuel. 1894. *History of Education in the Madras Presidency.* Madras: Srinivasa, Varadachari.
Savariroyan, D. 1900–1901. The Admixture of Aryan with Tamilian. *The Light of Truth or Siddhanta Deepika* 4: 104–8, 157–61, 218–20, 241–44, 269–71.
———. 1901a. Some Disputed Points Cleared. *The Light of Truth or Siddhanta Deepika* 5, no. 4: 78–81.
———. 1901b. Some Disputed Points Replied. *The Light of Truth or Siddhanta Deepika* 5, no. 12: 194–200.
———. 1907a. The Bharata Land or Dravidian India. *The Tamilian Antiquary* 1, no. 1: 1–28.
———. 1907b. On the Relation of the Pandavas to the Tamilian Kings. *The Tamilian Antiquary* 1, no. 1: 53–68.
Scafi, Alessandro. 1999. Mapping Eden: Cartographies of the Earthly Paradise. In *Mappings,* edited by D. Cosgrove. London: Reaktions Book.
Scharfe, Hartmut. 1973. Tolkāppiyam Studies. In *German Scholars on India.* Varanasi: Chowkambha Sanskrit Series Office.
Scholes, Robert. 1979. *Fabulation and Metafiction.* Urbana: University of Illinois Press.
Schroeder, Werner. 1984. *Man: His Origin, History, and Destiny.* Mount Shasta, CA: Ascended Master Teaching Foundation.
Schuchert, Charles. 1931. *Outlines of Historical Geology.* 2nd rev. ed. New York: John Wiley and Sons.
Schwartzberg, Joseph E. 1992. South Asian Cartography. In *Cartography in the Traditional Islamic and South Asian Societies,* edited by J. B. Harley and D. Woodward. Chicago: University of Chicago Press.
Sclater, Philip Lutley. 1864. The Mammals of Madagascar. *Quarterly Journal of Science* 1, no. 2: 213–19.
———. 1874. The Geographical Distribution of Mammals. *Science Lectures for the People,* 6th ser., 5: 67–84.
———. 1875. On the Present State of Our Knowledge of Geographical Zoology. Address Delivered to the Biological Section of the British Association, Bristol, August 25th, 1875. *Report of the British Association for the Advancement of Science* 2: 85–133.
Scott, G. Firth. 1898. *The Last Lemurian: A Westralian Romance.* London: James Bowden.
Scott, Robert. 1961. *Limuria: The Lesser Dependencies of Mauritius.* London: Oxford University Press.
Scott-Elliot, W. 1904. *The Lost Lemuria with Two Maps Showing Distribution of Land Areas at Different Periods.* London: Theosophical Publishing Society.
———. 1954. *The Story of Atlantis and The Lost Lemuria.* Reprint ed. London: Theosophical Publishing House.
Seetharaman, K. 1962. *Camūka A̱rivu Nūl: Aintām Vakuppu* [Social studies for class five]. 5th ed. Madras: Seshachala and Co.
Senghor, Leopold. 1977. Negritude and Dravidian Culture. In *Dravidians and Africans,* edited by K. P. Aravaanan. Madras: Tamil Kootam.
Sesha Aiyar, K. G. 1937. *Cera Kings of the Sangam Period.* London: Luzac.

Sesha Iyengar, T. R. 1933. *Dravidian India.* Madras: C. Coomaraswamy Naidu and Sons.

Seshagiri Sastri, M. 1897. *Essay on Tamil Literature.* Madras: S.P.C.K. Press.

Sethu Pillai, R. P. 1925. Tami̲lmo̲liyi̲n Perumaiyum Ata̲n Ta̲rkāla Nilaimaiyum [The greatness of the Tamil language and its present condition]. *O̲r̲umai* 5: 285–94.

———. 1950. *Kaṭa̲rkara̲iyil* [On the seashore]. Madras: Star Prachuram.

Shanmukhasundaram, M. 1939. *Iḷaiñar Intiyā Carittiram: Mutal Puttakam* [Indian history for the young: Part one]. Vol. 1. Madras: Aciriyar Nurpatippu Kalakam.

Shaver, Richard. 1948. *I Remember Lemuria and the Return of the Sathanas.* Evanston: Venture.

Shulman, David. 1980. *Tamil Temple Myths: Sacrifice and Divine Marriage in the South Indian àaiva Tradition.* Princeton, NJ: Princeton University Press.

———. 1988a. The Tamil Flood Myths and the Cankam Legend. In *The Flood Myth—Tamil,* edited by A. Dundes. Berkeley: University of California Press.

———. 1988b. Sage, Poet, and Hidden Wisdom in Medieval India. In *Cultural Traditions and Worlds of Knowledge: Explorations in the Sociology of Knowledge,* edited by S. N. Eisenstadt and I. F. Silber. Greenwich, CT: JAI Press.

Singaravelan, "Silambupani". 1981. *Aintām Ulakat Tami̲l Mānāṭu* [Fifth international Tamil conference]. Secundarabad: Privately printed.

Singhal, Jwala Prasad. 1963. *The Sphinx Speaks, or The Story of the Prehistoric Nations.* New Delhi: Sadgyan Sadan.

———. [1968?] *Forgotten Ancient Nations and Their Geography.* New Delhi: Sadgyan Sadan.

Sinnett, Alfred P. 1885. *Esoteric Buddhism.* Rev. ed. Boston: Houghton Mifflin.

———. 1954. Preface to *The Story of Atlantis and The Lost Lemuria,* by W. Scott-Elliot. Reprint ed. London: Theosophical Publishing House.

Sivagnana Yogi, Virudhai. 1913. *Tami̲lmo̲liyum Civane̲r̲iyum* [The Tamil language and Saivism]. Madurai: Tiruvitar Kalakam.

Sivagnanam, M. P. 1974. *E̲natu Pōrāṭṭam* [My struggle]. Madras: Inba Nilayam.

Sivaraja Pillai, K. N. 1932. *The Chronology of the Early Tamils.* Madras: University of Madras.

Sivasailam Pillai. 1951. *Putumu̲raik Ka̲lakat Tami̲lppāṭam (Muta̲r Puttakam: Mutar Parattirku)* [New Kazhagam Tamil reader for class six]. Rev. ed. Tirunelveli: South India Saiva Siddhanta Works Publishing Society.

Skaria, Ajay. 2000. *Hybrid Histories: Forests, Frontiers, and Wildness in Western India.* Delhi: Oxford University Press.

Smith, Alan G., and J. C. Briden. 1977. *Mesozoic and Cenozoic Paleocontinental Maps.* Cambridge: Cambridge University Press.

Smith, Alan G., David G. Smith, and Brian M. Funnell. 1994. *Atlas of Mesozoic and Cenozoic Coastlines.* Cambridge: Cambridge University Press.

Smith, Anthony D. 1986. Legends and Landscapes. In *The Ethnic Origins of Nations.* Oxford: Basil Blackwell.

Smith, Bernard. 1985. *European Vision and the South Pacific, 1768–1860: A Study in the History of Art and Ideas.* London: Yale University Press.

Smith, Geo. 1883. *The Geography of British India, Political and Physical.* London: John Murray.

Smith, Neil, and Anne Godlewska. 1994. Introduction: Critical Histories of Geography. In *Geography and Empire,* edited by Anne Godlewska and Neil Smith. Oxford: Blackwell.

Soja, Edward W. 1989. *Postmodern Geographies: The Reassertion of Space in Critical Social Theory.* London: Verso.

Somasundara Bharati, S. 1912. *Tamil Classics and Tamilakam.* Madras: N.p.

———. 1967. The Pre-deluge Pandinad and Her Southern Frontier. In *The Papers of Dr. Navalar Somasundara Bharatiar,* edited by S. Sambasivam. Madurai: Navalar Pusthaka Nilayam.

Sorkhabi, Rasoul B. 1996. What's in a Name: Gondwana or Gondwanaland? *Episodes* 19, no. 3: 82–84.

Spence, Lewis. 1933. *The Problem of Lemuria: The Sunken Continent of the Pacific.* London: Rider.

Sri, P. 1963. Kumarittuṟaiyil Cintaṉai Alaikaḷ [Waves of thought at Cape Kumari]. In *Potiyamum Imayamum* [Mount Potiyil and the Himalayas]. Madras: Amudha Nilayam.

Srinivasa Aiyangar, M. 1914. *Tamil Studies, Or Essays on the History of the Tamil People, Language, Religion, and Literature.* Madras: Guardian Press.

Srinivasa Iyengar, P. T. 1982. *History of the Tamils from the Earliest Times to 600 A.D.* Reprint ed. New Delhi: Asian Educational Services.

———. 1985. *Pre-Aryan Tamil Culture.* Reprint ed. New Delhi: Asian Educational Services.

Srinivasa Pillai, K. S. 1965. *Tamiḻ Varalāṟu (Muṟpākam)* [History of Tamil: First part]. 8th ed. Tanjavur: Verrivel Power Press.

Srinivasachari, C. S. 1938. Pre-Dravidian, Proto-Dravidian, and Dravidian. *Journal of the Bihar and Orissa Research Society* 24: 29–56.

Srinivasan, M. R. 1949. *Camūka Aṟivu Nūl* [Social Studies textbook]: *Book three for form three.* Madurai: N.p.

Srinivasan, R. 1956. *Tamiḻt Teṉṟal Vācakam (Mutaṟ Paṭivam: Ciṟappu Pakuti)* [Tamil Tenral Tamil reader for form three]. Kumbakonam: L. V. K. Publishing House.

St. Clair, David. 1972. *The Psychic World of California.* New York: Doubleday.

Starobinski, Jean. 1966. The Idea of Nostalgia. *Diogenes* 54: 81–103.

Steiger, Brad. 1978. *Worlds before Our Own.* New York: Berkley Publishing Corp.

Steiner, Rudolf. 1911. *The Submerged Continents of Atlantis and Lemuria: Their History and Civilization, Being Chapters from the Akashic Records.* Translated by Max Gysi. Chicago: Rajput Press.

———. 1959. *Cosmic Memory: Prehistory of Earth and Man.* Translated by Karl E. Zimmer. San Francisco: Harper and Row.

Stelle, Robert D. 1952. *The Sun Rises: The Story of the Formation of the First and Greatest Civilization on this Earth, as it actually occurred 78,000 Years Ago.* Ramona, CA: Lemurian Fellowship.

Stemman, Roy. 1976. *Atlantis and the Lost Lands: A New Library of the Supernatural.* London: Aldus Books.

Stewart, Susan. 1993. *On Longing: Narratives of the Miniature, the Gigantic, the Souvenir, the Collection.* Durham, NC: Duke University Press.

Stiebing, William H. 1984. *Ancient Astronauts, Cosmic Collisions, and Other Popular Theories about Man's Past.* Buffalo: Prometheus.

Stocking, George W., Jr. 1968. The Persistence of Polygenist Thought in Post-

Darwinian Anthropology. In *Race, Culture, and Evolution: Essays in the History of Anthropology*. Chicago: University of Chicago Press.

——— 1987. *Victorian Anthropology*. New York: Free Press.

Stoddart, David. 1986. *On Geography and Its History*. Oxford: Blackwell.

Stuart, H. A. 1893. *Census of India, 1891: Madras*. Madras: Government Press.

Sturge, P. H. 1897. *A Lecture on the Study of Indian History*. Madras: Commercial Press.

Subramania Aiyar, A. V. 1970. *Paṇṭait Tamiḻ Nāṭu* [Ancient Tamilnadu]. Tirunelveli: S. R. Subramania Pillai.

Subramania Pillai, K. 1949. *Ilakkiya Varalāṟu (Mutaṟ Pākam)*. 3rd rev. ed. Madras: Aciriyar Nurpatippu Kazhagam.

Subramania Sastri, S. 1915–16. Tamiḻnāṭukaḷ [Tamil territories]. *Centamiḻ* 14, no. 11–12: 417–38, 465–74.

Subrahmanian, N. 1996. *The Tamils: Their History, Culture, and Civilization*. Madras: Institute of Asian Studies.

Subramanian, Shaktidasan. 1939. *Tamiḻ Veṟi* [Passion for Tamil]. Madras: Sadhu Accukutam.

Subramania Sharma, N. R. 1922. *Centamiḻ Vācakat Tiraṭṭu* [Literary selections from Tamil poetry and prose]. Madurai: E. M. Gopalakrishna Kone.

Subramuniyaswami, Satguru Sivaya. 1998. *The Lemurian Scrolls: Angelic Prophecies Revealing Human Origins*. Kapaa, HI: Himalayan Academy Publications.

Suess, Eduard. 1904. *The Face of the Earth (Das Antlitz der Erde)*. Translated by Hertha B. C. Sollas. Vol. 1. Oxford: Clarendon Press.

Sullivan, Walter. 1974. *Continents in Motion: The New Earth Debate*. New York: McGraw.

Sundaram Pillai, P. 1895. *Some Milestones in the History of Tamil Literature Found In an Inquiry into the Age of Tiru Gnana Sambanda*. Madras: Addison.

Sundaravarada Acariyar, M. K. 1924. *Intutēcac Carittiram (Mutaṟ Puttakam)* [History of India, Part one]. 2nd ed. Madras: Indian Publishing House.

Suryanarayana Sastri, V. G. 1903. *Tamiḻmoḻiyiṉ Varalāṟu* [History of the Tamil language]. Madras: G. A. Natesan.

Suryanarayana Sastri, V. G., and T. Chelvakesavaraya Mudaliar. 1918. *Eighth Reader: Tamil for Form IV of High School* [Eṭṭam Vācaka Puttakam]. London: Macmillan.

———. 1904. *Ninth Reader: Tamil for Form V of High School* [Oṉpatām Vācaka Puttakam]. London: Macmillan.

Swaminatha Aiyar, U. V. 1978. *Caṅkakālatamiḻum Piṟkālattamiḻum* [Sangam Tamil and later-day Tamil]. Reprint ed. Madras: Dr. U. V. Swaminathaiyar Library.

———, ed. 1950. *Iḷaṅkōvaṭikaḷaruḷiceyta Cilappatikāra Mūlamum Arumpatavuraiyum Aṭiyārkkunallāruraiyum* [Ilango Adigal's *Cilapatikaram* with Adiyarkkunallar's Commentary]. Reprint ed. Madras: Dr. U. V. Swaminathaiyar Library.

Swaminathan, A. 1996. *Tamiḻaka Varalāṟu 1987 Varai* [History of the Tamil realm until 1987]. 13th ed. Madras: Deepa Patippakam.

Swinfen, Anne. 1984. *In Defence of Fantasy*. London: Routledge and Kegan.

Taffinder, Adelia H. 1908. A Fragment of the Ancient Continent of Lemuria. *Overland Monthly* 52: 163–67.

Tamby Pillai, V. J. 1907. The Solar and the Lunar Races of India. *The Tamilian Antiquary* 1, no. 1: 29–50.

Tamilnadu Textbook Society. 1977. *Intiya Varalāṟu: 6* [History of India for class six]. Madras: Tamilnadu Text Book Society.

———. 1981a. *Intiya Varalāṟu: 6* [History of India for class six]. Madras: Tamilnadu Text Book Society.
———. 1981b. *History of India (Standard VI)*. Madras: Tamilnadu Text Book Society.
———. 1982. *Tamiḻp Pāṭanūl: Āṟām Vakuppu* [Tamil reader for class six]. Madras: Tamilnadu Text Book Society.
———. 1996a. *Tamiḻp Pāṭanūl: Āṟām Vakuppu*. 2nd ed. Madras: Tamilnadu Text Book Society.
———. 1996b. *Tamiḻp Pāṭanūl: Ēḻām Vakuppu* [Tamil reader for class seven]. Madras: Tamilnadu Text Book Society.
———. 2001. *Tamiḻ 7* [Tamil reader for class seven]. Rev. ed. Madras: Tamilnadu Text Book Society.
Taussig, Michael. 1987. *Shamanism, Colonialism, and the Wild Man: A Study in Terror and Healing*. Chicago: University of Chicago Press.
Taylor, William. 1835. *Oriental Historical Manuscripts in the Tamil Language: Translated with Annotations*. 2 vols. Madras: N.p.
Tennent, James Emerson. 1977. *Ceylon: An Account of the Island, Physical, Historical and Topographical with Notices of its Natural History, Antiquities, and Productions*. Vol. 1. 6th rev. ed. Dehiwala, Sri Lanka: Tisara Prakasakayo Ltd.
Terdiman, Richard. 1985. *Discourse/Counter-Discourse: The Theory and Practice of Symbolic Resistance in Nineteenth-Century France*. Ithaca, NY: Cornell University Press.
Termier, Henri. 1960. *Atlas de Paléogéographie*. Paris: Masson.
Thanikachalam, K. 1992. *Tamiḻar Varalāṟum Ilaṅkai Iṭappeyer Āyvum* [History of Tamils and research on Sri Lanka as a place]. Madras: Saravana Patippakam.
Theunissen, Bert. 1989. *Eugene Dubois and the Ape-Man from Java: The History of the First "Missing Link" and its Discoverer*. Dordrecht: Kluwer Academic Publishers.
Thongchai Winichakul. 1994. *Siam Mapped: A History of the Geo-Body of a Nation*. Honolulu: University of Hawaii Press.
Thurston, Edgar. 1899. The Dravidian Problem. *Madras Government Museum Bulletin* 2, no. 3: 182–97.
Thurston, Edgar, and K. Rangachari. 1987. Introduction to *Castes and Tribes of Southern India*. New Delhi: Asian Educational Services.
Tirumalaikolunthu Pillay, S. A. 1900. A Short Sketch of Tamil Literature. *The Light of Truth, or Siddhanta Deepika* 4: 12–18, 85–89, 161–67, 198–201, 221–35, 245–54.
Tiruvenkadatayyangar, V. 1948. *Intu Tēca Carittiram: Mutaṟ Puttakam (Āṟām Vakuppu & Mutal Pāram)* [History of India, part one, for sixth standard and form one]. Reprint ed. Kumbakonam: V. S. Venkataraman.
Tiryakian, Edward A. 1974. Preliminary Considerations. In *On the Margin of the Visible: Sociology, the Esoteric, and the Visible*, edited by E. A. Tiryakian. New York: John Wiley and Sons.
Topinard, Paul. 1878. *Anthropology*. Translated by Robert T. H. Bartley. Library of Contemporary Science. London: Chapman and Hall.
Toussaint, Auguste. 1967. *History of the Indian Ocean*. Translated by June Guicharnaud. Chicago: University of Chicago Press.
Trevor-Roper, Hugh. 1983. The Invention of Tradition: The Highland Tradition of Scotland. In *The Invention of Tradition*, edited by Eric Hobsbawm and Terence Ranger. Cambridge: Cambridge University Press.

Trigger, Bruce G. 1989. *A History of Archaeological Thought.* Cambridge: Cambridge University Press.

Tuan, Yi-Fu. 1977. *Space and Place: The Perspective of Experience.* Minneapolis: University of Minnesota Press.

Tyson, Peter. 2000. *The Eighth Continent: Life, Death and Discovery in the Lost World of Madagascar.* New York: William Morrow.

Vaiyapuri Pillai, S. 1956. *History of the Tamil Language and Literature (Beginning to 1000 A.D.).* Madras: New Century Book House.

———. 1957. History of the Tamil Language and Literature. In *A Comprehensive History of India,* edited by K. A. Nilakanta Satri. Bombay: Orient Longmans.

van der Gracht, W. A. J. M. van Waterschoot, et. al. 1928. *Theory of Continental Drift: A Symposium on the Origin and Movement of Land Masses Both Inter-continental and Intra-continental, as Proposed by Alfred Wegener.* Tulsa: American Association of Petroleum Geologists.

van Oosterzee, Penny. 1997. *Where Worlds Collide: The Wallace Line.* Ithaca, NY: Cornell University Press.

van Steenis, C. G. G. J. 1962. The Land-Bridge Theory in Botany. *Blumea* 11, no. 2: 235–372.

Vasbinder, Samuel H. 1982. Aspects of Fantasy in Literary Myths about Lost Civilizations. In *The Aesthetics of Fantasy Literature and Art,* edited by R. C. Schlobin. Notre Dame: University of Notre Dame Press.

Velan, N. K. 1995. *Pūmiyiṉ Katai* [The story of earth]. 6th ed. Madras: South India Saiva Siddhanta Works Publishing Society.

Venkatachalam, K. 1956. *Cāmpā Tamiḻ Vācakam: Aintām Puttakam* [Samba Tamil reader: Book five]. Madras: Samba Publishing.

Venkatachalam Pillai, R. 1951. *Arunkalait Tamiḻ Vācakam (Mutal Pāram)* [Tamil reader for class six]. Madurai: E. M. Gopalakrishna Kon.

Venkataramayyar, U. S. 1934. *Centamiḻ Nūṟkōvai* [Garden of glorious Tamil]. Kumbakonam: G. V. K. Sami.

Venkatasami, Mayilai. C. 1983. *Maṟaintupōṉa Tamiḻ Nūlkaḷ* [Lost Tamil books]. Reprint ed. Chidambaram: Manivasagar Patippakam.

Vilvapati, K., et al. 1956. *Mūvar Tamiḻ Vācakam (Mūṉṟā Paṭivam: Iraṇṭām Pirivu): Ciṟappu Pakuti* [Moovar Tamil reader for class eight: Special section]. Reprint ed. Madras: Vardha Publishing House.

Vimalanandam, M. C. 1987. *Tamil Ilakkiya Varalāṟṟuk Kaḷañciyam* [Encyclopedia of Tamil literary history]. Madras: Aintinai Patippakam.

Vincent, Louis Claude. 1969. *Le Paradis Perdu de Mu.* Marsat: Editions de la Source.

Viswanathan, Gauri. 1989. *Masks of Conquest: Literary Study and British Rule in India.* New York: Columbia University Press.

———. 1998. Conversion, Theosophy, and Race Theory. In *Outside the Fold: Conversion, Modernity, and Belief.* Princeton, NJ: Princeton University Press.

———. 2000. The Ordinary Business of Occultism. *Critical Inquiry* 27, no. 1: 1–20.

Vitaliano, Dorothy B. 1973. *Legends of the Earth: Their Geologic Origins.* Bloomington: Indiana University Press.

Vivian, E. Charles. 1973. *City of Wonder.* Reprint ed. New York: Centaur Press.

von Ihering, H. 1907. *Archhelenis und Archinotis, Gesammelte Beitrage zur Geschichte der Neotropischen Region.* Leipzig: Engelmann.

Wadia, D. N. 1919. *Geology of India for Students.* London: Macmillan.
———. 1939. *Geology of India.* 2nd ed. London: Macmillan.
Wakelam, K. B. 1975. The Lost Continents of Atlantis and Lemuria. In *Seminar on Theosophy and Science (December 1975: Section III).* Madras: Theosophical Society.
Wallace, Alfred Russel. 1860. On the Zoological Geography of the Malay Archipelago. *Journal of the Proceedings of the Linnean Society of London (Zoology)* 4: 172–74.
———. 1876. *The Geographical Distribution of Animals with a Study of the Relations of Living and Extinct Faunas as Elucidating the Past Changes of the Earth's Surface.* 2 vols. London: Harper and Brothers.
———. 1877. The Comparative Antiquity of Continents, as indicated by the Distribution of Living and Extinct Animals. *Proceedings of the Royal Geographical Society,* o.s., 21, no. 6: 505–35.
———. 1880. *Island Life: Or, The Phenomena and Causes of Insular Faunas and Floras, Including a Revision and Attempted Solution of the Problem of Geological Climates.* London: Macmillan.
Wallis, Helen M., and Arthur H. Robinson, eds. 1987. *Cartographical Innovations: An International Handbook of Mapping Terms to 1900.* London: Map Collector Publications.
Walton, Bruce, ed. 1985. *Mount Shasta: Home of the Ancients, Subsurface Mysteries.* Mokelumne Hill, CA: Health Research.
Washington, Peter. 1995. *Madame Blavatsky's Baboon: A History of the Mystics, Mediums, and Misfits Who Brought Spiritualism to America.* New York: Schocken Books.
Wauchope, Robert. 1962. *Lost Tribes and Sunken Continents: Myth and Method in the Study of American Indians.* Chicago: University of Chicago Press.
Weber, Max. 1946. Science as Vocation. In *Max Weber: Essays in Sociology,* edited by H. H. Gerth and C. W. Mills. Oxford: Oxford University Press.
Weeks, David, and Jamie James. 1995. *Eccentrics: A Study of Sanity and Strangeness.* New York: Villard.
Wegener, Alfred L. 1924. *The Origin of Continents and Oceans.* Translated by J. G. A. Skerl. 3rd rev. ed. New York: E. P. Dutton.
———. 1966. *The Origin of Continents and Oceans.* Translated by J. Biram. 4th rev. ed. New York: Dover.
Weinstein, Fred. 1990. The Social Significance of Fantasy Thinking. In *History and Theory after the Fall: An Essay in Interpretation.* Chicago: University of Chicago Press.
Wellard, James Howard. 1975. *The Search for Lost Worlds: An Exploration of the Lands of Myth and Legend, including Atlantis, Sheba, and Avalon.* London: Pan Books.
Wells, H. G. 1932. *The Outline of History Being a Plain History of Mankind.* Vol. 1. Rev. ed. New York: Macmillan.
Whitehead, Harriet. 1974. Reasonably Fantastic: Some Perspectives on Scientology, Science Fiction, and Occultism. In *Religious Movements in Contemporary America,* edited by I. I. Zaretsky and M. P. Leone. Princeton, NJ: Princeton University Press.
Whitfield, Peter. 1994. *The Image of the World: Twenty Centuries of World Maps.* San Francisco: Pomegrante Artbooks in association with the British Library.
Wild, H. 1965. Additional Evidence for the Africa–Madagascar–India–Ceylon Land-Bridge Theory with special reference to the genera Anisopappus and Commiphora. *Webbia* 19, no. 2: 497–505.

Williams, Stephen. 1991. *Fantastic Archaeology: The Wild Side of North American Prehistory.* Philadelphia: University of Pennsylvania Press.

Willis, Bailey. 1932. Isthmian Links. *Bulletin of the Geological Society of America* 43: 917–52.

Wilson, Horace Hayman. 1838. *Historical Sketch of the Kingdom of Pandya, Southern Peninsula of India.* Madras: American Mission Press.

Winchester, Simon. 2001. *The Map that Changed the World: William Smith and the Birth of Modern Geology.* New York: HarperCollins.

Wiseman, J. D. H., and R. B. S. Sewell. 1937. The Floor of the Arabian Sea. *Geological Magazine* 74: 219–30.

Wood, D., and J. Fels. 1986. Designs on Signs: Myth and Meaning in Maps. *Cartographica* 23, no. 3: 54–103.

Wood, Robert M. 1985. *The Dark Side of the Earth.* London: George Allen and Unwin.

Wright, Claude Falls. 1894. *An Outline of the Principles of Modern Theosophy.* Boston: New England Theosophical Corporation.

Wright, John K. 1947. Terrae Incognitae: The Place of the Imagination in Geography. *Annals of the Association of American Geographers* 37: 1–15.

Yaeger, Patricia. 1996. Introduction: Narrating Space. In *The Geography of Identity,* edited by P. Yaeger. Ann Arbor: University of Michigan Press.

Yonemoto, Marcia. 1999. Maps and Metaphors of the "Small Eastern Sea" in Tokugawa Japan (1603–1868). *Geographical Review* 89, no. 2: 169–87.

Young, Iris M. 2000. House and Home: Feminist Variations on a Theme. In *Resistance, Flight, Creation: Feminist Enactments of French Philosophy,* edited by D. Olkowski. Ithaca, NY: Cornell University Press.

Young, Terence. 2001. Place Matters. *Annals of the Association of American Geographers* 91, no. 4: 681–82.

Zanger, Michael. 1992. *Mt. Shasta: History, Legend & Lore.* Berkeley, CA: Celestial Arts.

Zerubavel, Yael. 1995. *Recovered Roots: Collective Memory and the Making of Israeli National Tradition.* Chicago: University of Chicago Press.

Zitko, Howard John. 1941. *Unfurl the Banners: A Keynote Address at the Lemurian National Convention, Milwaukee, Wisconsin, August 9th, 1941.* Milwaukee: Lemurian Press.

Zvelebil, Kamil. 1973. The Earliest Account of the Tamil Academies. *Indo-Iranian Journal* 15, no. 2: 109–35.

———. 1992. *Companion Studies to the History of Tamil Literature.* Leiden: E. J. Brill.

INDEX

Compositor: BookMatters, Berkeley
Text: 10/12 Baskerville
Display: Baskerville
Printer and Binder: Maple-Vail Manufacturing Group

Milton Keynes UK
Ingram Content Group UK Ltd.
UKHW021659170824
447059UK00001B/29

9 780520 244405